IEE CIRCUITS, DEVICES AND SYSTEMS SERIES 12

Series Editors: Dr D. G. Haigh
Dr R. S. Soin
Dr J. Wood

Strained silicon heterostructures:

materials and devices

Other volumes in the Circuits, Devices and Systems series:

Strained silicon heterostructures:
materials and devices

Edited by
C. K. Maiti
N. B. Chakrabarti
and S. K. Ray

The Institution of Electrical Engineers

Published by: The Institution of Electrical Engineers, London,
United Kingdom

The Institution of Electrical Engineers,
Michael Faraday House,
Six Hills Way, Stevenage,
Herts. SG1 2AY, United Kingdom

British Library Cataloguing in Publication Data

Maiti, C.K.
    Strained silicon heterostructures: materials and devices.
    –(IEE circuits, devices and systems series; no. 12)
    1. Heterostructures 2. Silicon
    I. Title. II. Chakrabarti, N. B. III. Ray, S. K. IV. Institution of Electrical
    Engineeers
    620.1'93

ISBN 0 85296 778 0

Printed in England by Antony Rowe Ltd, Chippenham, Wiltshire

Contents

Preface

For the last several years, silicon–germanium (SiGe) strained layers have shown promising results for device applications. The driving forces have been to make new devices and the key to success has been the possibility of bandgap engineering in silicon-based materials. The development of powerful epitaxial growth techniques such as molecular beam epitaxy (MBE) and the ultra high vacuum chemical vapor deposition (UHVCVD) system has given rise to a new branch of engineering – namely bandgap engineering in group-IV alloys. The MBE and UHVCVD techniques have paved the way, not only for heterojunction bipolar transistors, field effect and modulation doped structures, but also for fascinating new quantum devices.

Advances in wireless communication and information technology require implementation of very high performance electronic systems. SiGe heterojunction bipolar transistors (HBTs) have emerged as a leading contender to satisfy these needs. The use of wide bandgap emitters allows design of HBTs with very high base doping levels, which significantly reduce base resistance and increase device speed. The low emitter–base turn-on voltage and device scaling to submicron dimensions significantly reduce power consumption in circuit operation. SiGe-HBTs are ideally suited for low voltage, low power wireless communications applications.

With the advent of manufacturable SiGe-HBTs and a greater architectural design flexibility, it is envisaged that SiGe technology will find applications in the 2–30 GHz range for next generation rf communication system applications. It is imperative that the essential details are systematically reviewed including materials growth, device fabrication and circuit applications, and this has been attempted in this book.

In Chapter 1, the advantages of heterojunction bipolar devices for wireless communication applications are introduced. As deposition of high quality films is the most important step for silicon-based heterostructure devices to ensure acceptable device performance, a review of the essential technologies for strained layer epitaxy of group-IV alloys is presented in Chapter 2.

In Chapter 3, SiGe and other group-IV alloy semiconductor properties such as bandgap, lattice constant, strained and unstrained layers, critical thickness, bandgap narrowing and discontinuities, electron and hole mobilities, and velocity overshoot are presented. The problems of growth of gate insulators for heterostructure devices compatible with silicon IC technology are not trivial and are discussed in Chapter 4.

The development of SiGe-HBTs is discussed in Chapter 5. Processes for the fabrication of SiGe-HBTs using heterojunction technologies are described. The chapter begins with a description of the general dc behavior of SiGe-HBTs and bandgap engineering to control the electron and hole currents in HBTs is introduced. Noise performance, rf circuits, SiGe-MMICs using SiGe-HBTs, and passive components are also discussed in this chapter.

The application of strained layers to heterostructure FETs is not as well developed as HBTs. The principles and performance of heterostructure field effect transistors using strained layers are discussed in Chapter 6. To a different class belong Si heterostructure MODFETs and BICFETs combining bipolar and FET operation in a single device. Unconventional structures such as BICFETs and resonant tunneling devices are treated in Chapter 7.

Chapter 8 describes how the growth of a strained Si layer on a relaxed SiGe buffer layer has led to higher values of electron mobility with significant enhancement in high frequency performance of MODFETs in Si technology.

Chapter 9 is concerned with the contact metallization on strained layers, and the fabrication and properties of metal–SiGe Schottky junctions. Silicon-based optoelectronics has recently made inroads in integrated optoelectronics and is another promising research field for group-IV alloy layers. Chapter 10 deals with the key devices in SiGe-based photonic integrated circuits.

This is the first book on Si heterostructures of which we are aware covering material properties of group-IV alloy layers such as SiGe, SiC, SiGeC and strained Si and their applications. The book treats SiGe, SiGeC and strained Si heterojunction devices, namely bipolar and field effect transistors, and other devices in a single text. The book is intended for use by senior undergraduate or graduate students in Electronic and Electrical Engineering, Applied Physics and Materials Sciences, and as a reference for engineers and scientists involved in semiconductor device research and development.

The book contains extensive references for those wishing to delve deeper into individual topics. Numerous up-to-date results from journals, conference proceedings, and books are cited. More than 250 figures and 25 tables are provided for illustration purposes. We hope that the book will serve as an useful reference and guide for silicon heterostructure device research and development and stimulate further work in this exciting and expanding field.

We would like to thank Dr. Robin Mellors-Bourne, Director of Publishing, and Dr. Roland Harwood, Commissioning Editor, for their support for this project. We also thank Ms. Diana Levy, Production Editor. It was due to her skill and efforts that the project could be completed in a relatively short time. The help of the production department in proof-reading the manuscript and indicating the deficiencies in many places is gratefully acknowledged. We are also thankful to Mr. Jonathan Simpson and Mr. Eric Willner of the IEE who were involved in the preliminary stages of the book.

Many people have contributed directly or indirectly towards this book and it would be impossible to find space to thank them all. Nevertheless, we would like to extend our special thanks to Dr. Deepak K. Nayak and Dr. Kalyan Mondal who have made particularly large contributions. A debt of gratitude is also owed to our past and present research students, who have greatly contributed to our work.

Finally, we must thank sincerely our families for their support and help during the preparation of this book.

C K Maiti
N B Chakrabarti
S K Ray
November 17 2000

Chapter 1

Introduction

The first transistor invented used elemental germanium as the semiconducting material. Silicon because of its inherent advantages, such as thermal stability, abundance and availability of a good oxide, soon replaced Ge as the substrate material. The transistors in the early phase of semiconductor device development were bipolar junction transistors (BJTs). Junction field effect transistors (FETs) were fabricated soon after, followed by metal oxide semiconductor field effect transistors (MOSFETs) in the late 1950s. Alloys of silicon and germanium were occasionally studied principally for optical applications. Silicon carbide has been used as a resistive element and sometimes utilized in transistors. A very different course is evident in electron devices made of III–V semiconductors. To realize a semiconductor out of metals and semimetals, one must use a compound. The first binary III–V semiconductor developed for device use was GaAs. GaAs diodes found applications as varactors, Gunn diodes and impact ionization avalanche transit time diodes (IMPATTs). The GaAs metal semiconductor field effect transistor (MESFET) was the first III–V compound semiconductor transistor. Meanwhile ternary compounds were developed and heterostructure diodes were fabricated which found a phenomenal success in diode lasers.

The concept of heterojunction devices is as old as the transistor itself which appeared in the late 1940s. William Shockley suggested in the patent granted to him in 1951 that improved unidirectional charge carrier injection could be obtained in bipolar devices by using a wide bandgap emitter [1]. The basic theory of operation of heterojunction transistors was later described in a classic paper by H. Kroemer during the 1950s [2]. Bandgap engineering with semiconductors of different bandgaps allows an additional degree of freedom

1

in the design of semiconductor devices, extending the device performance limits. However, fabrication techniques were not available for the realization of heterojunction transistors until 1969 when the first heterojunction transistors were manufactured using GaAs [3].

In semiconductor technology, heterostructures generally refer to a combination of two lattice matched semiconductors with different bandgaps. Heterostructures can cause changes in bandgap, effective mass, mobility and optical properties of the semiconductors. Lattice mismatched semiconductors can also be joined without defects at the interface if the mismatch between the two lattices is within a few percent and the epitaxial growth technique used can deposit very thin layers of the second semiconductor on the first.

Electron devices employing heterojunctions were first demonstrated in III–V material systems such as AlGaAs–GaAs or InGaAs–InP which are lattice matched. Epitaxial techniques such as liquid phase epitaxy, molecular beam epitaxy (MBE), or metal organic chemical vapor deposition have been developed to produce high quality interfaces at the heterojunctions [4]. Traditionally, compound semiconductors based on group III–V elements of the periodic table, like GaAs and InP, have utilized bandgap engineering to fabricate novel and high speed semiconductor devices such as heterostructure bipolar transistors (HBTs), modulation doped field effect transistors (MODFETs) and quantum well lasers. However, the cost and difficulty in depositing and processing III–V compound semiconductors have limited the use to specialized high speed and optical applications.

Different configurations of heterojunction bipolar transistors include a high bandgap emitter, both high bandgap emitter and collector, a high bandgap emitter with graded base, and a graded emitter. For a high bandgap emitter, the characteristic potential spike present between the emitter and base enhances the emitter injection efficiency and allows the designer to dope the base region more heavily to reduce overall base resistance. High frequency performance is improved since the cutoff frequency f_T is largely determined by the base transit time.

In the configuration where both emitter and collector are made of high bandgap materials, the injection of holes from the heavily doped base to the lightly doped collector is reduced and also lower collector doping results in a reduced collector junction capacitance. In this configuration, the emitter and collector are interchangeable, which simplifies the construction of logic circuits.

Silicon integrated circuits presently dominate the semiconductor industry. The two important devices used in Si technology are field effect and bipolar junction transistors. For digital circuit applications, complementary metal oxide semiconductor (CMOS) technology dominates because of its low power dissipation and high density of integration. CMOS has been the work horse for microprocessors and static random access memories. Bipolar transistors with their high transconductance have mostly been used in analog applications. The main drawback in bipolar digital circuits is the high power consumption. To improve single chip functionality, bipolar complementary metal oxide semiconductor (BiCMOS) processes have been developed to combine the advantages of CMOS and bipolar devices [5].

In conventional Si technology, great emphasis is now placed on decreasing the lateral and vertical dimensions of the individual transistors to reduce parasitic capacitances and resistances due to the "extrinsic" device structure. Advanced bipolar devices, with a very thin base achieved by ion implantation, are fabricated in double polysilicon, self-aligned processes using deep trench isolation. In an effort to further enhance the limits of Si technology, it now seems feasible to incorporate epitaxially grown materials which form a heterojunction with Si to improve device performance over homojunction devices.

Silicon, carbon, germanium and α-Sn are group-IV elements in the periodic table. As Si and Ge are completely miscible, $Si_{1-x}Ge_x$ films can be formed without many of the deposition and processing problems associated with compound semiconductors. As $Si_{1-x}Ge_x$ is grown on a Si substrate, existing Si processing equipment can be used and the mechanical and thermal stability of the Si substrate is retained. Although Si and Ge are chemically compatible, a $\sim 4.17\%$ difference exists between the lattice constants of the two elements. In order for a defect-free interface to be grown, the larger lattice spacing of Ge must compress in order to be accommodated in the smaller lattice constant of Si. The process by which this is accomplished is called "strained layer epitaxy". This strain can be used to adjust the energy-band structure, charge carrier transport and optical properties.

Early work on unstrained (relaxed) SiGe layers was done by several workers in the 1950s and a large part of the modern strained layer epitaxy is based on the results of these early measurements [6]. Carrier mobilities were measured by Levitas [7], Glicksman [8], and Busch and Vogt [9]. Braunstein *et al.* [10] made extensive measurements on fundamental optical absorption in the alloys while Dismukes *et al.* [11] made measurements on lattice constants.

Kasper and Herzog reported the growth of strained SiGe films in 1977 [12]. SiGe research may be considered to have started in the early 1980s at AT&T Bell Laboratories, IBM and Daimler-Benz (now Daimler-Chrysler) research laboratories. Initially, most SiGe alloys were grown using MBE that provided better performance and optical properties than single crystalline Si. The most significant step towards the commercialization of SiGe appeared in 1986 with the development of a new epitaxial growth process, known as ultrahigh vacuum chemical vapor deposition (UHVCVD), which allows deposition of an ultrathin, defect-free SiGe interface and could be more readily extended to industry. The first report on SiGe-HBTs appeared in 1987.

Recent progress in the heteroepitaxial deposition of group-IV alloys, specifically $Si_{1-x}Ge_x$ and its variants on Si, has led to the application of bandgap engineering techniques to silicon-based heterostructure bipolar and field effect devices. Currently, the majority of the SiGe hetero-bipolar devices are basically a Si-BJT with Ge selectively introduced in the base region. This means that existing Si fabrication processes can be easily modified to produce SiGe-HBTs. The only additional process is depositing Ge in the base region. However, stability is a major concern for the epitaxially grown SiGe layers as one cannot exceed the critical layer thickness for a given Ge mole fraction. Exceeding this limit leads to the formation of defects that can result in an unstable or metastable film and ultimately render the device useless.

The basic advantage of SiGe-HBTs lies in the ability to engineer the bandgap of Si and increase the transistor's inherent switching speed. Conventional Si has a fixed bandgap of 1.12 eV. By grading the Ge content across the transistor base region, however, the bandgap can be significantly and selectively reduced. Figure 1.1 shows the effects of this graded Ge profile by comparing: (i) the engineered bandgap profile that results in a ramped potential surface and (ii) the linear grading of Ge atoms across the base region to produce the engineered bandgap.

The addition of Ge to the base decreases the base bandgap and introduces a valence band discontinuity between the emitter and base. The decreased base bandgap leads to improved electron injection from emitter to base, an exponential increase in collector current and a much higher current gain for a given base doping level. Further, if a graded Ge profile is used in the base, the grading induces electric fields that reduce base transit time, improving the cutoff frequency. Another advantage of the grading in a SiGe-HBT is higher Early voltage and higher output resistance than in a Si-BJT [13]. Furthermore, in SiGe-HBTs, current gain, cutoff frequency and maximum oscillation frequency, f_{max} all increase at low temperature [14, 15]. This makes the SiGe-HBT a prime candidate for cryogenic applications which are usually reserved for Si CMOS and III–V compound semiconductors.

Two types of SiGe-HBTs with different Ge profiles in the base have been demonstrated: a graded-base (IBM) [16] and a true box-profile HBT (Daimler-Benz) [17]. In both processes, the addition of Ge in the base allows for a reduced base transit time for a given base sheet resistance, thus providing a

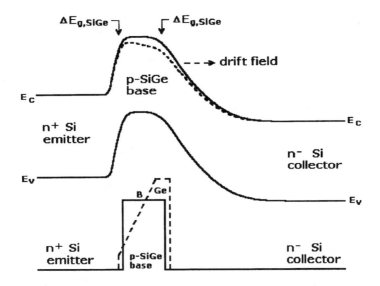

Figure 1.1.   (a) Band diagram of a SiGe-HBT and (b) graded Ge profile. Source: IBM website.

device with simultaneously high f_T and f_{max}, as shown in figure 1.2. The higher base doping concentration gives rise to a high Early voltage (due to less modulation of the space charge region into the neutral base) and a low noise figure (due to the low base resistance), which translate into performance advantages for rf applications.

In the last decade, the SiGe-BiCMOS technology has developed (see table 1.1) from a laboratory curiosity to a viable manufacturing technology (see figure 1.3) that replaces and extends the performance of silicon-based BiCMOS technology [18]. The impetus for this development has been due to the recent explosion in information technology requiring a very large bandwidth in network communication at speeds up to 40 Gbit/s and the rapid growth of global cellular and wireless applications (see figure 1.4).

The performance advantages of SiGe-HBTs over Si-BJTs for radio frequency integrated circuit (RFIC) applications are mainly the extremely high cutoff frequencies with reported f_T of 130 GHz at room temperature and 213 GHz at 77 K [19, 20] and f_{max} of 160 GHz [21], emitter coupled logic (ECL) gate delay of 6.7 ps [22] and a current mode logic gate delay down to 7.7 ps [23]. Figure 1.5 shows the lowest ECL gate delay measured using ring

Table 1.1. SiGe-HBTs: laboratory research to production. After S. Subbanna et al., IEEE ISSCC Tech. Dig., 66–67 (1999).

1982	UHVCVD growth technique developed
1986	UHVCVD epitaxial-base Si transistors fabricated
1987	First SiGe base HBTs fabricated using MBE
1988	Graded-base SiGe poly-emitter HBTs demonstrated – based on double-poly bipolar transistor process
1993	Analog LSI circuit (> 1 Gs/s, 12-bit DAC) demonstrated – based on CMOS-like base processing
1993	SiGe-HBTs with f_T > 100 GHz fabricated
1994	SiGe BiCMOS process in 200 mm Si-CMOS fabrication – aligned processing with CMOS base
1996	RF technology with passive components, models, design kit
1997	Full reliability – manufacturing qualification of BiCMOS rf technology
1998	Commercial products announced by IBM and Temic

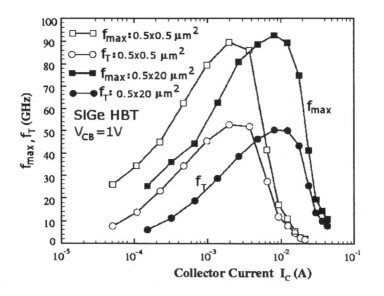

Figure 1.2. f_T and f_{max} of SiGe-HBTs. After S. P. Voinigescu et al., IEEE RFIC Symp. Dig., 131–134 (1999), copyright ©1999 IEEE.

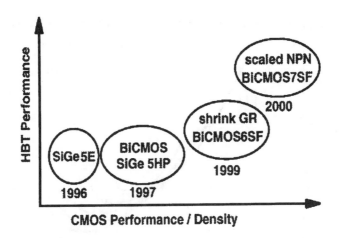

Figure 1.3. Performance evolution of SiGe technologies from bipolar field effect transistor 5E to 5, 6 and 7 generations of SiGe-BiCMOS. After S. Subbanna et al., IEEE IEDM Tech. Dig., 845–848 (1999), copyright ©1999 IEEE.

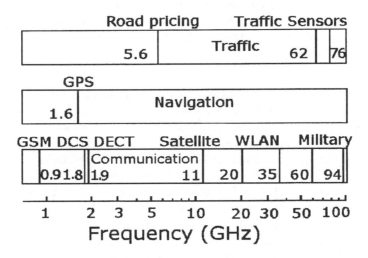

Figure 1.4. Selected high frequency applications and allocated frequency bands between 1 and 100 GHz. The three market segments labeled communication, traffic and navigation are expected to expand enormously in the next few years, mainly in the range up to about 10 GHz. After F. Schaffler, Thin Solid Films, **321**, 1–10 (1998).

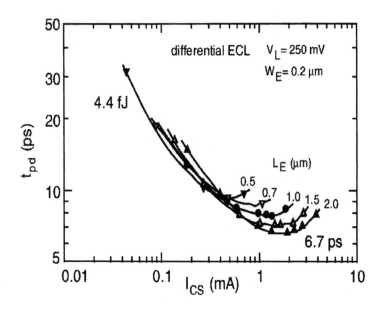

Figure 1.5. ECL gate delay measured using ring oscillators in SiGe-HBTs with an emitter area of 0.2 μm by 0.5, 0.7, 1, 1.5 and 2 μm at a single-ended logic swing voltage of 250 mV. After K. Washio et al., IEEE IEDM Tech. Dig., 557–560 (1999), copyright © 1999 IEEE.

oscillators using SiGe-HBTs having various emitter areas for a single-ended logic swing voltage of 250 mV. Noise measurements on SiGe-HBTs show that the low frequency ($1/f$) noise properties of SiGe devices are comparable to similar Si devices and are superior to both GaAs MESFETs and Si MOSFETs [24]. Since the base resistance can be made considerably lower for a SiGe-HBT compared to Si, one would expect that the high frequency thermal noise will also be reduced. Radiation hardness studies suggest that the SiGe-HBT is more resistant to certain types of radiation than its all-Si counterpart [25].

SiGe technology not only promises to significantly increase performance and lower power dissipation, but can also support the integration of radio frequency functions with digital circuits [26]. Currently, Si bipolar technologies are mostly used in analog and rf amplifier applications to about 3 GHz and in microwave oscillators up to approximately 20 GHz. Given the superior performance of the III–V compounds and the theoretical 50 GHz threshold for Si bipolar technology, one might expect Si to stay out of the high frequency microwave arena. However, SiGe-HBT technology, promising to multiply performance thresholds by two or three times that attainable with Si, is changing the view. Using bandgap engineering, SiGe may reach III–V type performance enhancements in high frequency integrated circuits, possibly up to 150 GHz or higher in Si technology (see table 1.2).

Table 1.2. Comparison of IBM SiGe-BiCMOS HBT with competing production technologies and products for low noise amplifier circuits. The IBM SiGe process offers a much lower cost than GaAs. After N. King and A. Victor, IBM Microelectronics, First Quarter, 5–8 (1999).

Parameter	Si bipolar (BiCMOS)	GaAs MESFET	GaAs HBT	GaAs HEMT	IBM SiGe BiCMOS
Gain (dB)	12	10	11	12	12
Noise figure (dB)	1.6	1.5	1.5	0.8	1.2
Out IP3 (dBm)	18	20	21	22	25
DC operating current at 2.7 V	15	15	10	20	6
Single positive supply voltage	yes	no	yes	no	yes
Integration	unlimited	rf/VHF	rf/VHF	rf	unlimited

As can be seen from the comparison shown in table 1.2, the SiGe-HBTs provide equal or better performance for low noise amplifier applications and SiGe-BiCMOS technologies allow for high levels of integration with a cost structure similar to standard BiCMOS Si. When used for wireless applications the SiGe process provides high performance HBT devices for almost 50% less power and at a much lower cost than GaAs [27]. Low noise amplifier circuits show performance that rivals products designed with GaAs, but with the features and capabilities of Si for less dc power. Figure 1.6 shows the breakdown voltage vs. peak cutoff frequency of various hetero-bipolar devices. As indicated in the figure the breakdown voltage for GaAs based devices is higher than that of SiGe, making SiGe devices attractive for low power high frequency applications. Furthermore, improved power added efficiency (PAE) at low dc voltages makes HBTs a good choice for power amplifier applications. Similar performance advantages have been demonstrated for personal digital communication, code division multiple access driver amplifiers and for oscillators, including voltage controlled oscillator (VCO) circuits [28]. Figure 1.7 shows the photograph of a die of Temic Semiconductors SiGe-based front-end integrated circuit (IC).

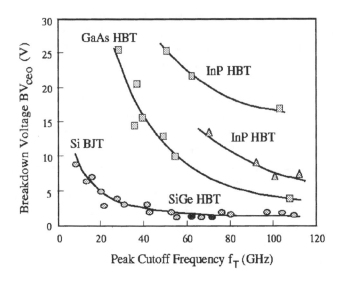

Figure 1.6. Breakdown voltage vs. peak cutoff frequency of various heterojunction bipolar transistors. Source: SiGeCom website.

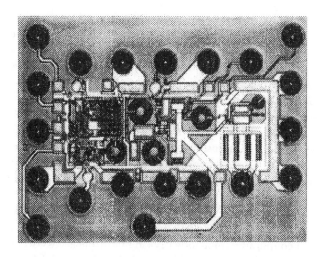

Figure 1.7. Die photograph of a SiGe front-end IC. After M. Bopp et al., IEEE ISSCC Conf. Tech. Dig., 68–69 (1999), copyright ©1999 IEEE.

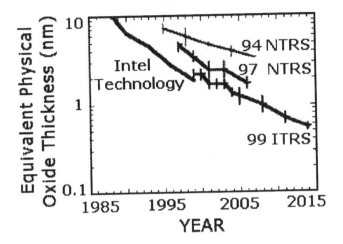

Figure 1.8. Trend of Si MOSFET oxide thickness in both research and production. ITRS roadmap predictions and Intel logic technology data for gate oxide thickness vs. time. After J. D. Plummer, IEEE DRC Dig., 3-6 (2000), copyright ©2000 IEEE.

Scaling trends in conventional Si-CMOS technology have seen an aggressive reduction in feature size, down to the sub-0.1 μm regime for faster devices with greater packing density, low power dissipation and reduced cost. Individual working transistors with gate lengths of 400–500 Å have already been demonstrated [29, 30]. The trend of Si-MOSFET gate oxide thickness in both research and production is shown in figure 1.8. The figure shows the International Technology Roadmap for Semiconductors (ITRS of Semiconductor Industry Association, SIA) roadmap of CMOS technology, predicting that the minimum oxide thickness will be around 1–1.5 nm. It is also predicted that minimum feature size will approach 10 nm by 2024, and for the most aggressively scaled dynamic random access memory (DRAM) the scale of integration will reach 64 Gbits in 2010. However, a number of device and technology issues, which might ultimately determine the limit of such scaling trends, have to be resolved [31].

The hole mobility in Si is much lower than that of an electron. In order to match the current drives of n-MOS and p-MOS in a CMOS technology, the area of the p-MOS device has to be about three times larger than its counterpart. This adversely affects the level of integration and device speed. Therefore, there has been an increased interest in band engineered field effect devices.

Si/SiGe and strained Si MOSFETs have been demonstrated to have better performance than conventional Si-MOSFETs [32–36]. Numerical simulations show that Si/SiGe MOSFETs continue to offer enhanced performance in the deep submicron regime as well [35, 37–39]. Velocity overshoot at the source end of a SiGe channel heterostructure field effect transistor (HFET) as compared to bulk Si channel MOSFETs has also been reported [37]. With such promises of high performance in the deep submicron regime, SiGe devices can be viewed as an alternative and supplement to scaling of Si-CMOS.

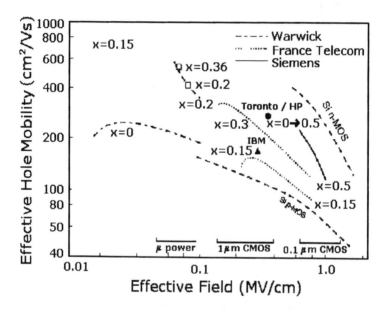

Figure 1.9. Reported experimental data on 300 K effective hole mobilities obtained in pseudomorphic Si/Si$_{1-x}$Ge$_x$/Si structures (see text) plotted against effective field (E$_{eff}$). The bars indicate the range of E$_{eff}$ values present in micropower, 1 and 0.1 μm CMOS technologies. After E. H. C. Parker and T. E. Whall, Solid-State Electron., 43, 1497–1506 (1999).

Figure 1.9 shows effective hole mobilities as a function of effective (vertical) field for fully pseudomorphic structures of nominally uniform alloy composition (except the Toronto/HP sample) produced by epitaxial growth, compared to mobilities observed at the unstrained Si/SiO$_2$ interface. Clearly, enhancements are observed through the use of SiGe. A significant improvement in mobility by a factor of two is seen in the Siemens device at high E$_{eff}$ values for $x = 0.5$.

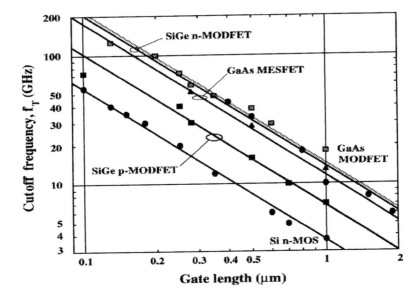

Figure 1.10. The cutoff frequency and delay time as a function of transistor gate length for SiGe-MODFETs compared to other technologies. After D. J. Paul, Advanced Materials, 11, 191–204 (1999).

Figure 1.10 shows the cutoff frequency as a function of gate length for a number of different transistor families [40–42]. These include n-MOS in conventional Si, p- and n-MODFETs in SiGe and MODFETs and MESFETs in GaAs. The plot shows only the results for $Si_{1-x}Ge_x$ devices with f_{max} comparable to or higher than f_T. The lines correspond to theoretical scaling of the device geometry and for small gate lengths the experimental points lie below these curves due to increased parasitics. The $Si_{1-x}Ge_x$ n-MODFET has a switching time comparable to n-MODFETs in GaAs/AlGaAs and better than GaAs MESFETs. The $Si_{1-x}Ge_x$ p-MODFET is even more impressive as the devices are faster than any other p-channel transistors reported [43].

Figure 1.11 shows a performance summary of some of the state-of-the-art devices compared to conventional technologies. As lower power solutions become available, the markets in the future are predicted to develop towards battery operated products. Battery powered portable electronic solutions are expected to become a major part of the total semiconductor market. The SiGe-HBT is already in commercial production and strained SiGe CMOS have a great potential to at least provide one extra generation of CMOS production.

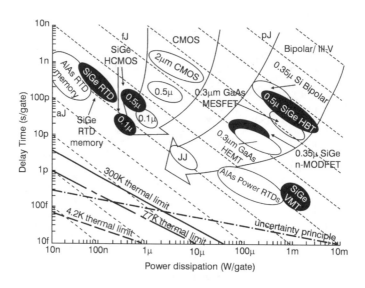

*Figure 1.11. Power delay products for different families of transistors. After D. J. Paul, Advanced Materials, **11**, 191–204 (1999).*

However, a low defect density virtual substrate or bulk $Si_{1-x}Ge_x$ substrate is needed for the implementation [43].

Increase of Ge content in the SiGe layer leads to a very thin film (limited by critical thickness criteria), large strain and reduced thermal stability of the layers, thereby limiting the application of binary alloys. Incorporation of carbon in a silicon–germanium alloy gives an additional degree of freedom in device design. By incorporating smaller-sized C atoms substitutionally in SiGe layers to form a ternary $Si_{1-x-y}Ge_xC_y$, the strain can be compensated, extending the SiGe-based heterostructures to allow more flexible device design. This new material can alleviate some of the constraints on strained $Si_{1-x}Ge_x$ and may help to open up new fields of device applications for heteroepitaxial Si-based systems [44–47]. When compared with SiGe technologies, the addition of carbon offers a significantly greater flexibility in process design and a greater latitude in processing margins.

An application of C-containing Si and SiGe layers is to increase the thickness, stability and Ge content of $Si_{1-x}Ge_x$ for p-channel FETs and npn-HBTs. One can also use strained $Si_{1-y}C_y$ on Si instead of strained Si on a relaxed buffer for n-channel FETs and pnp-HBTs. Design of new buffer layers with $Si_{1-x-y}Ge_xC_y$ is needed for virtual substrates for hetero-FETs. These help achieve increased performance and process margins for HBTs, suppress transient enhanced diffusion of boron and reduce undoped SiGe spacer layer thickness. Another application is in strain symmetrization for superlattices on Si(001) for optical applications.

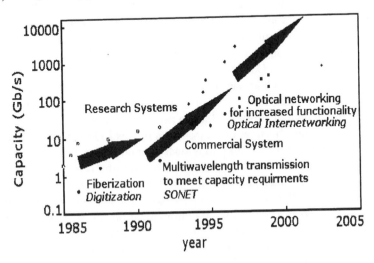

*Figure 1.12. Chronology on the progress of single fiber capacity from research (•) and commercial systems. After L. Lunardi, Solid-State Electron., **43**, 1627–1632 (1999).*

The last few years have seen the evolution of large capacity optical networks, due to the ever increasing demand for more bandwidth. Ultrahigh speed monolithic ICs will be the key components in future optical fiber communication systems that operate at over 10 Gb/s. The progress reported in the last few years on transmission systems beyond 10 Gb/s capacity is shown in figure 1.12. A SiGe base bipolar transistor is a very attractive candidate because of its low ECL gate delay (less than 7 ps) and a cutoff frequency of over 100 GHz [22].

In the effort of fabricating Si based optoelectronic devices, the inherent disadvantage of the indirect bandgap of Si can in some respects be overcome by the introduction of a SiGe heterosystem which allows application-specific tailoring of the electronic and optical material properties. The evolution of Si alloy based heterostructures over the last decade has led to dramatic progress in the capability of infrared ($> 1.2 \ \mu$m) light to be waveguided, detected, emitted, modulated and switched in Si. Light emission from nanocrystallite films of group-IV alloys has been observed utilizing low dimensional quantum confined electronic states. Much of this effort was inspired by the discovery of visible room temperature photoluminescence in porous Si. Since then many other methods, like folded Brillouin zone in strained layer superlattices, nanocrystals embedded in an oxide matrix and implanted defect induced transitions have been investigated for light emission of group-IV alloys.

An important advantage of using SiGe absorbing material is tunability over the 1.1–1.6 μm range. Enhancement of absorption by using strained layer multiple quantum wells (MQWs) and optical cavities has been demonstrated. Perhaps the greatest advantage is compatibility with Si technology from the point of view of integration. Several Si/SiGe optical components such as detectors grown on a Si substrate for the near (~ 1.3 μm) and the mid-infrared (3–5 μm) region have been reported (see for a review Reference 48). In addition other passive optical device functions such as modulators and interferometers with SiGe waveguides have been realized on Si substrates which opens the way for a monolithic integration of Si based optical devices integrated with Si electronic circuits. These devices are envisaged for different application fields such as inter- and intrachip optical communication on Si IC chips and high efficiency thermal imaging detectors for different applications.

The main technological steps for fabrication of silicon ICs are well known [4]. These include the growth of a substrate and buffer layers, epitaxy for the active regions, doping of the active layers, formation of buried layers and wells, formation of conducting layers to access the internal terminals (base and collector in bipolar and source and drain in FET), deposition/growth of dielectrics for isolation of individual device units, gate insulator for MOS, passivation, lithography, deposition of gate materials for MOS, formation of contact regions, metallization for interconnection. Fortunately it is possible to carry out the processes using Si, polysilicon, silicon oxide/nitride and silicides. In modern IC technology both vertical and lateral dimensions have become very small. The base in bipolars has a typical thickness less than 500 Å and the active layer in MOSFETs is less than 200 Å. The supply layer in a modulation doped FET is a few monolayers and the oxide thickness of the MOS gate is typically 100 Å. The precision and control necessary are therefore very high.

It is important to retain the integrity of the carefully engineered structures during subsequent thermal steps. This requires processes with a small value of the product of the diffusion constant of the materials and the time of processing which determines the spread by diffusion. The problem of thermal redistribution is more acute in Si/SiGe heterostructures. Thermal processes modify the alloy composition profile and the dopant profile. A consequence of this redistribution is to change the character and position of the silicon/alloy interface.

Attention must also be called to the special issues in heterostructure devices and their evaluation. Heterostructure MOSFETs utilize a buried channel defined by quantum wells. But surface channels are often found to coexist. This makes it difficult to extract the device parameters and evaluate intrinsic performance. Intrinsic device parameters used for MOS evaluation are the saturation and subthreshold behavior, low field mobility, source end velocity and low and high V_{DS} transconductance. The measurement of intrinsic mobility and transconductance requires an accurate estimation of the source and drain resistance. Extrinsic parameters are more dependable as evaluators. It is perhaps more justified to assess the claims of superior performance on the basis of indices like inverter delay and transistor cutoff frequency, noting, however, that these depend not only on device transconductance, transit time and capacitances but also on inseparable parasitics.

The above brief survey has shown that group-IV heterostructures composed of compounds of silicon, germanium and carbon promise enhancement of performance over bulk Si devices and also new applications. The principal problem encountered in realizing these potentials concerns the quality of the films which must often be crystalline to be grown. Fabrication of high quality films is recognized to be the most important step for silicon-based heterostructure devices to ensure acceptable device performance. Different techniques developed over the last two decades for growing $Si_{1-x}Ge_x$ and C-containing SiGe films are discussed in Chapter 2. The other limitation is the substantial lattice mismatch between the group-IV elements and therefore the thin films required are necessarily strained. The strain in the film limits the permissible thickness of the strained film and at the same time causes important modifications in the band and transport properties. Chapter 3 provides the background for understanding the limitations mentioned above and the material and electronic properties and their directionality.

For a device to provide satisfactory performance it is not enough that the active layers are of high quality. One must ensure that all the interfaces are defect free, all the contacts for accessing and controlling the channels, the gate insulator in MOSFET and all the passivating insulators have high quality. Even with the long experience of Si IC technology, the problems of growth of the gate insulator for heterostructure devices compatible with Si IC technology are not trivial and require careful detailed investigation. These are discussed in Chapter 4.

Silicon transistors are commonly classified as bipolar and unipolar. Devices widely used in Si ICs are surface p- and n-channel MOSFETs with current flow parallel to the interface, and npn-BJTs with current flowing normal to the interface and lateral pnp structures. These forms can be readily modified by exchanging active surface Si and base regions by silicon–germanium alloy. Many modifications of these basic structures have been developed over last fifteen years for obtaining enhanced performance. To a different class belong Si heterostructure MODFETs and bipolar inversion channel field effect transistors (BICFETs) which are based on the availability of heterojunctions. The development of SiGe-HBTs is discussed in Chapter 5.

The performance of silicon-based heterojunction transistors depends on the direction of current flow, the orientation of film growth, and the polarity and magnitude of the strain in the films determining the band splitting and band alignment. One thus has vertical or lateral transistors and active layers under compressive or tensile strain. In the case of MOSFETs, another important consideration is the location of the channel (buried or surface) and its composition (strained Si or SiGe). A strained SiGe channel suffers from the handicap of alloy scattering. Advanced silicon-based MODFETs make use of modulation doping for reducing impurity scattering. One thus notices that there are wide classes of heterostructure devices. The principles and performance of SiGe channel HFET devices using relatively simple technology are discussed in Chapter 6.

Heterostructure devices may be built using crystalline, polycrystalline or amorphous alloys. Though most such devices employ crystalline material and utilize strain in pseudomorphic structures, amorphous and polycrystalline IV–IV alloys have numerous low-cost applications and are useful when the thickness of the alloy layer must exceed several microns. Amorphous silicon–germanium is a useful component in efficient tandem solar cells because of a better utilization of the solar spectrum by the alloy. Another application is the use of SiGeC alloys in low-cost avalanche photodiodes (APDs) with responsivity tuned to the visible light emitting diodes (LEDs). Applications of poly-SiGe in thin film transistors and as gate material in p-MOSFETs are also documented in Chapter 6.

Unconventional structures such as BICFETs combining bipolar and FET operation and resonant tunneling devices are treated in Chapter 7. It is well known that the best device performance is achievable in MODFETs which locate the high mobility channel as far away as possible from surfaces and interfaces and the ionized impurities. Chapter 8 discusses the physical realization and performance of p- and n-channel MODFETs.

Chapter 9 is concerned with metallization of SiGe alloy layers and fabrication and properties of metal–SiGe junctions. Applications of elemental

Si and Ge as photodiodes, phototransistors and avalanche photodiodes are well established. Charge coupled devices (CCDs) using Si MOSFET technology are the recognized mainstay of infrared imaging. Crystalline and hydrogenated amorphous Si have found wide applications in photovoltaics. Development of group-IV alloys and Si/SiGe heterostructures together with advances in Si ICs has led to extension of the range of wavelength and also applications. Use of multiple quantum wells and microcavities to enhance absorption, new structures of separated carrier generation and multiplication detectors in the wavelength region of interest in fiber optics and metal-semiconductor-metal (MSM) photodiodes are some of the recent developments of great interest.

Integrated optoelectronics used to mean combination of III–V compound semiconductor devices and guiding and modulating structures realized with lithium niobate/tantalate and other electro-optic and acousto-optic dielectrics. Silicon based optoelectronics has recently made inroads in integrated optoelectronics. Waveguides have been realized in Si/SiO$_2$. Si/SiGe and other silicon-based structures and active devices, such as heterojunction photodiodes and phototransistors, have been integrated with waveguides. These developments and their potential in silicon-based optical systems are discussed in Chapter 10.

1.1   References

1 W. Shockley, "US Patent 2569347," 1951.

2 H. Kroemer, "Theory of a wide-gap emitter for transistors," *Proc. IRE*, vol. 45, pp. 1535–1537, 1957.

3 D. K. Jadus and D. L. Feucht, "The Realization of a GaAs-Ge Wide Band Gap Emitter Transistor," *IEEE Trans. Electron Dev.*, vol. ED-16, pp. 102–107, 1969.

4 S. M. Sze, *VLSI Technology*. McGraw Hill Book Co., Singapore, 2nd ed., 1988.

5 N. B. Chakrabarti, *Introduction to Integrated Circuit Logic and Memory*. Oxford and IBH Limited, New Delhi, 1994.

6 S. C. Jain, *Germanium–Silicon Strained Layers and Heterostructures*. Academic Press Inc., New York, 1994.

7 A. Levitas, "Electrical Properties of Germanium–Silicon Alloys," *Phys. Rev.*, vol. 99, pp. 1810–1814, 1955.

8 M. Glicksman, "Mobility of Electrons in Germanium–Silicon Alloys," *Phys. Rev.*, vol. 111, pp. 125–128, 1958.

9 Von G. Busch and O. Vogt, "Electrische Leitfahigkeit und Halleffekt von GeSi Legierungen," *Helv. Phys. Acta*, vol. 33, pp. 437–459, 1960.

10 R. Braunstein, A. R. Moore and F. Herman, "Intrinsic optical absorption in germanium-silicon alloys," *Phys. Rev.*, vol. 109, pp. 695–710, 1958.

11 J. P. Dismukes, L. Ekstrom and H. J. Paff, "Lattice Parameter and Density in Germanium Silicon Alloys," *J. Phys. Chem.*, vol. 68, pp. 3021–3027, 1964.

12 E. Kasper and H. J. Herzog, "Elastic Strain and Misfit Dislocation Density in $Si_{0.92}Ge_{0.08}$ Films on Silicon Substrates," *Thin Solid Films*, vol. 44, pp. 357–370, 1977.

13 E. J. Prinz and J. C. Sturm, "Current gain–Early voltage products in heterojunction bipolar transistors with nonuniform base bandgaps," *IEEE Electron Dev. Lett.*, vol. 12, pp. 691–693, 1991.

14 J. D. Cressler, J. H. Comfort, E. F. Crabbe, G. L. Patton, J. M. C. Stork, J. Y.-C. Sun and B. S. Meyerson, "On the profile design and optimization of epitaxial Si- and SiGe base bipolar technology for 77 K applications - Part I : Transistor DC design considerations," *IEEE Trans. Electron Dev.*, vol. 40, pp. 525–541, 1993.

15 J. D. Cressler, J. H. Comfort, E. F. Crabbe, G. L. Patton, J. M. C. Stork, J. Y.-C. Sun and B. S. Meyerson, "On the profile design and optimization of epitaxial Si- and SiGe base bipolar technology for 77 K applications - Part II: Circuit performance issues," *IEEE Trans. Electron Dev.*, vol. 40, pp. 542–556, 1993.

16 G. L. Patton, J. H. Comfort, B. S. Meyerson, E. F. Crabbe, G. J. Scilla, E. De Fresart, J. M. C. Stork, J. Y.-C. Sun, D. L. Harame and J. N. Burghartz, "75-GHz f_T SiGe base heterojunction bipolar transistors," *IEEE Electron Dev. Lett.*, vol. EDL-11, pp. 171–173, 1990.

17 E. Kasper, A. Gruhle and H. Kibbel, "High Speed SiGe-HBT with Very Low Sheet Resistivity," in *IEEE IEDM Tech. Dig.*, pp. 79–82, 1993.

18 S. Subbanna, G. Freeman, D. Ahlgren, D. Greenberg, D. Harame, J. Dunn, D. Herman, B. Meyerson, Y. Greshishchev, P. Schvan, D.

Thornberry, G. Sakamoto and R. Tayrani, "Integration and Design Issues in Combining Very-High-Speed Silicon–Germanium Bipolar Transistors and ULSI CMOS for System-on-a-Chip Applications," in *IEEE IEDM Tech. Dig.*, pp. 845–848, 1999.

19 K. Oda, E. Ohue, M. Tanabe, H. Shimamoto and K. Washio, "DC and AC Performances in Selectively Grown SiGe-Base HBTs," *IEICE Trans. Electron.*, vol. E82-C, pp. 2013–2020, 1999.

20 N. Zerounian, F. Aniel, R. Adde and A. Gruhle, "SiGe heterojunction bipolar transistor with 213 GHz f_T at 77 K," *Electronics Lett.*, vol. 36, pp. 1076–1078, 2000.

21 A. Schuppen, U. Erben, A. Gruhle, H. Kibbel, H. Schumacher and U. Konig, "Enhanced SiGe Heterojunction Bipolar Transistors with 160 GHz-f_{max}," in *IEEE IEDM Tech. Dig.*, pp. 743–746, 1995.

22 K. Washio, M. Kondo, E. Ohue, K. Oda, R. Hayami, M. Tanabe, H. Shimamoto and T. Harada, "A 0.2-μm Self-Aligned SiGe HBT Featuring 107-GHz f_{max} and 6.7-ps ECL," in *IEEE IEDM Tech. Dig.*, pp. 557–560, 1999.

23 E. Ohue, K. Oda, R. Hayami and K. Washio, "A 7.7-ps CML using selective-epitaxial SiGe HBTs," in *IEEE BCTM Proc.*, pp. 97–100, 1998.

24 L. S. Vempati, J. D. Cressler, J. A. Babcock, R. C. Jaeger and D. Harame, "Low-Frequency Noise in UHV/CVD Epitaxial Si and SiGe Bipolar Transistors," *IEEE J. Solid-State Circuits*, vol. 31, pp. 1458–1467, 1996.

25 J. A. Babcock, J. D. Cressler, L. S. Vempati, S. D. Clark, R. C. Jaeger and D. L. Harame, "Ionizing Radiation Tolerance of High-Performance SiGe HBTs Grown by UHV/CVD," *IEEE Trans. Nuclear Sci.*, vol. 42, pp. 1558–1566, 1995.

26 S. Subbanna, J. Johnson, G. Freeman, R. Volant, R. Groves, D. Herman and B. Meyerson, "Prospects of silicon-germanium-based technology for very high-speed circuits," in *IEEE MTT-S Dig.*, pp. 361–364, 2000.

27 N. King and A. Victor, "Enhanced Wireless Circuit Performance with Silicon Germanium Technology," *IBM MicroNews*, vol. 5, pp. 5–7, 1999.

28 B. S. Meyerson, "Silicon:germanium-based mixed-signal technology for optimization of wired and wireless telecommunications," *IBM J. Res. Develop.*, vol. 44, pp. 391–407, 2000.

29 A. Hori, H. Nakaoka, H. Umimoto, K. Yamashita, M. Takase, N. Shimizu, B. Mizuno and S. Odanaka, "A 0.05 μm-CMOS with ultra shallow source/drain junctions fabricated by 5 keV ion implantation and rapid thermal annealing," in *IEEE IEDM Tech. Dig.*, pp. 485–488, 1994.

30 M. Ono, M. Saito, T. Yoshitomi, C. Fiegna, T. Ohguro and H. Iwai, "A 40 nm gate length n-MOSFET," *IEEE Trans. Electron Dev.*, vol. 42, pp. 1822–1830, 1995.

31 Y. Taur, D. A. Buchanan, C. Wei D. J. Frank, K. E. Ismail, L. Shih-Hsien, G. A. Sai-Halasz,· R. G. Viswanathan, H.-J. C Wann, S. J. Wind and H.-S. Wong, "CMOS scaling into the nanometer regime," *Proc. IEEE*, vol. 85, pp. 486–504, 1997.

32 D. K. Nayak, J. C. S. Woo, J. S. Park, K. L. Wang and K. P. MacWilliams, "Enhancement-Mode Quantum-Well Ge$_x$Si$_{1-x}$ PMOS," *IEEE Electron Dev. Lett.*, vol. EDL-12, pp. 154–156, 1991.

33 P. M. Garone, V. Venkataraman and J. C. Sturm, "Mobility Enhancement and Quantum Mechanical Modeling in Ge$_x$Si$_{1-x}$ Channel MOSFETs From 90 to 300 K," in *IEEE IEDM Tech. Dig.*, pp. 29–32, 1991.

34 K. L. Wang, S. G. Thomas and M. O. Tann, "SiGe Band Engineering for MOS, CMOS and Quantum Effect Devices," *J. Mater. Sci.: Mater. Electron.*, vol. 6, pp. 311–324, 1995.

35 S. Verdonckt-Vandebroek, F. Crabbe, B. S. Meyerson, D. L. Harame, P. J. Restle, J. M. C. Stork and J. B. Johnson, "SiGe-Channel Heterojunction p-MOSFETs," *IEEE Trans. Electron Dev.*, vol. 41, pp. 90–102, 1994.

36 C. K. Maiti, L. K. Bera, S. S. Dey, D. K. Nayak and N. B. Chakrabarti, "Hole mobility enhancement in strained Si p-MOSFETs under high vertical fields," *Solid-State Electron.*, vol. 41, pp. 1863–1869, 1997.

37 A. G. O'Neill and D. A. Antoniadis, "Deep Submicron CMOS Based on Silicon Germanium Technology," *IEEE Trans. Electron Dev.*, vol. 43, pp. 911–918, 1996.

38 G. A. Armstrong and Chinmay K. Maiti, "Strained Si Channel Heterojunction p-MOSFETs," *Solid-State Electron.*, vol. 42, pp. 487–498, 1998.

39 G. Halkias and A. Vegiri, "Device Parameter Optimization of Strained Si Channel SiGe/Si n-MODFETs Using a One-Dimensional Charge Control Model," *IEEE Trans. Electron Dev.*, vol. 45, pp. 2430–2436, 1998.

40 K. Ismail, "Si/SiGe High-Speed Field-Effect Transistors," in *IEEE IEDM Tech. Dig.*, pp. 509–512, 1995.

41 M. Arafa, K. Ismail, J. O. Chu, B. S. Meyerson and I. Adesida, "A 70-GHz f_T low operating bias self-aligned p-type SiGe MODFET," *IEEE Electron Dev. Lett.*, vol. 17, pp. 586–588, 1996.

42 R. Hagelauer, T. Ostermann, U. Konig, M. Gluck and G. Hock, "Performance estimation of Si/SiGe hetero-CMOS circuits," *Electronics Lett.*, vol. 33, pp. 208–210, 1997.

43 D. J. Paul, "Silicon–Germanium Strained Layer Materials in Microelectronics," *Advanced Materials*, vol. 11, pp. 191–204, 1999.

44 H. J. Osten, *Carbon-Containing Layers on Silicon - Growth, Properties and Applications.* Trans-Tech Publications, Switzerland, 1999.

45 A. Gruhle, H. Kibbel and U. Konig, "The reduction of base dopant outdiffusion in SiGe heterojunction bipolar transistors by carbon doping," *Appl. Phys. Lett.*, vol. 75, pp. 1311–1313, 1999.

46 H. J. Osten, D. Knoll, B. Heinemann and P. Schley, "Increasing Process Margin in SiGe Heterojunction Bipolar Technology by Adding Carbon," *IEEE Trans. Elec. Dev.*, vol. 46, pp. 1910–1912, 1999.

47 S. John, S. K. Ray, E. Quinones, S. K. Oswal and S. K. Banerjee, "Heterostructure P-channel metal–oxide–semiconductor transistor utilizing a $Si_{1-x-y}Ge_xC_y$ channel," *Appl. Phys. Lett.*, vol. 74, pp. 847–849, 1999.

48 H. Presting, "Near and mid infrared silicon/germanium based photodetection," *Thin Solid Films*, vol. 321, pp. 186–195, 1998.

Chapter 2

Strained Layer Epitaxy

The advances in crystal growth technologies, such as molecular beam epitaxy, gas-source molecular beam epitaxy (GSMBE), organometallic vapor phase epitaxy (OMVPE) and chemical vapor deposition (CVD) have enabled ultrathin epitaxial semiconductor layers to be routinely grown with both monolayer precision in thickness and composition control to about 1 at.%. Major advances have been achieved in binary SiGe alloys where the built-in strain and the composition of a pseudomorphic $Si_{1-x}Ge_x$ layer on a Si substrate affect the band structure, energy gap as well as the band offset significantly. High performance devices and circuits based on $Si/Si_{1-x}Ge_x$ heterostructures have been demonstrated. There is now a substantial literature on various aspects of $Si_{1-x}Ge_x$ epitaxial growth and devices and a number of review articles treating various aspects [1–4].

Many methods have been used for deposition of epitaxial Si and alloys incorporating Ge, C and Sn on Si substrates. These can be broadly categorized into physical vapor deposition and chemical vapor deposition methods. The main physical vapor deposition method is MBE which is widely used because of its excellent control over thickness and composition of layers. Chemical vapor deposition methods are now available for the growth of very high quality strained layers. Notable among them are: limited reaction processing CVD (LRPCVD) [5], rapid thermal chemical vapor deposition (RTCVD) and low temperature ultra high vacuum chemical vapor deposition (UHVCVD) [6]. Gibbons and his group at Stanford were one of the first to demonstrate high quality $Si_{1-x}Ge_x$ on Si using LRPCVD. The lamp-heated limited reaction processing reactor (LRP) laid the groundwork for other lamp-heated systems at Princeton University and AT&T Bell Laboratories. The UHVCVD reactor pioneered by Meyerson and his coworkers at IBM appeared at nearly the same

24

time as LRPCVD. Combining a standard diffusion furnace with ultra high vacuum, they have made a very significant impact in growing high quality alloy layers at low temperature for the fabrication of SiGe-HBTs. An excellent review of the UHVCVD technique and of the devices fabricated using this method of growth has been published [1]. Other CVD techniques have also been used to grow device quality strained alloy layers. Results of $Si_{1-x}Ge_x$ depositions at atmospheric pressure by ASM, the only commercial entry, and another group at IBM have been published [7–9]. These atmospheric pressure CVD results appear to be very promising for widespread application of $Si_{1-x}Ge_x$ on Si heterostructures in a production environment.

Carbon has a very low bulk solubility in Si (10^{-4} at.% at its melting point) and an even lower solubility in Ge. It is known that the incorporation of elements into Si at concentrations far in excess of the bulk solubility limit is possible by solid phase epitaxy (SPE). SPE thus provides another possible synthesis route for forming metastable $Si_{1-y}Ge_y$ or $Si_{1-x-y}Ge_xC_y$ layers. In this technique, amorphous mixtures of Si/Ge/C obtained mainly by ion implantation are recrystallized with a well defined temperature profile. Good quality epitaxial layers over a relatively large range of compositions can be obtained. However, it is known that carbon inhibits the kinetics of Si SPE such that higher temperatures will be required. SPE based on the preamorphization by ion implantation encounters the problem that implantation profiles are often not sharp enough to guarantee well localized buried layers.

In this chapter we discuss the technology of growth of group-IV alloy films and their characterization. We shall examine the deposition of heteroepitaxial films in greater depth using various reactors. Focus is placed on systems that have successfully demonstrated devices. Atmospheric pressure systems are also covered since they have a very great potential of widespread commercial use. As the reactor configurations differ substantially, the advantages and disadvantages of each system are compared. Wafer cleaning methods, reaction kinetics such as constituent incorporation control, dopant control, and selective deposition are examined. Characterization of strained epitaxial films using Rutherford backscattering spectroscopy analysis (RBS), x-ray photoelectron spectroscopy (XPS), spectroscopic ellipsometry (SE), high resolution x-ray diffractometry (HXRD) and atomic force microscopy (AFM) is discussed.

2.1 Film Deposition

In this section, we discuss the different approaches that have been used for surface preparation for low temperature epitaxy and the conditions that must be satisfied for pseudomorphic growth. The important requirements for high-quality epitaxial growth include: an atomically clean substrate surface, contamination free (ultrahigh vacuum or inert) environment, source purity, a source–substrate geometry that allows uniform deposition, uniform substrate temperature, absence of particulate and undesired impurity atoms, compatibility with analytical tools to allow *in situ* monitoring of growth quality and control of layer composition and thickness.

Perhaps the single most important issue in silicon-based epitaxy is wafer preparation and *in situ* cleaning prior to epitaxial growth. Epitaxial layer growth is degraded by submonolayer surface coverage of oxygen and carbon. Poor surface cleaning results in defects at the epitaxial interface that are independent of the lattice mismatch between Si and the grown film. Improper surface preparation prior to epitaxial growth and/or particulates from the reactor determine the number of threading dislocations in an as-deposited film. To prevent reoxidation or contamination of the cleaned surface, at least the final stages of the cleaning process are preferably done *in situ* under UHV conditions. Because a native oxide forms on Si shortly after exposure to oxygen, even at room temperature, *in situ* cleaning prior to epitaxial growth is needed.

In general, the requirements for cleaning elemental group-IV substrates are more stringent than those for cleaning III–V compound semiconductor substrates. Conventional Si substrate cleaning techniques fall into three general classes: (i) a combination of RCA and Piranha cleans leading to the growth of a relatively volatile surface SiO_x ($x \approx 1$) layer, followed by *in situ* desorption of the volatile oxide at a much lower temperature than for the native SiO_2 oxide, (ii) *in situ* removal of the surface and near-surface region by sputtering with inert (usually Ar) ions, followed by a thermal anneal to remove sputtering damage and (iii) *in situ* back-etching of the substrate under chlorine gas to remove several hundred angstroms of material. Conventional Si homoepitaxial reactors use an *in situ* high temperature hydrogen or hydrogen chloride ambient to ensure that the surface is free of oxide prior to epitaxial growth.

The standard wet cleaning process (RCA) used for many years prior to epitaxial growth consists of the standard clean step 1 (SC1) which is a 10 min treatment in a boiling solution of 1 H_2O_2 : 1 NH_4OH : 5 H_2O followed by standard clean step 2 (SC2) in 1 H_2O_2 : 1 HCl : 5 H_2O solution. The self-passivating oxide film left is stripped by dipping in a dilute HF etch (40:1 H_2O : HF(48%)) and a thorough rinse in deionized water. This leaves some amount of silicon oxide on the surface in addition to other species such as hydrogen, fluorine and hydroxyl groups. The most straightforward way to clean the surface is by thermal desorption of the oxide layer in the deposition chamber. If the oxide is thick enough to be stoichiometric SiO_2 the overall reaction is

$$SiO_2 + Si \rightarrow SiO(g) \qquad (2.1)$$

Alternative techniques for surface preparation that have been developed more recently involve the creation of a hydrogen-terminated surface. This is done by simple dilute HF dip (not followed by a deionized water rinse) immediately prior to loading the wafer into the deposition system. Experimentally, it is found that such a surface can be exposed to the laboratory ambient without significant oxidation over a period of hours. In this case the dangling bonds on the surface are nearly all terminated with hydrogen, which considerably reduces oxidation of the surface. The hydrogen desorbs from the surface at moderate temperatures ($\sim$ 400–500 ^{o}C) and thus growth can commence as soon as the wafer is raised to the deposition temperature.

Data shown in figure 2.1 are useful to place an explicit lower bound on the operating temperature required to maintain an oxide-free Si surface by preventing reoxidation in the presence of residual oxygen and water vapor. In classical Si epitaxy, high temperature ($T \geq 1000$ ^{o}C) prebakes of Si wafers prior to epitaxy served to remove silicon oxides present initially as well as to maintain a clean, bare Si surface in the presence of significant, fixed levels of contaminating species in reactors. Oxygen is somewhat less reactive than water vapor, requiring a relatively low temperature to maintain a clean Si surface. Several approaches to the cleaning problem have been made in the low temperature deposition of $Si_{1-x}Ge_x$ on Si: retaining the high temperature step and using an ultra high vacuum to desorb oxide, using a lamp-heated system to rapidly change from the cleaning temperature to the deposition temperature, using ion bombardment to physically remove the oxide, or using the unique properties of Si wafers after dipping in dilute hydrofluoric acid. ·

We now turn to other possible contaminants such as carbon and oxygen in the growing films. Low bulk concentrations of oxygen, carbon and heavy metals are desired for high minority carrier lifetimes. Carbon can come from various sources. It is present in all but the most carefully controlled chemical cleaning solutions. Even if such control is maintained, an atomically clean surface will immediately react with carbon in ambient air. *Ex situ* cleaned samples are, therefore, generally terminated with the formation of a completely inert chemical oxide that can be readily reduced and desorbed by heating in vacuum. Carbon may also come from the decomposition of oils used in certain types of vacuum pumps, although oil-free pumps and closed-cycle He cryopumps are mostly used in MBE systems. In gas-source MBE or MBE/CVD systems ion pumps are used. However, ion pumps

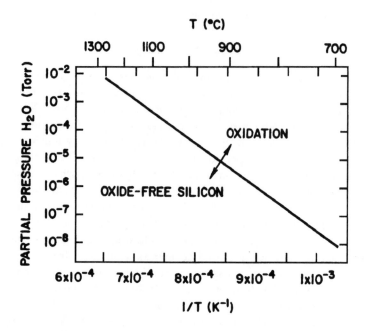

Figure 2.1. Chemical equilibrium data for the maintenance of an oxide-free Si surface in the presence of water vapor. After B. S. Meyerson, Proc. IEEE, 80, 1592–1608 (1992), copyright ©1992 IEEE.

cannot handle significant gas loads, and the accumulation of typical gas source chemicals within cryopumps poses both toxic hazards and the possibility of ignition and/or explosion. Gas-source MBE systems must therefore use nonaccumulating pumps such as turbomolecular pumps backed by roots and rotary. The challenge is thus to develop hydrocarbon trapping techniques and chemically resistant oil-free pumps. Having a low base pressure, however, reduces the oxygen and carbon contamination in the bulk and, if the base pressure is low enough, the formation of a native oxide is also prevented. Using a loadlock during the wafer load and unload is an additional method of keeping the deposition chamber free of oxygen and carbon from the atmosphere. Carbon can be removed by heating the substrate to extremely high temperatures ($\sim$ 1150 ^{o}C) when it dissolves into the Si. It is desirable to prevent carbon from getting on the surface to begin with, and this is largely decided by the cleaning reagent purity.

2.2 Growth Kinetics

In this section, we discuss the present understanding of the surface chemistry of common reactant gases used in low temperature epitaxial growth using the chemical vapor deposition method which is a commercial technique in Si and group-IV alloy technology. The deposition chemistry is simple in solid source MBE since a chemical reaction does not take place. However, complex surface reactions and gas chemistry are involved in both CVD based and gas-source MBE processes. In CVD, however, the growth rate strongly decreases with temperature. A high temperature is required for the precursor gases to decompose, while a low temperature is necessary to prevent dislocation formation and islanding during the growth.

Although most of the early work on SiGe alloy films was carried out by MBE, CVD is more acceptable in production and thus there has been considerable work on the development of CVD processes for group-IV alloy films. An extremely broad range of deposition conditions and reactor designs has been employed. For high temperature (above 1000 °C) and at atmospheric pressure, the deposition/growth rate is determined mainly by mass transport through the gas phase, a process which is not very sensitive to temperature. At lower temperatures ($\sim$ 750–1000 °C) and pressures (0.001–0.1 Torr), the chemical reaction becomes slower and the transport properties of the species in the gas phase change and become a function of complicated diffusion phenomena and chemical reaction rates. In the lower pressure range (mTorr range) the process can become more complicated. A knowledge of growth kinetics and surface reactions becomes very important in low temperature growth. Many reports have emerged describing the deposition kinetics primarily from a phenomenological point of view. In the following, we discuss the general principles of low temperature epitaxial growth and discuss the present understanding of the surface chemistry and growth kinetics of common reactant gases used in low temperature epitaxial growth.

2.2.1 Hydrogen

Hydrogen is generally present in almost all epitaxial processes because it is commonly used as a carrier gas and also the decomposition of many of the precursor gases produces hydrogen. Thus the adsorption and desorption reactions of hydrogen are of considerable importance as these block surface sites and influence the growth rate. Hydrogen desorption from the Si <100> surface has been studied extensively. Early work showed that dissociative adsorption of molecular hydrogen was inefficient. The important features of thermal desorption spectra were reported by Cheng and Yates [10]. Adsorption of atomic hydrogen at 210 K results in a saturation coverage of 1.9 monolayer and formation of a (1×1) surface reconstruction. A thermal desorption spectrum taken from this surface shows three hydrogen desorption peaks that are identified as the β_1, β_2 and β_3 peaks. These are attributed to desorption from monohydride, dihydride, and trihydride species, respectively. However, at 600 K, the saturation coverage is 1.0 monolayer and the (2×1) surface reconstruction is found, with only the monohydride (β_1) peak observed during thermal desorption measurements indicating that in the ideal case there are no reactive surface sites and surface reactions cannot take place.

Thus desorption from the monohydride is of dominant importance in determining growth rates. It might be expected that the surface desorption reaction would be

$$2\mathrm{SiH}* \rightarrow 2\mathrm{Si}* + \mathrm{H}_2 \qquad (2.2)$$

where $*$ represents a surface site and Si$*$ represents the species Si bonded to a surface site. Adsorbed H_2 blocks the surface sites and thus influences the growth rate and the adsorption and desorption reactions of H_2 are of considerable importance. Although the adsorption of molecular hydrogen is weak and is difficult to study, it has been demonstrated that Ge enhances the desorption rate of H_2 considerably [11]. Garone *et al.* [12] have proposed an empirical relation for the hydrogen desorption rate.

$$R_{\mathrm{desorption}}(\mathrm{H}) = \nu N \phi [x e^{-E_{\mathrm{dg}}/kT} + (1-x)e^{-E_{\mathrm{ds}}/kT}] \qquad (2.3)$$

where ν is the frequency factor, N is the number of sites/cm^2, ϕ is the fraction of filled sites, x is the Ge fraction in the film, and E_{dg} and E_{ds} are the activation energies of H_2 desorption from Ge and Si, respectively.

2.2.2   Silane

The film growth kinetics at low temperatures and especially at very low pressures can become quite complicated due to the varying hydrogen surface coverage. Although there is a vast amount of literature and some differences in conclusions, most studies are based on phenomenological observations (H_2 evolution and/or growth rate) and have been interpreted in terms of a hypothetical reaction sequence. Mostly this sequence consists of some adsorption reaction

$$\mathrm{SiH}_4 + * \rightarrow \mathrm{SiH}_4* \qquad (2.4)$$

which is followed by the intermediate formation of several metastable silicon-containing species: SiH, SiH$_2$, SiH$_3$ etc. Finally a deposition reaction takes place resulting in solid Si and molecular H_2 which has to be evaporated:

$$SiH_4* \rightarrow \text{film} + 2H_2 \uparrow \qquad (2.5)$$

This sequence can be further refined by adding various intermediate reaction steps. In order to understand unambiguously the underlying fundamental reaction steps, one needs to identify and quantify the intermediate reaction products upon adsorption of silane onto the Si surface, preferably in conditions of controllable surface coverage, i.e., under ultrahigh vacuum. The decomposition of silane on $< 100 >$–(2×1) and $< 111 >$–(7×7) Si surfaces for a wide range of temperature has been studied in detail by Gates *et al.* [13]. The authors used static secondary ion mass spectroscopy (SSIMS) and thermal desorption spectroscopy to identify the SiH_x species during adsorption of silane. The reactive sticking probability of silane to the surface was measured using temperature programmed desorption (TPD) of H_2 as a function of H coverage. According to these findings, SiH_4 adsorbs dissociatively, consuming two active sites:

$$2SiH_4(g) + 2* \rightarrow 2H* + 2SiH_3* \qquad (2.6)$$

This is followed by a number of steps in which SiH_x fragments are consecutively dehydrogenated until the Si adatom is incorporated into the growing film, releasing the active site and desorbing H_2:

$$2SiH* \rightarrow H_2(g) + \text{film} \qquad (2.7)$$

Adsorption at 150 K showed a sticking coefficient of roughly 10^{-2} and a saturation coverage of roughly 0.2 SiH_4 molecules/surface atom. However, at high temperature a sticking coefficient of $\approx 4 \times 10^{-3}$ was extracted from fitting CVD growth rates [14]. SSIMS showed that the predominant species on the surface were SiH and SiH_3. The elementary mechanism as proposed by Gates *et al.* [13], however, has been contested by Hirose *et al.* [15]. From TPD measurements, the authors concluded that silane adsorbs dissociatively, not into H* and SiH_3*, but into 4 monohydride units.

2.2.3   Silane–Germane

In order to grow SiGe alloy layers, GeH_4 is added to the commonly used Si source SiH_4. Important questions concerning the growth of alloy layers are: (i) the incorporation ratio of Ge, (ii) the influence of GeH_4 on growth rate and the temperature dependence of growth rate and (iii) the influence of GeH_4 on dopant incorporation.

Germane is chemically similar to silane and as a result similar behavior might be expected. It has been reported [16] that germane dissociation on the $Si < 100 >-(2\times1)$ surface proceeds according to the reaction

$$GeH_4 + 2* \rightarrow GeH_3* + H* \qquad (2.8)$$

From growth studies, it has been shown that the sticking coefficient of GeH_4 is lower than that of SiH_4 [11, 17, 18]. The subsequent regeneration of active sites proceeds by desorption of hydrogen as discussed earlier. The Ge incorporation ratio is given by $x_{Ge}/[p_{GeH_4}/(p_{GeH_4} + p_{SiH_4})]$ where x_{Ge} is the atomic fraction of Ge in the layer and p_{GeH_4} and p_{SiH_4} are the respective partial pressures. According to most authors the ratio is 3 and is independent of temperature between 580 and 700 °C [18, 19]. The influence of GeH_4 on the growth rate appears to be rather complicated. At a lower Ge fraction, the growth rate is enhanced considerably, whereas at higher Ge contents it decreases again. The peak of highest growth rate shifts from $x = 0.2$ at 500 °C towards lower x with increasing temperature, vanishing completely at 750 °C [11, 19, 20]. An explanation for this behavior at low x has been suggested by Meyerson *et al.* [11], whereas the phenomenon at high x has been explained by Robbins *et al.* [19]. The rate limiting is due to H_2 desorption, which is assumed to proceed more easily from a Ge site at the surface than from Si, reducing the activation energy at low x. At higher x, though, the surface contains more and more Ge, and therefore less H. So, the rate control shifts to dissociative MH_4 adsorption (where M = Si or Ge).

2.2.4 Dichlorosilane

The use of dichlorosilane has so far been limited to systems operating at or near atmospheric pressure. There is an extensive literature on the modeling of Si epitaxy at high temperature using dichlorosilane where gas phase reactions and boundary layer transport are dominant. Comparatively few attempts have been made to relate the effect of germane on growth kinetics and the underlying surface reactions in the low temperature regime [12].

Dichlorosilane (SiH_2Cl_2) based atmospheric pressure CVD (APCVD) at temperatures below 800 °C has been used for Si epitaxy. Unfortunately, there are no detailed surface science studies of the decomposition of dichlorosilane on Si. However, there has been some work on decomposition of dichlorosilane and other chlorine-containing species on $Si < 111 >-(7\times7)$ [10]. During undoped growth, the most abundant species at the Si surface is Cl, a reaction by-product that has to desorb as HCl (or $SiCl_2$, but this does not contribute to the deposition). Growth rates for dichlorosilane are generally lower than for silane at low temperatures, and this can be explained by the fact that predominantly hydrogen desorption occurs at low temperatures, leaving a surface that becomes increasingly saturated with chlorine [21, 22]. Dichlorosilane has an initial reactive sticking coefficient of approximately 0.36 at low temperatures, decreasing to 0.0025 at 850 K [23]. The decrease in sticking coefficient with temperature suggests adsorption mediated by a precursor state

$$SiH_2Cl_2(g) \Leftrightarrow SiH_2Cl_2(ad) \qquad (2.9)$$

followed by dissociative chemisorption with the overall reaction [23]

$$SiH_2Cl_2(ad) + 4* \rightarrow SiCl* + Cl* + 2H* \qquad (2.10)$$

It has become clear that there are important differences between dichlorosilane and silane with respect to doping and surface preparation. The role of hydrogen is different in the case of growth with a dichlorosilane source. The incorporation of dopants, especially n-type dopants, is very different when dichlorosilane is used. All dopant species explored (diborane, phosphine and arsine) increase the growth rate at low temperatures. A better doping control has been obtained in LRP/RTCVD systems using PH_3 in a SiH_2Cl_2 ambient at temperatures above 800 °C and in APCVD for AsH_3 as well as for PH_3. This is so because Cl desorption can be accelerated due to formation of volatile AsCl or PCl components and the growth rate increases. This reaction assists also in the desorption of As or P making much sharper dopant transitions possible.

In the case of growth from either silane or dichlorosilane, the growth rate is limited at low temperatures by the desorption of hydrogen or HCl from the wafer surface which opens up new active sites. As both desorption processes are thermally activated, the growth rate strongly decreases with decreasing temperature. Ge increases the desorption rate but the growth rate is nevertheless eventually limited by the availability of active sites for adsorption. So the minimum practical growth temperature for CVD will always be greater than that for MBE.

Selective epitaxy is a growth process in which deposition occurs only on the bare Si surface in an oxide-patterned substrate. This is a unique feature of the CVD techniques. It has attracted considerable interest due to its potential application in the self-aligned processing of ultra-high speed devices. Selective epitaxial growth (SEG) in Si has been widely demonstrated in a $SiH_2Cl_2/HCl/H_2$ based chemistry at high temperatures. With the recent advances in low temperature epitaxy, SEG has now been achieved at a growth temperature much below 1000 °C using chlorinated hydrides as well as chlorine-free hydrides such as disilane, silane and germane. Selectivity is affected by the oxide material. Kato *et al.* [24] have observed differences in the Si nucleation rate on the thermally grown oxide and undoped and doped silicate glasses under the same deposition conditions. Doping during the deposition can also influence the selectivity [25].

2.2.5   Ge Incorporation and Abruptness

As the Ge sticking coefficient is unity, the composition of thick layers is determined by the ratio of Ge and Si fluxes. However, when thin layers (tens of angstroms) are grown, surface segregation may cause the Ge profile to depart from that expected as Ge starts to float to the surface by changing sites with surface Si atoms tending towards the energetically favorable sites. Surface segregation is driven by the reduced free energy when Ge is on surface sites. Although Ge surface segregation can be detected by a variety of techniques, SIMS depth profiling is perhaps the best as it is sensitive to small concentrations and provides a large dynamic range of concentration variation.

Figure 2.2 shows SIMS profiles obtained when a sheet of Ge atoms is deposited and then capped with Si [26]. In this case growth was at 450 °C at a rate of 1 Å/s. In SIMS, the slope of the trailing edge of the profile (starting from the surface) is degraded by ion knock-on mixing, so an ideally abrupt profile would have a sharper slope on the leading edge. Instead, a smaller slope is observed on the leading edge. Fukatsu *et al.* [26] have modeled the observed behavior using a two-site model for the segregation process. The behavior of the decay length is illustrated in figure 2.3 using model parameters extracted from the measurements.

At low temperatures the rate of atomic interchange between surface and subsurface sites is reduced, leading to a sharp reduction in the decay length. The decay length is at a maximum for a growth temperature of 400 °C and decreases again at high temperatures as the additional thermal energy increases the probability that Ge will occupy high energy sites (equilibrium segregation). The observed decay lengths are large enough to influence the growth of quantum wells and especially short-period superlattices. Figure 2.3 shows that highly abrupt transitions require growth at very low temperatures, where other issues such as limited epitaxial layer thickness and defects become important. An alternative approach is to use a surfactant at a somewhat higher growth temperature. Segregation of the surfactant species is more energetically favorable than Ge and thus Ge is driven to subsurface sites. Surfactants that have been explored include Bi, Sn, As and Sb, with the best results obtained with Sb [27]. Tsai *et al.* [28] has proposed a model to estimate the interfacial abruptness of the Si/SiGe heterojunction grown by UHVCVD.

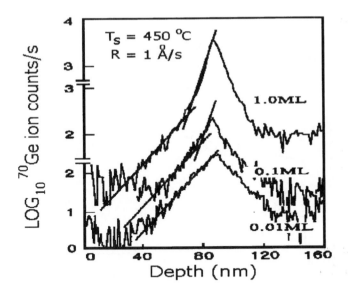

Figure 2.2. SIMS profile showing Ge segregation during MBE growth. A sheet of Ge atoms is capped with Si at 450 °C (growth rate 1 Å/s). The slope of the trailing edge (to the right) should ideally be abrupt but is degraded by ion knock-on mixing. The leading edge slope is even more gradual, indicating that segregation of Ge has occurred. After S. Fukatsu et al., Surf. Sci., 267, 79–82 (1992).

2.2.6   Carbon-Containing Alloys

In the case of MBE, sublimation from a graphite filament is used as the elemental carbon source, whereas gaseous sources for CVD include hydrocarbons and metallo-organics. Binary $Si_{1-y}C_y$ films have been deposited using hydrocarbons and organometallic sources. Propane (C_3H_8) has been used but carbon incorporation efficiency has been found to be very low (about 0.002) [29]. The carbon incorporation efficiency has been found to be insufficient (about 0.025) in the case of growth of ternary SiGeC using ethylene (C_2H_4) [30]. This has been attributed to the fact that hydrocarbons contain many strong C–H and C–C bonds that need high decomposition temperatures, which in turn prevent efficient C incorporation. Regolini *et al.* [31] used organosilicon sources like tetraethylsilane ($Si(CH_3CH_2)_4$) and tetramethylsilane ($Si(CH_3)_4$) and were able to incorporate up to 0.7 at.% of

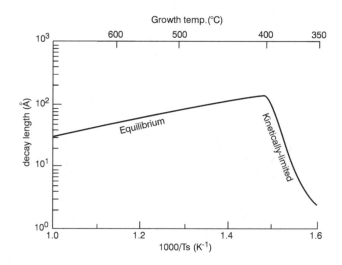

*Figure 2.3. Decay length for Ge segregation as a function of growth temperature (growth rate 1 Å/s). The decay length is calculated using the two-site model with parameters extracted from SIMS measurements. The decay length is less than 10 Å only for very low growth temperatures (< 375 °C). After S. Fukatsu et al., Surf. Sci., **267**, 79–82 (1992).*

C in SiGeC epilayers. Other organosilicon sources such as tetrasilylmethane ($C(SiH_3)_4$) which has no C–C and C–H bonds and methylsilane (SiH_3CH_3) have been used for the deposition of SiGeC layers [32–34].

Methylsilane (SiH_3CH_3) is the simplest organosilicon compound without any C–C bond and it is possible to incorporate C into the lattice without breaking the preformed Si–C bond [29]. Thus it is possible to carry out the epitaxy at lower temperatures than those needed with hydrocarbon sources. In addition, as methylsilane is in the gas phase at room temperature and can be diluted in H_2, it is compatible with the $SiH_4/GeH_4/H_2$ system. Use of silane as the Si source is preferred to dichlorosilane (DCS) for $Si_{1-y}C_y$ growth at low temperature. Typical growths result in a ratio of less than one in five substitutional carbon atoms in the $Si_{1-y}C_y$ layer for samples with carbon concentrations between 0.3 and 2.6 at.%. Limited decomposition of DCS restricts its use to temperatures above 675 °C for Si and $Si_{1-y}C_y$ growth. For simplicity the chemical reactions involving binary $Si_{1-y}C_y$ growth [35] using methylsilane are discussed here. Since Ge does not interfere with C incorporation, the proposition should also be valid for $Si_{1-x-y}Ge_xC_y$ growth.

The decomposition mechanisms generally invoked for explaining Si growth from SiH_4 are

$$SiH_4(g) + 2* \rightarrow H + SiH_3 \tag{2.11}$$

$$SiH_3 + * \rightarrow H + SiH_2 \tag{2.12}$$

$$2SiH_2 \rightarrow H_2(g) + 2SiH \tag{2.13}$$

$$2SiH \rightarrow H_2(g) + 2* + \text{film} \tag{2.14}$$

Available sites (dangling bonds) on the surface are represented by an $*$. They are consumed in reactions (2.11) and (2.12), and regenerated in reaction (2.14). The latter is the rate-limiting step for low temperature growth. The species SiH_2 (silylene) is very reactive. In the SiH_4/SiH_3CH_3 system, the following reaction may occur:

$$SiH_2 + SiH_3CH_3(g) \rightarrow SiH_4(g) + SiCH_4 \tag{2.15}$$

in which molecular SiH_4 is regenerated. $SiCH_4$ is the main radical needed for C incorporation. $SiCH_4$ can diffuse across the surface over H covered sites and, like SiH_2, it can be inserted into the Si–H or C–H surface bond. It is plausible that the growth rate is determined by reactions (2.11)–(2.14), while C incorporation is controlled by reaction (2.15). For the reactions given above, the probability of obtaining C radicals by thermal pyrolysis of SiH_3CH_3 is only small. If SiH_4 is present, the partial pressure p_{SiH_4} is too low or $p_{SiH_3CH_3}$ is too high, the reaction between SiH_2 and SiH_3CH_3 is very fast. C radicals which are produced too rapidly cannot be frozen into substitutional sites and act as contaminants. This results in twinning or even amorphous growth. Lower growth temperature and higher silane partial pressure reduce the mobility of C on the surface during growth, increasing the probability of C incorporation into substitutional sites. By increasing p_{SiH_4}, an equilibrium between the production of C radicals and the incorporation of substitutional C is established. In addition to improving the fraction of C which is substitutional on the lattice, an increase in silane partial pressure also improves the surface morphology for layers with higher carbon contents. Finally, if p_{SiH_4} increases beyond a certain value, three-dimensional and/or polycrystalline growth will occur as in the case of Si deposition from SiH_4.

2.3 Deposition Techniques

In this section we shall examine the deposition of heteroepitaxial films in greater depth using various reactors. Focus is placed on systems that have successfully demonstrated devices. High quality pseudomorphic $Si_{1-x}Ge_x$, $Si_{1-y}C_y$ and $Si_{1-x-y}Ge_xC_y$ alloy layers with a carbon concentration up to 7% have been prepared by MBE [36] and other techniques. MBE is preferred for research work because of its flexibility, but chemical vapor deposition systems are more convenient for production consistency.

2.3.1 Molecular Beam Epitaxy

Molecular beam epitaxy, a physical deposition process, uses an ultrahigh vacuum chamber with a base pressure of about 10^{-10} Torr. Cryopumps provide an oil-free evacuation system. A modern MBE system used for Si heteroepitaxy is shown in figure 2.4. A load-lock is used for wafer loading to prevent exposing the main deposition chamber to the atmosphere. Bean and his coworkers [37] used an argon sputter cleaning cycle to etch 10 nm from the surface of the wafer. The etch was followed by an 850 °C anneal before lowering to the deposition temperature, between 500 and 750 °C. It has been demonstrated that sputter cleaning leads to degradation in the minority carrier lifetime by heavy metal contamination sputtered from the chamber onto the surface of the wafer [38]. Because of the ultra high vacuum conditions, medium temperature ($<$ 850 °C) bakeouts may be sufficient to cause native oxide and other contaminants to desorb from the surface of the wafer [39]. Most MBE systems retain some type of high temperature cleaning or anneal cycle. Fortunately, the resistively heated substrate can be lowered to the deposition temperature without concern about surface recontamination because of the very low partial pressures of oxygen and carbon.

The deposition kinetics are simple in MBE since a chemical reaction does not take place. Deposition occurs when extremely pure elemental sources are heated and the evaporated molecules travel towards the surface of a heated wafer. The heated substrate provides the surface mobility necessary to epitaxially align the impinging molecules. Deposition rate is controlled by the flux of the evaporated molecules and the substrate temperature. The chamber pressure increases to about 5×10^{-8} Torr during depositions so the evaporated species still have a very long mean free path, hence the terminology molecular beam. However, deposition uniformity is a limitation in MBE. Rotating the substrate may circumvent the problem, but large wafers ($>$ 125 mm) may present an insurmountable problem from a uniformity ($\pm$ 5%) standpoint.

Solid-source MBE has two distinct advantages: good control of layer thickness and composition, and growth at low temperature. However, the major disadvantage is the formation of a particulate, defect producing mechanism in the grown films. Gas-source MBE may be considered a

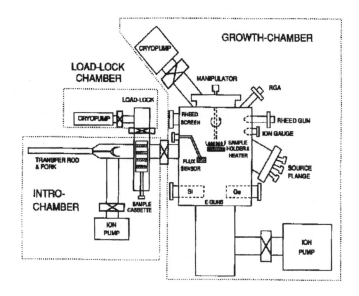

Figure 2.4.   Molecular beam epitaxy system optimized for heteroepitaxy. Wafers up to 150 mm diameter are introduced from the sample cassette holding up to 25 wafers. After S. S. Rhee, Studies of SiGe Tunneling Heterostructures Grown by Si Molecular Beam Epitaxy, *PhD Thesis, Univ. Calif. at Los Angeles, 1991.*

simple derivative of the solid-source MBE. Use of gas sources addresses three weaknesses of conventional MBE. First, the limited capacity of Knudsen evaporation cells requires replenishment of source materials at regular intervals leading to nonstoichiometric growth due to the shift in evaporation rates. Frequent vacuum breaks not only decrease chamber up-time but can result in a significant overall increase of vacuum impurity levels. The use of external gas cylinders in gas-source MBE eliminates this problem.

Gas sources have the second advantage that they overcome difficulties in handling pyrophoric species such as white phosphorus. Phosphine, although highly toxic, is not spontaneously flammable and with proper handling will not accumulate in the MBE system. Lastly, with proper choice of gases (reaction chemistry) and growth temperatures, the reactive species from gas sources may decompose at the heated substrate surface alone. There is little or no wall deposition, and the elimination of such accumulations largely eliminates particulate contamination due to flaking. Gas-source MBE may be described as a hybrid MBE/CVD system, but the deposition pressure is an order of magnitude or more below other CVD systems. At these deposition pressures, gas phase equilibrium may not be achieved, and CVD kinetics may not apply.

As mentioned earlier, in spite of widespread use, MBE suffers from prohibitive initial cost due to ultra high vacuum components and high system down-time. Replenishing the sources requires a system vent followed by a few days of chamber bakeouts to restore the base pressure. Questions still remain about using MBE for VLSI processing probably due to longstanding biases on cost, throughput and wafer size. These concerns may be unwarranted, but for Si alloys MBE appears to remain a tool for research rather than production.

2.3.2   UHVCVD

Ultra high vacuum chemical vapor deposition addresses the batch processing concerns raised by MBE while retaining the excellent control of dopant and compositional profiles The reactor consists of a diffusion furnace under ultra high vacuum, as shown in figure 2.5. Since the base pressure of 10^{-9} Torr is comparable to MBE, the advantages of low contamination and prevention of native oxide after loading are maintained. A load-lock is also used to prevent exposing the deposition chamber to the atmosphere. The control of the wafer temperature in diffusion furnaces is extremely good. As a result, a surface rate-limited reaction results in a very uniform layer under the assumption of little gas depletion.

2.3.3   RTCVD and LRPCVD

Rapid thermal chemical vapor deposition is a processing technique that combines radiant heated lamps and a cold wall CVD chamber. The technique allows one to clean wafers *in situ* at a high temperature and grow epitaxial layers at a lower temperature immediately thereafter. While rapid thermal processing (RTP) generally refers to any process in which the substrate is ramped to a desired temperature in a matter of seconds, limited reaction processing refers to the use of rapid temperature cycles to control thermally driven surface reactions. Both are single wafer processes with the wafer residing on low thermal mass quartz pins inside a quartz deposition chamber. Figure 2.6 illustrates the four steps of a typical single limited reaction process cycle: (i) the wafer is held at a low temperature, T_i, and a desired reactive gas is flowed over the substrate, (ii) using a high-intensity radiant source, the substrate is heated at a rate of the order of 100–400 ^{o}C per second to a temperature, T_f, and a chemical reaction between the substrate surface

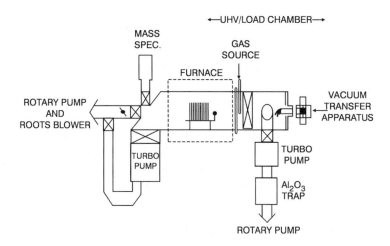

Figure 2.5. Schematic of a UHVCVD system. Base pressure in the UHV segment of the apparatus is 10^{-9} Torr, while operating pressures during film deposition are 10^{-3} Torr. After B. S. Meyerson, Proc. IEEE, **80***, 1592–1608 (1992), copyright ⓒ1992 IEEE.*

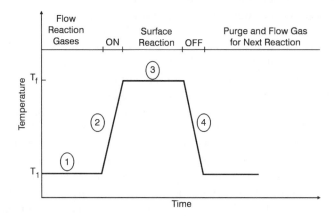

Figure 2.6. Typical temperature and process profile of a limited reaction processing cycle reactor.

and the reactive gas is induced, (iii) the reaction is allowed to proceed at an appropriate temperature for a predetermined, precisely controlled period, and (iv) the reaction is stopped by turning off the radiant source and allowing the wafer to rapidly cool to T_i.

Limited reaction processing CVD for Si homoepitaxy and $Si_{1-x}Ge_x$ heteroepitaxy was initiated by Gibbons and his coworkers at Stanford [40]. The base pressure is about 1 mTorr which is much higher than the systems described earlier. The wafer is lamp-heated, allowing rapid changes in wafer temperature (typically 100 ^oC/s). As a result of the rapid temperature transitions, the high temperature *in situ* cleaning step can occur in a short time, thus reducing the total thermal budget compared to commercial epitaxial deposition systems. Furthermore, a sharp doping profile and compositional transitions are made by using the lamps as a thermal switch to control the reaction. The gas flows are established at low temperature and the lamps are turned on to initiate the deposition. Use of a load-lock and a gas purifier leads to a very low residual oxygen and carbon content in the $Si_{1-x}Ge_x$ films and contamination problems may be avoided [41].

Many other researchers have used similar configurations to the Stanford group with continuous improvements in base pressures. A new name, rapid thermal CVD instead of LRPCVD, has been adopted because of the use of gas switching rather than lamp heating to control the reaction. Lower base pressure (10^{-7} Torr) with greater temperature and gas flow uniformity was examined by Green and his coworkers at AT&T Bell Laboratories [42]. Lamp-heated systems retain the high temperature pre-epitaxial cleans with hydrogen or hydrogen chloride but reduce the total time spent at high temperature. Typical *in situ* cleaning cycles are 1000 ^oC for 15 s in hydrogen. This high temperature hydrogen clean can reduce interfacial doses of both oxygen and carbon to below 5×10^{13} cm^{-3} from 5×10^{17} cm^{-3} at a growth temperature of 750 ^oC [43]. Severe undercutting of the oxide appears when this cleaning method is used on patterned oxide wafers [44]. The process of a low temperature cleaning, using a remote fluorine-based plasma, has as yet only achieved oxygen and carbon doses of 5×10^{14} cm^{-2} and 3×10^{15} cm^{-2}, respectively. It is clear from the above discussions that the characteristics of rapid thermal grown layers are:

(i) abrupt composition profiles: since the substrate is hot only during layer growth, the thermal exposure of the substrate for any growth temperature is inherently minimized (like RTP), reducing the broadening of interfaces by diffusion and intermixing. This allows the growth of layers at fairly high temperatures while maintaining abrupt composition profiles.

(ii) growth of ultrathin layers: the accuracy and reproducibility of substrate temperature vs. time profiles produced by lamp heating allows excellent control of the extent of surface reactions. This ability translates into the controlled growth of very thin layers of semiconductors and insulators.

(iii) *in situ* multilayer processing by changing the reactive gases between high temperature cycles, while the substrate is cool: multiple thin layers of different composition can be grown without removing the substrate from the processing chamber.

2.3.4   Atmospheric CVD

Atmospheric pressure reactors hold great promise for widespread commercial use for heteroepitaxy on Si. $Si_{1-x}Ge_x$ layers have been deposited in a commercial (ASM), single wafer, lamp-heated Si epitaxial reactor at atmospheric pressure [45–47]. IBM researchers [7–9] have deposited Si and $Si_{1-x}Ge_x$ at atmospheric pressure using SiH_2Cl_2 and GeH_4. Gas purifiers at point-of-use, ultra-clean gas handling and load-locks were used to reduce the oxygen and carbon incorporation. The IBM system uses a silicon carbide susceptor, whereas the ASM system uses a quartz support plate. The former group deposited smooth $Si_{1-x}Ge_x$ layers with up to 44% Ge at 550 °C, leading to the conclusion that the chlorine-based gas chemistry suppresses islanding at high Ge concentrations. Chemical vapor deposition of epitaxial SiGe films from SiH_4–GeH_4–HCl–H_2 gas mixtures in an atmospheric pressure CVD process has also been reported [48]. Layer deposition has been carried out in a horizontally arranged, induction heated and air-cooled conventional epitaxy reactor.

Selective deposition of epilayers allows fabrication of high density circuits and reduces parasitic capacitances. Therefore, deposition of $Si_{1-x}Ge_x$ layers on patterned wafers has been attempted [49, 50]. The pattern sensitivity of the thickness of selectively deposited $Si_{1-x}Ge_x$ has been found to depend on the total system operating pressure. It has been observed that pattern sensitivity is much greater for atmospheric pressure deposition than for reduced-pressure deposition.

2.3.5   Remote Plasma CVD

Plasma enhanced CVD is another low temperature growth technique used in Si processing. However, plasma induced damage is unavoidable when the substrates are immersed in a plasma. An attractive solution is to use a remote plasma in which the excited gaseous species are transferred to the substrate located away from the plasma generation zone. Remote plasma CVD (RPCVD) is a low temperature process (150–450 °C) and has been successfully employed in *in situ* remote hydrogen plasma clean of Si surfaces, prior to growth of Si homoepitaxy and $Si_{1-x}Ge_x$ heteroepitaxy. A typical growth process employs an *ex situ* wet chemical clean, an *in situ* remote hydrogen plasma clean, followed by a remote argon plasma dissociation of silane and germane to generate the precursors for epitaxial growth. Boron doping concentrations as high as 10^{21} cm^{-3} have been achieved in low temperature epitaxial films by introducing B_2H_6/He during growth. The growth rate of epitaxial Si can be varied from 0.4 Å/min to 50 Å/min by controlling the rf power. The wide range of controllable growth rates makes RPCVD an excellent tool for applications ranging from superlattice structures to more conventional Si epitaxy. Defect densities below the detection limits of TEM ($\sim 10^5$ cm^{-2} or less) have been reported. The RPCVD process also exploits the hydrogen passivation effect at temperatures below 500 °C to minimize the adsorption of C and O during growth. Low oxygen contents of $\sim 3 \times 10^{18}$ cm^{-3} have been achieved by RPCVD.

A plasma source is also useful for the growth of Si alloy epitaxial layers in a low pressure CVD (LPCVD) or a very low pressure CVD (VLPCVD) system. Unlike MBE or UHVCVD, the base pressure in VLPCVD is not low enough to prevent the formation of oxide in the reaction chamber. Therefore *in situ* plasma cleaning techniques are needed to prepare the surface for epitaxial deposition. The use of quartz halogen lamps as a radiant heat source is similar to LRPCVD/RTCVD systems, except a high thermal mass susceptor is used to eliminate temperature differences across the wafer. Low temperature *in situ* sputter cleaning of patterned oxide wafers offers advantages over the HF dip used in UHVCVD and the high temperature hydrogen clean used in LRPCVD/RTCVD. *In situ* cleaning does not require a hydrophobic surface and does not undercut the oxide. An optimized argon sputter cleaning process for temperatures greater than 750 °C in a VLPCVD reactor has been described by Comfort and Garverick and their coworkers [51, 52] and the development of sputter cleans at 750 °C and below has been reported by Yew and Reif [53]. Deposition of *in situ* doped n- and p-type layers of up to 10^{20} cm^{-3} dopant concentrations and selective epitaxial layers have been reported [54–56].

2.3.6 Solid Phase Epitaxy

From the viewpoint of the compatibility with present Si integrated circuit fabrication processes, it may be difficult and extremely costly to merge MBE or UHVCVD deposition technologies within a standard commercial process. An alternative approach to forming heterolayers is to implant high dose Ge or C ions on a Si substrate followed by recrystallization using solid phase epitaxy

[57–59]. The method is very attractive for the growth of C-containing alloys as implantation allows the incorporation of elements into Si at concentrations far in excess of their solubility limit. Typical experimental steps of solid phase epitaxy involve depositing a film of SiGe epilayer on a Si substrate, amorphizing from the free surface to a certain depth and then annealing so that the amorphized layer can regrow epitaxially on the underlying substrate. When only the epilayer, i.e., SiGe, is amorphized, solid phase epitaxy is termed unstrained or stress-relaxed SPE, whereas it is called strained SPE when the underlying Si layer is also amorphized. This method is fully compatible with any standard Si IC manufacturing process and is relatively simple compared to the expensive MBE or UHVCVD method.

Solid phase epitaxial growth of SiGe alloy using Ge ion implantation and prolonged furnace anneal has been reported [60–63]. Annealing of implanted surface alloy layers induces the formation of dense concentrations of end-of-range defects. Annealing of buried layers also results in the generation of stacking faults within the region of the alloy layer having the highest (peak) Ge concentrations. Such defects cause serious degradation of the electrical properties in functioning devices. The amorphization of deep layers obtained by high energy Si pre-implants on cooled substrates can be a suitable way to confine the end-of-range damage well below the active region of the device. On the other hand, it is known that carbon implanted in Si with various dopants strongly depresses the formation of secondary defects produced in subsequent high temperature annealing. Boron can behave similarly, acting both as a dopant and as a source for strain reduction.

Rapid thermal anneal (RTA) which is now used extensively in recent VLSI processing is used for SPE to reduce the thermal budget during annealing. A combination of low temperature thermal anneal with RTA or sequential RTA can also be used to optimize the electrical activation with reduced defect concentration. There are at present two proposed models for the SPE mechanism [64]. These are the dangling bond model and the diffusion model. Despite the significant advances made in SPE, there still exists an anomaly that neither model can explain with regard to the composition dependence of the activation energy of the growth. As the composition x in $Si_{1-x}Ge_x$ increases, the activation energy goes through a maximum instead of showing a monotonic change from bulk Si to bulk Ge.

The demand for alternative technologies to either combine with or replace the existing technologies persists. Low temperature techniques for material deposition/growth are being developed. Among these are plasma-CVD of $Si_{1-x}Ge_x$. Researchers from Stanford University have applied pulsed laser induced epitaxy (PLIE) and gas immersion laser doping (GILD) to the fabrication of layers in the Si–$Si_{1-x}Ge_x$ system and have also demonstrated that the fabrication of an epitaxial structure for HBT is feasible.

2.4   Doping of Alloy Layers

In Si homoepitaxy, emphasis is placed on obtaining a high growth rate to solve the problems of throughput and on reducing autodoping from deposition onto highly doped layers. In low temperature Si and $Si_{1-x}Ge_x$ epitaxy, autodoping is not a problem and desired layer thicknesses are on the order of 100 nm or less. Precise control of the Ge and dopant concentration profiles becomes more important than high growth rates. Certain device applications can benefit from bandgap grading, so control of Ge incorporation down to 1–2% is desirable. *In situ* doping is a necessity to get the profiles without implantation. Control of *in situ* doping profiles down to 50 nm and formation of dopant peaks below the surface are extremely desirable for precise vertical dopant profiles and lower junction capacitance. Ion implantation cannot achieve these types of profiles. High and moderate levels of dopants of both types are needed to form device structures. Quick transitions from high to low and low to high dopant concentrations are also desired for the formation of lightly doped spacers.

The ability to *in situ* dope eliminates many of the implantation and anneal steps in a device fabrication process. The goal is to obtain good control of the dopant incorporation in the alloy layers. The ability to control p- and n-type dopants is necessary in the design of the vertical profile of transistors. In addition, the transition from high to low and low to high concentrations warrants investigation because lightly doped spacers may be useful in improving breakdown voltage and reducing tunneling current.

2.4.1   Doping in MBE Growth

At present, boron is commonly used for p-doping as it has a larger equilibrium solid solubility and smaller diffusion coefficient. Common boron sources used in MBE may be classified as the molecular sources (B_2O_3, HBO_2 and boric acid) and elemental boron. The sticking coefficient of all of these source species is unity and boron does not desorb from the surface at growth temperatures. Molecular sources are easier to work with as the cell temperatures required are reasonably low (typically, 700–1150 °C for B_2O_3). However, these compounds evaporate molecularly and as a result oxygen is deposited on the wafer surface along with the desired boron. Oxygen concentrations are roughly equal to the boron concentration unless the deposition temperature is high enough. Device designs generally require relatively high boron concentration and low growth temperatures that can lead to the incorporation of considerable amounts of oxygen. As a result, many researchers have shifted to elemental boron sources.

Evaporation of elemental boron requires a special crucible design capable of withstanding temperatures of 1300–2000 °C. The deposition of heavily boron-doped $Si_{1-x}Ge_x$ by MBE has been studied by Lin *et al.* [65]. They found that $Si_{0.70}Ge_{0.30}$ layers doped to 2×10^{20} cm^{-3} with an HBO$_2$ source were highly defective, while similar doping levels could be achieved with an elemental boron source without defects. Layers with boron concentration of 5×10^{20} cm^{-3} and good surface morphology were grown at 350 °C. Boron concentrations at this level are sufficient for virtually all devices. However, the abruptness of the boron profile is also important. Like Ge, boron tends to segregate to the surface. Jorke and Kibbel [66] have reported decay lengths of about 10 Å for boron-doped Si and $Si_{0.80}Ge_{0.20}$ grown at low temperatures (450 °C). At higher growth temperatures the decay length increases, with evidence for a transition from kinetically limited to equilibrium segregation at about 600 °C. Thus δ-doped regions must be grown by depositing boron and capping it at low temperatures. Kuo *et al.* [67] have shown that *in situ* boron diffusion is primarily a function of Ge content and does not show strong dependence on biaxially strained $Si_{1-x}Ge_x$ films.

In situ n-type doping during MBE growth is problematic because of low sticking coefficients and surface segregation. Thus a limited range of dopant concentrations is available. Low energy implantation during deposition seems to solve these problems, but increases the complexity and cost of MBE. One of the attractive n-type dopants for precise control of the junction depth is antimony. In contrast to the behavior of boron and Ge, the behavior of antimony is complex. The sticking coefficient varies from near unity at 600 °C to about 10^{-3} at 700 °C, decreasing further at higher substrate temperatures [68]. Again in contrast to Ge and boron, antimony desorbs from the surface above about 675 °C [69]. Finally, the incorporation into the substrate is very low and temperature dependent. Due to the poor incorporation, the maximum doping concentration that can be achieved at moderate growth temperatures is quite small. To obtain abrupt transitions and controllable high concentrations two approaches have been used. At low temperatures (325 °C) segregation is considerably reduced and concentrations up to 5×10^{19} cm^{-3} have been obtained together with δ-doped regions with 50 Å width.

The inability to *in situ* dope n-type dopants and to deposit selective layers has been surmounted by using gas-source MBE [39, 70–72]. Surface morphology and poor incorporation of AsH$_3$ *in situ* doping of Si have motivated the investigation of PH$_3$. Hirayama and Tatsumi [73], using PH$_3$ in a gas-source MBE reactor, found a similar decrease in growth rate and degradation of surface morphology with increasing PH_3 gas concentration. In both cases of AsH_3 and PH_3 *in situ* doping of Si homoepitaxy, the adsorption of the dopant radical is cited as the deposition rate-limiting step instead of the SiH_2 radical. This change in the surface rate-limiting step appears to affect the surface morphology. Although not fully understood, GeH_4 probably plays a role in improving the surface morphology during AsH_3 doping.

2.4.2   Doping in CVD Growth

Doping in CVD is achieved simply by adding the appropriate dopant gaseous species, e.g., AsH_3, PH_3 and B_2H_6, to the source gas stream during growth. The transition abruptness of the doping profile is determined by the gas switching system used in the reactor and the incorporation kinetics of the dopant species at the growth front.

p-type doping by adding B_2H_6 to the source gas is a rather straightforward in low temperature CVD. Decomposition studies of diborane on $< 100 >$ Si were reported by Yu *et al.* [74]. Diborane is an unusual molecule in which two of the hydrogen atoms are bonded to two boron atoms and diborane is termed an electron-deficient molecule. Upon heating, the decomposition product is H and boron does not desorb from the surface at least up to 850 °C. Surface hydrogen was shown to block adsorption of diborane. Studies of UHVCVD doping suggest that boron incorporation requires two active sites which is consistent with the reaction

$$B_2H_6 + 2* \rightarrow 2BH_3* \tag{2.16}$$

This is followed by the overall reaction

$$BH_3* \rightarrow B* + 3H* \tag{2.17}$$

and finally desorption of hydrogen takes place. It is seen that germane, silane and diborane all appear to require two sites for initial adsorption. As a consequence, the relative incorporation rates will remain roughly constant as the temperature (and thus hydrogen coverage) change. The above studies show that the SiGe layer is only very slightly influenced by the addition of B_2H_6: it is increased by about 10% for 5×10^{19} cm^{-3} of B. It has been shown that the initial sticking coefficient of diborane is approximately twice as large as that of silane [75, 76]. LRP/RTCVD, UHVCVD and other forms of low temperature CVD [77, 78] appear to perform equally well concerning sharpness of dopant transition and boron incorporation. The boron content in the grown films follows very closely the gas phase B_2H_6 concentration over several orders of magnitude [1].

Typically, n-type doping is obtained by adding AsH_3 or PH_3 to the source gas. In contrast to p-type doping, n-type doping behaves rather differently in various CVD systems. In the case of Si, Yu *et al.* [74] found that the sticking coefficient of phosphine is nearly one. The saturation coverage on $< 100 >$ Si at 400 ^oC corresponds to one phosphine molecule per four surface atoms, suggesting an initial reaction given by

$$PH_3 + 4* \rightarrow P* + 3H* \qquad (2.18)$$

So far the reactions are similar to those of other hydrides; but there are important differences. At substrate temperatures above 550 ^oC desorption of phosphorus can occur, with both P and P_2 species detected by mass spectrometry [79]. This desorption reaction is second order in phosphorus surface coverage [80], i.e.

$$2P* \rightarrow 2* + P_2 \qquad (2.19)$$

Phosphorus usually forms three bonds; consequently when it is incorporated into a Si dimer at the surface, one active site is removed from the surface. At temperatures above 400 ^oC, H desorbs and the surface becomes fully covered with P or As. This results in a blocking of all active sites, reducing the growth rate considerably. Doping with phosphorus has been studied by Jang *et al.* [81] for VLPCVD growth from silane and germane (deposition pressure approximately 13 mTorr). For SiGe growth, it was shown that phosphine had a smaller effect on the growth rate, phosphorus was more effectively incorporated, and retrograde doping was possible.

2.5 Binary $Si_{1-x}Ge_x$ Alloy Films

Most of the early work on binary $Si_{1-x}Ge_x$ alloy films was performed using MBE whereas growth using CVD systems started much later. Several excellent reviews on SiGe alloy layers have appeared in the literature [1, 82, 83]. In the following, we briefly discuss the development and general characteristics of SiGe films grown using various techniques.

Kasper and Herzog [84] were the first to attempt the growth of $Si_{1-x}Ge_x$ on Si at 750 ^oC using MBE. At this high temperature, three-dimensional growth and islanding were observed. Using MBE, the concept of strained layer epitaxy [85] was applied again to Si/SiGe material system by Bean *et al.* [86] in 1984 at a low temperature (550 ^oC). In the case of strained layer epitaxy, lattice mismatched layers can be grown without misfit dislocations if the thickness is kept below a critical thickness. At this low temperature, it was found that the critical thickness for the epilayer was several times higher than the value that can be obtained from the mechanical equilibrium theory [87]. The key feature of this low temperature technique is that two-dimensional growth of strained $Si_{1-x}Ge_x$ containing a high concentration of Ge is possible as shown in the atomic force micrograph (figure 2.7).

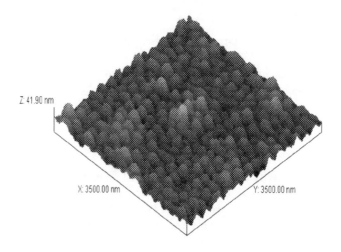

Figure 2.7. AFM surface image of a strained $Si_{0.74}Ge_{0.26}$ film grown on Si using gas-source molecular beam epitaxy.

For pseudomorphic growth of $Si_{0.5}Ge_{0.5}$ on Si, the growth temperature must be lower than 550 °C. Bean and his coworkers [37, 86] found that the maximum Ge incorporation before the occurrence of non-planar growth depends on the deposition temperature (figure 2.8). At 750 °C the maximum Ge mole fraction is 10%, whereas at 550 °C 100% Ge is attainable. It is speculated that higher deposition temperatures resulted in surface mobility high enough to cause surface tension problems and islanded growth. Deposition rates of up to 600 nm/min for Si are possible. However, typical $Si_{1-x}Ge_x$ deposition rates are in the range of 30 nm/min for greater profile control [37]. Also extremely abrupt compositional profile control is possible by the use of mechanical shutters. At 630 °C, the $Si_{1-x}Ge_x$ growth rate decreases with increasing partial pressure of GeH_4.

Zhang *et al.* [88] have demonstrated excellent SiGe film quality using Si_2H_6 and Ge as sources and reported the growth process, layer morphology, structural quality and growth kinetics. The variations of alloy composition with change in Si_2H_6 flow rate and Ge flux for the two growth techniques

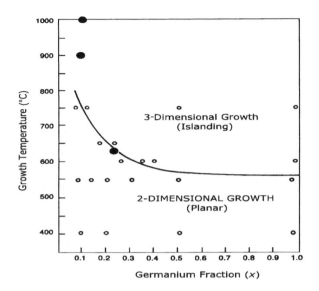

*Figure 2.8. Plot of growth regimes vs. growth temperature and Ge mole fraction (x) for $Si_{1-x}Ge_x$ grown on Si substrates. The solid line and the open circles refer to MBE, while the solid circles refer to LRP samples which exhibited planar growth. After J. C. Bean et al., Appl. Phys. Lett., **44**, 102–104 (1984).*

are shown in figures 2.9a and 2.9b, respectively. In the first, the Ge cell temperature is held constant and in the second the Si_2H_6 flow rate is held constant. As expected, the Ge content in the alloy increases with decrease of Si_2H_6 flow rate and with increase of Ge flux.

Figures 2.10a and 2.10b show the variation of growth rate with Ge content in the alloy for the two methods of growth. In the first case, when the Ge cell temperature, or flux, is held constant, the growth rate monotonically decreases with Ge content in the alloy. However, when the composition is varied by varying the Ge flux, as shown in figure 2.10b, the growth rate first decreases and then increases sharply.

The growth behavior indicated by figure 2.10a is easily understood. As the Si_2H_6 flow rate is decreased, the Ge content in the alloy increases, but the growth rate decreases. A similar trend has been observed as reported by Mokler *et al.* [89] during growth of $Si_{1-x}Ge_x/Si$ from gaseous Si_2H_6 and GeH_4. The growth phenomenon represented by the data of figure 2.10b is more

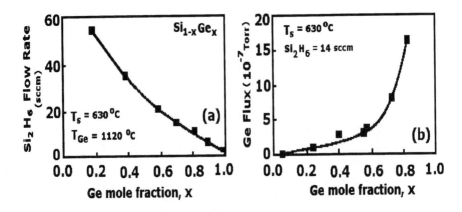

Figure 2.9. Dependence of Ge content in $Si_{1-x}Ge_x$ alloys for two methods of growth: (a) the Ge cell temperature is held constant at $1120\,°C$ and the Si_2H_6 flow rate is varied and (b) the Si_2H_6 flow rate is held constant at 14 sccm and the Ge cell temperature is varied. The squares are experimental data and the solid lines are a guide for the eye. After F. C. Zhang et al., Appl. Phys. Lett., 67, 85–87 (1995).

interesting. Here the Si_2H_6 flow rate is maintained constant and the different alloy compositions are obtained by varying the Ge flux. One would expect that the growth rate would monotonically increase with Ge composition. However, it is noticed that the growth rate initially decreases up to $x \sim 0.5$ and then increases. A curve fitting indicates that the measured data could be the sum of two curves, one decreasing slowly with x (shown in figure 2.10b by the dashed line) and the other increasing with x (the dot-dashed line). The latter represents the increase of growth rate with increasing Ge flux. The first curve, however, would suggest that there is actually a surface phenomenon that tends to decrease the growth rate with increasing Ge flux.

The UHVCVD system, which was developed by Meyerson [6], uses a base pressure of 10^{-9} Torr, a growth pressure of 10^{-6} Torr and a growth temperature as low as 550 °C. This type of UHVCVD is ideally suited for a high throughput manufacturing environment. This system, however, has a few limitations. As the source gas flow is increased (at constant temperature) to increase the Ge concentration during growth of a $Si_{1-x}Ge_x$ layer, the growth rate increases. The deposition pressure is about 1–2 mTorr with deposition rates around 1–2 nm/min. Because of the low deposition pressure, gas depletion does not occur as in standard low pressure CVD polysilicon reactors at 0.4 Torr. *In situ*

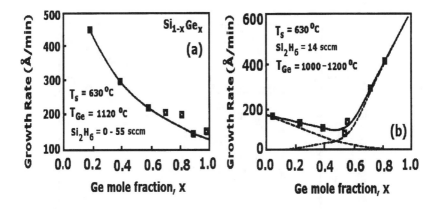

Figure 2.10. $Si_{1-x}Ge_x$ *layer growth rates as a function of Ge fraction for two methods of growth: (a) the Ge cell temperature is held constant at 1120 °C and the Si_2H_6 flow rate is varied and (b) the Si_2H_6 flow rate is held constant at 14 sccm and the Ge cell temperature is varied. The squares are experimental data and the solid lines are a guide for the eye. In (b), the dashed line indicates the decreasing curve, the dot-dashed line, the increasing curve. After F. C. Zhang et al., Appl. Phys. Lett.,* **67***, 85–87 (1995).*

doping of both types is possible [1] and boron dopant content in the film is linear as shown in figure 2.11. The addition of PH_3 to SiH_4 produces n-type doping. Although dopant control is similar to that of boron, the dynamic range over which P content remains stable is far more limited (as shown in figure 2.12) and allows the incorporation of up to $\sim 1 \times 10^{20}$ cm^{-3}. Apparently no problems with islanded growth at high Ge concentrations appear, because the deposition temperature is below 550 °C. The addition of GeH_4 enhances the growth rate of $Si_{1-x}Ge_x$ over Si homoepitaxy at 550 °C [11]. The catalytic effect of GeH_4 in removing hydrogen from the surface is cited as the reason for the increased $Si_{1-x}Ge_x$ growth rate. Ge incorporation in alloy layers is roughly linear in source Ge content, while the behavior of growth rate is somewhat more complex, as shown in figure 2.13. Although blanket depositions are trivial, problems appear with patterned oxide depositions because the HF dip cleaning technique is hydrophilic on oxide regions. Polysilicon layers on top of the oxide may eliminate this problem at the expense of greater process complexity. This problem may ultimately limit the manufacturability of the UHVCVD technique. Selective Si epitaxy was examined by researchers at Tohoku University [90] and selective $Si_{1-x}Ge_x$ deposition was studied by

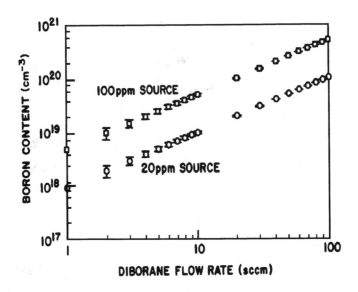

*Figure 2.11.   Dopant accuracy for two dopant source concentrations. Limitations are flow controller absolute accuracy and offset, as well as temperature induced fluctuations. After B. S. Meyerson, Proc. IEEE, **80**, 1592–1608 (1992), copyright ©1992 IEEE.*

Greve and Racanelli [91, 92]. The authors observed that GeH_4 enhanced the nucleation time of polycrystalline material on oxide regions of patterned wafers.

High quality Si and SiGe films have been obtained using the RTCVD technique [42, 93, 94]. This is a film growth technique that results from the combination of rapid thermal annealing and chemical vapor deposition. In principle, growth of the epilayer can be controlled by gas switching, lamp switching, or both. The base pressure used in this technique (10^{-7} Torr) is much lower than that used in the LRP epitaxial growth technique. A wide range of growth temperature (600–900 °C) can be used. Using this growth technique, good quality epitaxial Si grown at 600–900 °C [42] and SiGe grown at 900 °C [94] have been demonstrated.

$Si_{1-x}Ge_x$ layers need to be deposited at lower temperatures to avoid relaxation and the three-dimensional growth problems. So the deposition temperature was reduced down to 625 °C for $Si_{1-x}Ge_x$ and increased back to 850 °C for Si cap layers. Fortunately, the $Si_{1-x}Ge_x$ growth rate increases with

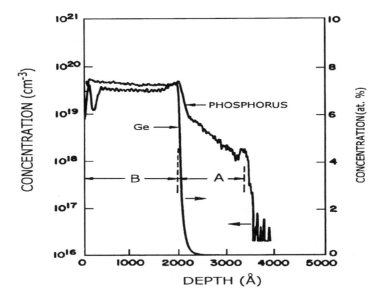

Figure 2.12. Film phosphorus content, by SIMS, as influenced by the addition of germane during UHVCVD film growth. Silane and phosphine flows were constant throughout growth (regions A and B), while germane was added at a constant flow in region B. Note the discontinuous jump in, and subsequent stability of, phosphorus incorporation in the presence of Ge.   Spreading resistance showed the phosphorus to be fully activated. After B. S. Meyerson, Proc. IEEE, **80***, 1592–1608 (1992), copyright* ©*1992 IEEE.*

the addition of GeH_4 [12, 95, 96]. The $Si_{1-x}Ge_x$ growth rate enhancement with increasing GeH_4 is similar to the UHVCVD result.   The maximum incorporation before three-dimensional growth from MBE work does not seem to apply to LRPCVD/RTCVD.

One of the major problems of reduced temperature growth is an increased oxygen incorporation in the $Si_{1-x}Ge_x$ layers. The contamination source was traced to high oxygen content present in SiH_2Cl_2. SiH_4 does not have the same problem because gas with much higher purity is available [93]. The oxygen incorporation problem diminishes with the use of a load-lock and point-of-use filtration of SiH_2Cl_2.

Atmospheric pressure chemical vapor deposition has also been used to grow good quality epitaxial Si and SiGe films [8]. This technique offers temperature flexibility over a wide range, excellent temperature uniformity in

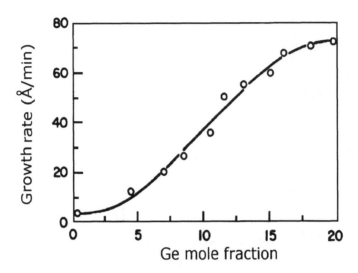

*Figure 2.13. Film growth rate (solid line) during SiGe UHVCVD as a function of film Ge content. A 25× increase in growth rate occurs for the addition of 20% Ge to the solid film. After B. S. Meyerson, Proc. IEEE, **80**, 1592–1608 (1992), copyright ©1992 IEEE.*

the commercial process and a simple conventional flow-through design which uses no vacuum pumps. An ultra clean APCVD process has been used to grow films at low temperatures, 550–850 °C, by rigorously minimizing oxygen and moisture in the growth environment by using load-lock, point-of-use gas purification and ultra clean gas handling [8]. $Si_{1-x}Ge_x$ layers were deposited over the temperature range from 600 to 900 °C. At 625 °C the deposition rate of the $Si_{1-x}Ge_x$ alloy was found to increase with GeH_4 partial pressure; the amount of Ge in the alloy increased rapidly with GeH_4 partial pressure at low GeH_4 partial pressures and appeared to approach saturation at higher GeH_4 partial pressures. The deposition rate depends weakly on the SiH_2Cl_2 partial pressure, with a slight decrease in deposition rate with increasing SiH_2Cl_2 partial pressure at higher GeH_4 partial pressures. The deposition kinetics appear similar to the LRP/RTCVD system since SiH_2Cl_2 and GeH_4 are used. When the deposition rate of the Si component over the temperature range from 600 to 900 °C was normalized by the Ge content in the alloy, an Arrhenius type behavior was observed with an apparent activation energy of about 1.9–

2.0 eV suggesting that the deposition process for Si is limited by reaction kinetics, rather than by mass transport. The Ge component of the deposition rate behaves quite differently as shown in figure 2.14. At low (600–650 °C) temperatures, the deposition rate displays an Arrhenius behavior with an apparent activation energy (0.7 eV) much lower than that of the Si component. At higher temperatures, the deposition rate of the Ge component depends only weakly on temperature. Remote plasma enhanced chemical vapor deposition has also been used for Si and $Si_{1-x}Ge_x$ epitaxy [97].

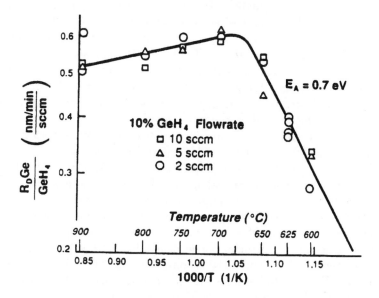

Figure 2.14. Ge component of deposition rate normalized by GeH₄ flow. After T. I. Kamins and D. J. Meyer, Appl. Phys. Lett., **59**, *178–180 (1991).*

2.6 Binary $Si_{1-y}C_y$ Alloy Films

The growth of $Si_{1-y}C_y$ epitaxial films on Si(100) presents a formidable challenge as: (i) the lattice mismatch of Si and C is –52%, (ii) C has a very low solubility in Si below 10^{-6} at 1400 K, (iii) C contamination of the Si surface during epitaxy disrupts the process, and (iv) silicon carbide exists as a stable phase out of several polytypes. The above problems can be avoided by growing the films in a condition far away from thermodynamic equilibrium. The typical conditions for the growth of good quality $Si_{1-y}C_y$ epitaxial films on Si are a low growth temperature, typically between 450 and 550 $^{\circ}$C, and use of an extremely pure elemental C beam.

Posthill and his coworkers [98] studied the feasibility of deposition of dilute $Si_{1-y}C_y$ epitaxial films on Si(100) using remote plasma-enhanced CVD. Carbon incorporation up to 3 at.% was achieved at a growth temperature of 725 $^{\circ}$C. The layers were characterized by x-ray diffraction and transmission electron microscopy (TEM). No evidence for the formation of silicon carbide was found.

Extensive work by IBM researchers on the growth of $Si_{1-y}C_y$ alloys on Si has been done using MBE. Iyer and his coworkers [99, 100] and Eberl and his coworkers [101] synthesized pseudomorphic $Si_{1-y}C_y$ alloys on Si(100) using solid source MBE. The layers were characterized by x-ray diffraction, secondary ion mass spectroscopy, TEM and Raman spectroscopy. They demonstrated that good quality layers with a few at.% of carbon ($y \leq 0.05$) can be grown by MBE if low growth temperatures, 500–600 $^{\circ}$C, and a growth rate of 0.2 nm/s are used. Amorphous growth occurs for lower substrate temperatures or higher carbon concentration.

The thermal stability of $Si_{1-y}C_y$/Si strained layer superlattices was studied by Goorsky *et al.* [102]. The superlattices were grown by MBE as described above, with three concentrations of C: 0.003, 0.008 and 0.013. The superlattices were stable on annealing for 2 h at temperatures up to 800 $^{\circ}$C. Between 800 and 900 $^{\circ}$C strain relaxation occurred by interdiffusion of C and Si at the interfaces. At 1000 $^{\circ}$C and above, precipitation of silicon carbide was observed. An x-ray diffraction rocking curve for a ten-period $Si/Si_{1-y}C_y$ superlattice taken at a glancing angle of incidence [100] is shown in figure 2.15. Each period of the superlattice consisted of a 320 Å Si layer and a 110 Å $C_{0.005}Si_{0.995}$ layer. TEM studies showed no evidence of dislocations or other defects. For 5% C concentration, the layers grown at 450 $^{\circ}$C were amorphous. At higher temperatures, precipitates of silicon carbide were formed.

One possible application of $Si_{1-y}C_y$ alloys is the fabrication of symmetrically strained SiC/SiGe superlattices on a Si substrate. Eberl and his coworkers [103] reported fabrication of $C_{0.01}Si_{0.99}/Ge_{0.1}Si_{0.90}$ superlattices on Si. The strain in the superlattices alternated between tensile and compressive in the individual $Si_{1-y}C_y$ and $Si_{1-x}Ge_x$ alloy layers, respectively. The individual layer thicknesses were 10 and 13.5 nm, respectively. X-ray rocking curves were measured and average strain in the superlattice was estimated. The strain was considerably less than that expected in a superlattice in which a $C_{0.01}Si_{0.99}$ layer is replaced by a Si layer. It is clear from these results that perfect symmetrically strained layer superlattices can be fabricated by adjusting the relative Ge and C content and/or by changing the thickness of the individual layers.

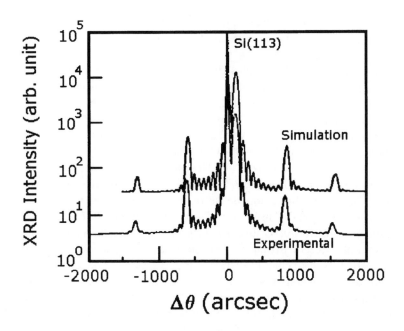

Figure 2.15. X-ray diffraction rocking curve for a $Si/Si_{1-y}C_y$ superlattice. The carbon content in the $Si_{1-y}C_y$ film is 0.005. After S. S. Iyer et al., Microelectron. Eng., 19, 351–356 (1992).

Faschinger *et al.* [104] have reported the electrical properties of undoped and Sb-doped $Si_{1-y}C_y$ alloys on high resistivity (1000 Ω cm) Si grown by MBE at 500 °C. The layers could be doped with Sb at low growth temperatures (at 350 °C). However, a significant Sb segregation takes place to the surface and leads to doping of the subsequently deposited layers. Modulation-doped $Si/Si_{1-y}C_y/Si$ structures with the modulation doping in the Si layer (to avoid unintentional doping of the $Si_{1-y}C_y$ channel and Si spacer layer) exhibited enhanced electron mobilities with peak values of about 10,000 cm^2/Vs, indicating that $Si_{1-y}C_y$ is strained on Si and forms an electron channel.

Growth of dilute SiC epitaxial layers on Si using RTCVD have been reported by several workers [105, 106]. Ray and his coworkers [106] have reported the growth of polycrystalline and epitaxial layers of metastable $Si_{1-y}C_y$ alloys using silane and propane at different temperatures with varying amounts of carbon ($y = 0.01$, 0.02 and 0.05) using rapid thermal chemical vapor deposition. Stoichiometric SiC layers were also grown for comparison by increasing the propane to silane flow ratio. Following usual cleaning schedules, p-type Si < 100 > wafers were pretreated with HF (2:1) diluted in deionized water vapor just before loading into the chamber. The authors have also reported solid phase epitaxial growth of $Si_{1-y}C_y$. Epitaxial regrowth by rapid thermal annealing of C^+-implanted Si and SiGe layers was used in the study of the formation of metastable phases. CO was used as the source gas for C implantation. The ion dose was chosen to incorporate 1%, 2% and 5% C into Si with two different energies. All the samples were rapid thermally annealed in nitrogen (1.2 liter/min) at 1046 °C for 30 s.

Fourier transform infrared (FTIR) spectroscopic measurements were carried out on all RTCVD and SPE grown layers to study the nature of carbon incorporation. The absorption peaks due to the localized vibrational mode (LVM) of substitutional C in Si at 16.7 μm (wavenumber 607.5 cm^{-1}) and the strong phonon absorption mode in silicon carbide at 12.6 μm (wave number 794 cm^{-1}) were monitored. Strane and his coworkers [107] reported the synthesis and detailed structural characteristics of $Si_{1-y}C_y$ strained layers on Si by solid phase epitaxy. Carbon was introduced into preamorphized Si by multiple energy implantation to form a uniform C concentration. Preamorphization was done using 60, 30 and 10 keV Si$^+$ ions with doses of 3 × 10^{15}, 6 × 10^{14} and 5 × 10^{14} cm^{-2}. This was followed by C implantation with doses of 2.34 × 10^{14}, 9.78 × 10^{14} and 4.44 × 10^{14} cm^{-2}, respectively, with corresponding energies of 25, 12.5 and 5 keV. Simulations showed that this procedure yielded a flat C profile. A 30 min anneal at 450 °C was used to remove end-of-range defects. A 30 min anneal at 700 °C was used for the solid phase epitaxial growth. High quality layers could be fabricated with C concentrations up to 1 at.% by this technique.

Fourier transform infrared spectroscopy measurements showed that the carbon occupies substitutional lattice sites. Film stability was studied in the temperature range 810–925 °C and the changes in strain were measured by double-crystal x-ray rocking curves. Carbon and oxygen concentrations were measured by SIMS and the microstructure by TEM. On annealing the layers for 240 min at 875 °C, the 607 cm^{-1} local mode due to substitutional C decreased and a broad peak at 810 cm^{-1} was observed. On annealing the layers for 1200 min at the same temperature, the local mode disappeared and the broad peak became stronger indicating the precipitation of silicon carbide.

2.7   Ternary $Si_{1-x-y}Ge_xC_y$ Alloy Films

The growth of SiGeC alloys on Si is a challenging task in the sense of finding suitable source materials and growth conditions for a ternary alloy. The major concern is to increase the carbon incorporation rate by avoiding SiC precipitation. Major published work indicates the use of MBE, CVD and solid phase epitaxy while a limited number of publications have appeared where some novel technique was used. In MBE, a solid elemental carbon (graphite) filament sublimation cell is heated to over 3000 °C in order to sublime C atoms from the filament. Acetylene has also been used as C source. Vegard's law predicts that for a complete strain compensation in the pseudomorphic $Si_{1-x-y}Ge_xC_y$ alloys, the ratio of Ge to C concentration should be about 8, whereas experimental values obtained by different authors vary between 8 and 12. In the following, we discuss various techniques used for SiGeC film growth, and the strain and stability of $Si_{1-x-y}Ge_xC_y$ layers grown on Si.

To participate in strain compensation, carbon must occupy substitutional sites within the SiGe lattice. Because of the low solubility of carbon within Si, low temperature growth techniques are required. Epitaxial growth of $Si_{1-x-y}Ge_xC_y$ has been demonstrated using both RTCVD [35] and MBE [108]. In both cases it is found that carbon concentrations of up to 2% can be incorporated substitutionally.

John *et al.* [109] have reported deposition of $Si_{1-x-y}Ge_xC_y$ layers in a cold-wall UHVCVD chamber. Briefly, the system consists of two chambers, a load-lock which permits the loading of three 4 inch wafers and a deposition chamber with a base pressure of approximately 1×10^{-10} Torr. Disilane (100%), germane (10% in helium) and methylsilane (20% in helium) were delivered to the deposition chamber using mass flow controllers. The substrate was heated using infrared radiation through a graphite chuck coated with pyrolytic boron nitride and the temperature was maintained using feedback control through a pyrometer. Ge mole fractions ranging from 0% to 40% were studied with carbon concentrations varying from 2×10^{19} to 2×10^{21} cm^{-3}.

The gas flows were adjusted to obtain the appropriate concentrations of Ge and C in the film. Except for those experiments in which the effect of pressure was being studied, the total pressure during the deposition was kept constant in the range 1–10 mTorr. The crystallinity of the samples was determined using reflection high energy electron diffraction (RHEED) immediately after deposition. The typical energy of the incident electrons was approximately 15 keV and a spot size of approximately 1 mm was used.

Figure 2.16 shows a SIMS profile of a typical $Si_{1-x-y}Ge_xC_y$ film. This spectrum was obtained using Cs^+ as the primary ion beam and a quadruple ion detector. This particular sample was grown at 550 °C and consists of a 400 Å Si buffer layer, followed by a 300 Å $Si_{1-x-y}Ge_xC_y$ layer and a 100 Å Si capping layer. The $Si_{1-x-y}Ge_xC_y$ layer was grown using a germane flow of 1.3 sccm, a methylsilane flow of 0.171 sccm and a disilane flow of 1.2 sccm.

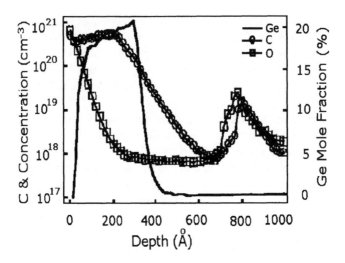

Figure 2.16.   SIMS profile of a typical $Si_{1-x-y}Ge_xC_y$ film grown using UHVCVD. After S. John et al., Proc. Mat. Res. Soc. Symp., 275–279 (1996).

The epi-substrate interface is marked by the presence of a peak in the C and O profiles. Within the film though, the oxygen levels approach the SIMS detection limits. Because the analysis was conducted on a very thin film, the surface carbon contamination interferes with the data within the first 100 Å, preventing the observation of the fall-off of the carbon layer at the capping–$Si/Si_{1-x-y}Ge_xC_y$ interface. Other analyses with thicker capping layers have shown that C does indeed fall off quickly to background levels. Within the $Si_{1-x-y}Ge_xC_y$ layer, both Ge and C are observed to incorporate well without any evidence of surface segregation. The tail in the C profile leading into the buffer layer is attributed to SIMS limitations as both the Ge and C flows were initiated at approximately the same depth (±10 Å).

SIMS analysis was conducted to determine the incorporation of C for various concentrations of Ge and results are presented in figure 2.17. All these films were grown at 550 °C. Depending on the carbon incorporation, the disilane flow was modulated between 1 and 20 sccm. The germane flow was varied from 0.15 to 1.5 sccm to obtain the appropriate mole fraction of Ge. This shows the controllability of the process to incorporate from 0.03% to 4.6% of C in $Si_{1-x-y}Ge_xC_y$ films. The data also show that the incorporation of carbon is independent of the flow of germane or the incorporation of Ge.

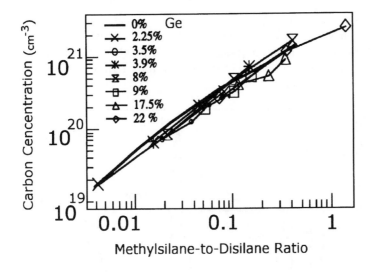

Figure 2.17. Carbon incorporation as a function of $CH_3SiH_3:Si_2H_4$ flow ratio.
After S. John et al., Proc. Mat. Res. Soc. Symp., 275–279 (1996).

Experiments were conducted to determine the effect of temperature (from 500 to 600 ^oC) upon the incorporation of carbon and Ge. For Ge incorporation, the authors have not observed any effect of temperature, within the limits of SIMS repeatability, and the incorporation was found to be dependent only upon the GeH_4 and Si_2H_6 flow ratio. Carbon incorporation, on the other hand, was found to be very dependent upon temperature. From 500 to 550 ^oC a drop in the C incorporation by about 33% and from 550 ^oC to 600 ^oC a drop in the incorporation of about 60% was reported. Preliminary studies indicated that there is no pressure dependence of incorporation for either Ge or C in the pressure range 1–100 mTorr.

Figure 2.18 shows two overlapping XRD (004) rocking curves from two samples. One of the two samples is a 5000 Å $Si_{1-x}Ge_x$ layer grown upon Si. The full width at half maximum (FWHM) of the rocking curve is found to be 40 arcsec. The separation of the film and substrate peak is found to be 400 arcsec from which it was determined that the Ge concentration is 3.9%. The second sample is a 4000 Å thick $Si_{1-x-y}Ge_xC_y$ layer with the same concentration of Ge. Carbon incorporated is enough not only to completely compensate the compressive strain caused by Ge, but to actually introduce tensile strain into the grown $Si_{1-x-y}Ge_xC_y$ layer. The FWHM of this film is 57 arcsec. Using Vegard's law and assuming complete substitutional incorporation the carbon concentration was found to be 0.68%. These XRD spectra demonstrate the ability of the UHVCVD techniques to form high quality epitaxial $Si_{1-x-y}Ge_xC_y$ layers ranging from compressively strained to fully strain-compensated layers to tensile strained layers.

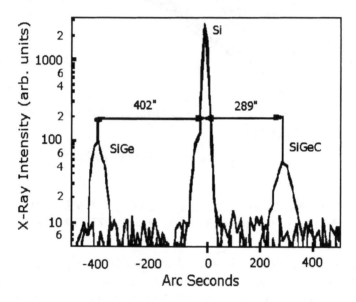

Figure 2.18.   Overlapping XRD rocking curves from $Si_{0.96}Ge_{0.04}$ and $Si_{0.9532}Ge_{004}C_{0.0068}$ samples. After S. John et al., Proc. Mat. Res. Soc. Symp., 275–279 (1996).

AFM under tapping mode was used as an *ex situ* method to analyze the surface morphology of epitaxial $Si_{1-x-y}Ge_xC_y$ layers as a function of Ge and C concentration. For the deposition of these layers, the disilane flow rate was varied between 1 and 5 sccm. Germane was varied between 1 and 3 sccm, and methylsilane was varied between 0.16 and 4 sccm. These flow ratios led to Ge mole fractions of up to 40% and carbon concentrations of up to 4%. AFM analysis of the surface indicates that the roughness is a function of both the carbon concentration and the film thickness. For high Ge concentrations with thickness beyond the critical thickness (of $Si_{1-x}Ge_x$), carbon is found to be useful in reducing the surface roughness of the film. Thus the surface morphology confirms the strain compensation provided by carbon which is also observed using XRD. For films below the critical thickness, as the carbon concentration is increased, three-dimensional islanding is observed by RHEED and AFM, degrading the epitaxial quality of the material.

Eberl and his coworkers [108] have grown $Si_{1-x-y}Ge_xC_y$ layers with 25% (x) Ge and several different C (y) concentrations by MBE using solid sources at low temperature (400–550 °C). The thickness of the layers varied between 200 and 250 nm. Both capped and uncapped layers were studied. In the layers containing 25% Ge and no carbon ($y = 0.0$), the strain was measured using an x-ray diffraction technique and was found to be 1% as expected. The strain decreased to 0.8% for $y = 0.0089$ and to 0.4% for $y = 0.019$ as shown in figure 2.19 showing the effect of carbon addition on partial strain compensation. The Raman spectra studies showed a substitutional carbon vibration mode at 600 cm^{-1}. No silicon carbide precipitation was observed from TEM and Raman studies. Several other authors [110–114] have studied growth, strain and stability of $Si_{1-x-y}Ge_xC_y$ (with up to 2% carbon) layers grown on Si using MBE. Osten *et al.* [110] have made more extensive studies of the ternary alloy layers. They varied Ge concentration between 0 and 100% and C concentration between 0 and 2.5%. Layers were grown by MBE using co-evaporation of Ge, Si and C at 500 °C. First a 60 nm thick Si buffer was deposited. Then they deposited an approximately 50 nm thick SiGe layer and opened the carbon source shutter for the same length of deposition time.

In principle, a high-dose Ge and/or C implantation process is advantageous over the MBE technique due to its compatibility with Si integrated circuit processing. However, the implantation techniques have to overcome two defect formation mechanisms related to solid phase epitaxy growth: end-of-range loops and strain-induced defects. $Si_{1-x-y}Ge_xC_y$ strained layers fabricated by solid phase epitaxy have been studied by several authors [115–117]. Lu and Cheung [117] have fabricated near-surface SiGeC alloys with Ge peak composition up to 16 at.% using high-dose (3.6×10^{16} cm^{-2}, 50 keV) Ge ion implantation and subsequent C^+ ion implantation. Solid phase epitaxial growth was accomplished by furnace annealing between 700 and 1100 °C for 30 min in N_2 ambient. RBS channeling spectra and cross-sectional transmission electron microscopy (XTEM) studies showed that high quality $Si_{0.91}Ge_{0.08}C_{0.01}$ crystals were formed at the Si surface, while $Si_{0.82}Ge_{0.16}C_{0.02}$ had extended defects. Strane and his coworkers [116] synthesized SiGeC epitaxial layers with 0.7 and 1.4 at.% carbon by solid phase epitaxial regrowth at 700 °C of multiple energy carbon implants into preamorphized $Si_{0.86}Ge_{0.14}$ layers on Si substrates. The pseudomorphic 160 nm $Si_{0.86}Ge_{0.14}$ layers on Si(100) were grown using the UHVCVD technique.

Ternary strain-compensated $Si_{1-x-y}Ge_xC_y$ alloys were prepared by

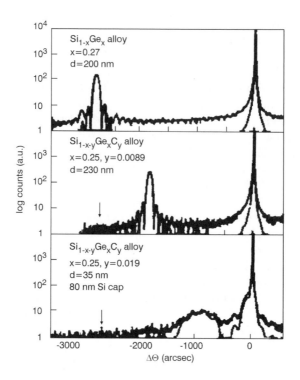

*Figure 2.19.   X-ray rocking curve data and corresponding simulations for (a) $Si_{1-x}Ge_x$, and (b) and (c) $Si_{1-x-y}Ge_xC_y$ layers with different carbon concentrations. After K. Eberl et al., Appl. Phys. Lett., **60**, 3033–3035 (1992).*

Menendez *et al.* [118] by implanting C ions into a 100 nm thick $Si_{0.923}Ge_{0.077}$ layer and subsequent annealing. X-ray diffraction rocking curves showed that the $(Si_{0.923}Ge_{0.077})_{0.991}C_{0.009}$ layers were cubic with a Si-like lattice constant. These measurements confirmed that C caused tensile strain and compensated the compressive strain in the SiGe layers in accordance with Vegard's law. Similar results (strain compensation by C) in the layers grown by solid phase epitaxy were observed by Im and his coworkers [115] These results show that the in-plane macroscopic strains caused by Ge and C cancel each other at a Ge:C ratio of about 10. Epitaxial regrowth by rapid thermal annealing of C^+-implanted MBE grown $Si_{0.82}Ge_{0.18}$ layers was used in a study of the formation of metastable phases by Maiti [119] CO was used as the source gas for C implantation. Following Vegard's law 2% C was implanted into MBE grown $Si_{0.82}Ge_{0.18}$ films for complete strain compensation.

Due to several advantages of rapid thermal chemical vapor deposition such as compatibility with Si processing, low thermal budget, precise control of surface reactions, it has been extensively used for growing epitaxial SiGeC layers by many workers [29, 31, 35, 120–127]. In most cases, dichlorosilane, germane and methylsilane are used as the precursors of Si, Ge and C, respectively, at a growth temperature between 550 and 650 ^{o}C. Mi and his coworkers [29, 31] have studied the carbon incorporation and strain in the epitaxial $Si_{1-x-y}Ge_xC_y$/Si heterostructures grown by rapid thermal vapor deposition using methylsilane (SiH_3CH_3) in the SiH_4/GeH_4/$SiCH_6$/H_2 system in the temperature range 550–600 ^{o}C at 1.5 Torr [29]. Defect-free alloy layers with compositions of up to 30 at.% Ge and 2.2 at.% C were reported. Figure 2.20 shows the variation of the measured C concentration and strain as a function of methylsilane concentration. The C incorporation efficiency is high and substitutional carbon incorporation of 2.2 at.% is possible.

Regolini *et al.* [31, 124] investigated the structure of $Si_{1-x-y}Ge_xC_y$ strained layers grown by rapid thermal chemical vapor deposition. Layers grown at 650 ^{o}C were of good quality. At lower temperatures, the growth became very slow. At higher temperatures, some of the C atoms were incorporated either as clusters or as SiC. Infrared measurements confirmed that most of the C atoms were substitutional. No dislocations were present in equally thick samples which did not contain C. The authors characterized the layers and measured Ge and C concentrations as well as strain using x-ray diffraction and RBS techniques. Epitaxial layers 4000 Å thick with $x = 0.20$ were compensated by up to 1% of substitutional C as indicated by the FTIR peak at 605 cm^{-1}. Compensation of strain by adding carbon and the validity of Vegard's law were confirmed. Carbon and Ge distributions in C-rich $Si_{1-x-y}Ge_xC_y$ layers were studied by Guedj *et al.* [127] by using high-resolution electron microscopy (HREM), XRD and Raman spectroscopy.

Epitaxial Si/SiGeC/Si heterostructures have been grown using LRPCVD by Rim *et al.* [120]. Dichlorosilane, germane and ethylene were used, respectively, as Si, Ge and carbon precursors in a hydrogen ambient at about 10 Torr. Diborane was used for *in situ* p-type doping. About 6000 Å of p-Si epilayer was grown at 1000 ^{o}C on which SiGeC layers were grown at either 500 or 600 ^{o}C and was capped by a Si layer grown at 700 ^{o}C. Samples were characterized for crystalline quality by x-ray diffraction, SIMS and FTIR. Film thickness and Ge composition were determined by RBS.

Atmospheric pressure chemical vapor deposition has also been employed [30, 128] for the growth of $Si_{1-x-y}Ge_xC_y$ layers using SiH_2Cl_2 as Si source,

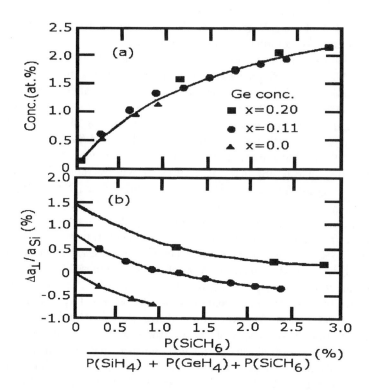

*Figure 2.20. Carbon incorporation in SiGeC films as a function of SiCH₆ content for different Ge concentrations. After J. Mi et al., Appl. Phys. Lett., **67**, 259–261 (1995).*

GeH$_4$ as the Ge source and C$_2$H$_2$ or C$_2$H$_4$ as the C source. Processing conditions are very similar to other CVD techniques and growth temperature was 625 °C. The crystallinity, composition and microstructure of the films were determined using RBS with channeling, SIMS and XTEM. It was found that crystallinity of the films is very sensitive to the flow rate of C$_2$H$_4$. Films containing up to 2% C was epitaxial with good crystallinity and very few interfacial defects. Sb ion implantation and annealing of SiGeC films grown by CVD techniques have been studied by Garcia and his coworkers [129]. Implantation was done with 1×10^{16} ions/cm^2 of 200 keV ions at 77 K in order to completely amorphize the layer. The samples were annealed under different conditions and were characterized using RBS with channeling. It was found that the regrowth kinetics were significantly different from those of SiGe alloys. A greater thermal budget is required for regrowth, and annealing at higher temperatures leads to an outward carbon diffusion and formation of a carbon depleted region.

Novel deposition chemistry has been used to obtain very high carbon concentrations in pseudomorphic $Si_{1-x-y}Ge_xC_y$ layers [32, 130–132]. Todd *et al.* [130] reported heteroepitaxial growth of SiGeC on Si(100) using a novel silicon–carbon precursor $C(SiH_3)_4$, with mixtures of SiH_4 and GeH_4, and a UHVCVD system. Film thicknesses of 100 to 110 nm with 4–6 at.% C were obtained. Growth of ternary $Si_{1-x-y}Ge_xC_y$ films from a single source precursor $Ge(SiMe_3)_4$ has been reported by Chiu *et al.* [131] using low-pressure chemical vapor deposition at temperatures of 600–700 ^{o}C. This study shows that it is possible to use organometallic compounds containing different numbers of Si, Ge and C atoms as single-source precursors to adjust the composition of $Si_{1-x-y}Ge_xC_y$ films grown by CVD. Recently, Akane *et al.* [132] have grown epitaxial $Si_{1-x-y}Ge_xC_y$ layers on Ge(001) substrates by metallo-organic molecular beam epitaxy (MOMBE) using disilane and monomethylgermane (CH_6Ge). No SiC formation or polycrystallization was observed by XRD. Kouvetakis *et al.* [32] have reported a wide range of compositions of high purity polycrystalline SiGeC diamond-structured materials using UHVCVD techniques and reactions of molecular carbon sources. Preformed Ge–C bonds (CH_3GeH_3) and Si–C bonds ($C(SiH_3)_4$) were used and films with composition $Si_{0.37}Ge_{0.13}C_{0.50}$ were deposited on Si by thermal decomposition of $Ge[Si(CH_3)_3]_4$ at 650–700 ^{o}C.

2.8 $Ge_{1-y}C_y$ and Related Alloy Films

Metastable $Ge_{1-y}C_y$ alloys have been prepared by MBE under non-equilibrium conditions at relatively low growth temperatures with a C concentration of 1% [133–136]. The bulk Ge layer's growth on Si shows a typical Stranski-Krastanov (SK) mode i.e., the growth starts in a two-dimensional mode up to a certain thickness and transition occurs to a three-dimensional island mode. The use of surfactant can inhibit three-dimensional island formation. Osten *et al.* [133, 134] applied the concept of Sb-mediated growth for the preparation of two-dimensionally grown $Ge_{1-y}C_y$ layers on Si. First a Si buffer layer was grown followed by the deposition of one monolayer of antimony. A 300 Å $Ge_{1-y}C_y$ layer was then deposited with a growth rate of 0.125 Å/s. The substrate temperature was 500 ^{o}C during the whole growth process. The amount of C coevaporated with Ge was adjusted to attain a total C concentration of 1%. The reflection high energy electron diffraction technique was used to monitor the in-plane lattice spacing of the layer during the growth. Lattice constants perpendicular to the interface ($a_\perp$) and parallel to the interface ($a_\parallel$) were also measured using x-ray techniques and strain in the layers was determined. Kolodzey *et al.* [135] have grown crystalline germanium–carbon alloys having a cubic diamond lattice on Si(100) using MBE at 600 °C. Measurements on thick relaxed alloy layers showed that up to 3 at.% C was incorporated, which reduced the lattice constant and increased the energy gap (0.875 eV) compared to bulk Ge.

One approach for realization of a direct energy-gap group-IV alloy system involves alloying Sn with Si or Ge to form epitaxially stabilized diamond cubic Sn_xGe_{1-x}/Ge and Sn_xSi_{1-x}/Si heterostructures. Growth of Sn-based heterostructures is challenged by the large lattice mismatch between α-Sn and Si (19.5%), very low solid solubility of Sn in crystalline Si (5×10^{19} cm^{-3}), and pronounced Sn segregation to the surface during growth at ordinary Si epitaxy temperatures ($T > 400$ °C). Growth conditions similar to MBE on Sb δ-doped layers in Si [137] had to be used for the growth incorporating Sn at high fractions while maintaining a high epitaxial quality. Min and Atwater [138] have fabricated ultrathin, coherently strained Sn/Si and Sn_xSi_{1-x}/Si alloy quantum well structures with substitutional Sn incorporation far in excess of the equilibrium solubility limit via substrate temperature and growth flux modulations in MBE. Sn/Si single and multiple quantum wells with Sn coverage up to 1.3 ML, $Sn_{0.05}Si_{0.95}$/Si multiple quantum wells of up to 2.0 nm and $Sn_{0.16}Si_{0.84}$/Si multiple quantum wells up to 1.1 nm were determined to be pseudomorphic.

He *et al.* [139] have grown strain compensated layers of $Si_y(Sn_xC_{1-x})_{1-y}$ alloy films on Si(100) with composition of Sn and C greatly exceeding their normal equilibrium solubility in Si. Amorphous SiSnC alloys were deposited by molecular beam deposition from solid sources followed by thermal annealing. *In situ* monitoring of crystallization rate was done using time-resolved reflectivity. Solid phase epitaxy for $Si_{0.098}Sn_{0.01}C_{0.01}$ occurs at a rate about 20 times slower than of bulk Si. The film was found to be dislocation free with good substitutionality of Sn and C.

2.9   Strained Si on Relaxed SiGe

In the following, we discuss the present status of growth of strained Si on relaxed SiGe buffer layers on Si using various techniques. It is now known that the problem of high threading dislocation densities in relaxed layers may be avoided by using a series of low mismatched interfaces and increasing the Ge concentration in steps (step grading) or linearly with a relatively high growth temperature [11, 140–143]. Because of gradual increase of the lattice mismatch in such a buffer, the misfit dislocation network is distributed over the range of compositional grading rather than being concentrated at the interface to the Si substrate. The greatly improved buffer quality via the compositional grading lowers the threading dislocation density by three orders of magnitude and results in a much improved electron mobility at low temperatures. Strained layer epitaxial growth on patterned substrates has been attempted [144] which can reduce epilayer threading dislocation densities by up to two orders of magnitude. Powell *et al.* [145] have proposed a method for producing an almost dislocation-free relaxed SiGe buffer layer on thin silicon-on-insulator (SOI) substrates. In this growth process, the SiGe epitaxial layer relaxes without the generation of threading dislocations within the SiGe layer rather in the bottom ultrathin SOI substrate with a superficial Si thickness less than the SiGe layer thickness.

Experimental studies for the last few years on strained SiGe materials have resulted in a significant progress in the understanding of strain relaxation kinetics and optimization of graded buffer layers with respect to relaxation and surface morphology [146–150]. These parameters are of crucial importance as they are interdependent and are affected by growth temperature, grading rate and composition. It appears that the competition between dislocation nucleation and propagation determines the final threading dislocation density in the film. The compositional grading is believed to promote propagation while suppressing nucleation of dislocations and leads to reduced amounts of surface strain, thus allowing higher growth temperatures [151, 152]. In fact, the use of a compositionally graded, relaxed, $Si_{1-x}Ge_x$ buffer layer has been advocated as "virtual substrate" allowing the strain in the film to be tailored at will (for a detailed discussion on strain adjustment in SiGe buffer layers, see for example an excellent review by Schaffler [153]).

Many methods exist for the deposition of strained Si on thick relaxed $Si_{1-x}Ge_x$ films on Si. Gibbons *et al.* [40, 95] at Stanford were one of the first groups to demonstrate high quality relaxed $Si_{1-x}Ge_x$ on Si. The lamp-heated limited reaction processing reactor has been used to grow linearly graded SiGe buffer layers and strained Si. High quality, epitaxial, relaxed $Si_{1-x}Ge_x$ layers have been grown by rapid thermal processing chemical vapor deposition by Jung *et al.* [154]. Further improvements of the relaxed buffer (step-graded) layer formation using APCVD with intermediate *in situ* annealing at high temperature have been reported by Kissinger *et al.* [155]. Threading dislocation densities as low as 100 cm^{-2} were found indicating that most of the misfit dislocations extended throughout the wafer.

High quality completely lattice-relaxed SiGe buffer layers have been grown on Si(001) using MBE in the temperature range 750–900 ^{o}C with compositional grading of the order of 10% μm^{-1} or less with final Ge concentrations of about 30%. Xie *et al.* [156] have grown compositionally graded relaxed Si$_{1-x}$Ge$_x$ buffer layers on Si with various composition gradients and temperatures. The authors reported a threading dislocation density in the range of 10^5–10^6 cm^{-2} in fully relaxed SiGe buffer layers grown using both MBE and rapid thermal chemical vapor deposition [157]. Gas-source molecular beam epitaxy [158, 159] has also been successfully employed for the growth of high quality completely lattice-relaxed step-graded SiGe buffer layers on Si(001) in the temperature range 750–800 °C. A more abrupt compositional transience of the SiGe/Si interface is expected in gas source MBE grown quantum wells, owing to reduced Ge segregation at the heterointerface [160] compared with in those grown by solid-source MBE where Ge segregation has been recognized as an important issue [161].

Another advantage of GSMBE is that uniform thickness and composition can be obtained without sample rotation. However, GSMBE is associated with autodoping of dopant gas impurities, which would affect the device characteristics. The characteristic cross-hatch surface roughness and the underlying strain fields of the misfit array can overlap, blocking threading dislocation glide leading to dislocation pile-ups during the growth of thick and high Ge content buffer layers. A method of controlling threading dislocation density in relaxed graded buffers containing 50 to 100% Ge has been developed by Currie *et al.* [162] using chemical-mechanical polishing (CMP) though the release of immobile dislocations located in dislocation pile-ups.

The growth of only a micron thick relaxed buffer layer is possible [163] using step-wise graded buffer based on a combination of Si$_{1-x}$Ge$_x$ and Si$_{1-x-y}$Ge$_x$C$_y$. The buffer concept is based on the fact that the addition of C to a SiGe layer not only reduces the strain, but also stabilizes the layer. Due to the very strong local strain field around the individual carbon atoms, dislocation glide requires a higher energy in SiGeC than in unperturbed, strain equivalent SiGe on Si. Using this method, a threading dislocation density below 10^5 cm^{-2} was obtained for a 73% relaxed homogeneous Si$_{0.7}$Ge$_{0.3}$ layer on top of the buffer structure. A stepped Si$_{1-x}$Ge$_x$ buffer with the identical thickness and strain profile grown at the same temperature shows a threading dislocation density above 10^7 cm^{-2}. The ternary SiGeC material should therefore be considered like a new material with its own strain degree and relaxation behavior rather than like a SiGe film with artificially reduced strain.

Structural characterization and film quality of strained Si layers are usually carried out by transmission electron microscopy (both plan-view and cross-section) for the determination of defects/dislocations within the layers and thickness of the layers and energy dispersive spectroscopy (EDS) utilizing a scanning TEM for film composition. Rutherford backscattering spectroscopy also yields similar information. High resolution x-ray diffraction is typically used to determine strain. In this technique, the lattice constant perpendicular to the sample surface, $a_\perp$, is determined. To extract the Ge content and strain state of the film, it is also necessary to find $a_\parallel$. This is achieved by measuring the Ge content by some other method (e.g., RBS) and measuring $a_\parallel$ by the grazing incidence x-ray technique. An alternative method that measures the strain in the film directly is Raman spectroscopy. Sputter depth profiling by secondary ion mass spectroscopy is used to study the chemical composition, e.g., Ge profile. However, this method has a limiting depth resolution of about 3 nm and cannot give a reasonable profile for ultrathin films.

Fitzgerald *et al.* [140, 164] have used x-ray diffraction and conventional plan-view and cross-sectional transmission electron microscopy for the determination of strain relaxation. Compositionally graded (10% Ge/μm) $Si_{1-x}Ge_x$ films grown at 900 °C with both MBE and rapid thermal chemical vapor deposition reveal that for $0.10 < x < 0.53$ the layers are totally relaxed. $Si_{1-x}Ge_x$ cap layers grown on these graded layers are free of threading dislocations when examined with plan-view and cross-sectional transmission electron microscopy. Electron beam induced current (EBIC) images were used to count the threading dislocation densities. The dislocation densities measured by EBIC for MBE grown samples with final Ge concentrations of 23, 32 and 50% were found to be $4.4 \times 10^5 \pm 5 \times 10^4$, $1.7 \times 10^6 \pm 1.5 \times 10^5$ and $3.0 \times 10^6 \pm 2 \times 10^6$ cm^{-2}, respectively.

2.10 Poly-SiGe Films

Doped polycrystalline silicon (poly-Si) is commonly used as gate material in MOS structures and as contact material for bipolar transistors. Poly-SiGe film has emerged as a new material for advanced CMOS technology and low temperature thin film transistor (TFT) fabrication for large area display electronics. Commonly used source gases for deposition of poly-SiGe are SiH_4 and GeH_4 in the temperature range 400–600 °C.

Rapid thermal CVD has been employed to grow poly-SiGe films [5]. However, these films suffer from higher oxygen content and difficulty of nucleation on oxide. VLPCVD has been found to yield high quality films on oxidized Si substrates at a lower temperature (400–600 °C) and at pressures < 4 mTorr using an rf power of 4 W only. Compared to LPCVD, the plasma enhanced deposition technique gives higher growth rate, smaller grain sizes, direct deposition on oxide, and improved structural properties such as smoother surface and a more columnar grain structure.

King and Saraswat [165] have used the conventional LPCVD technique to deposit polycrystalline $Si_{1-x}Ge_x$ films of different compositions at 625 ^{o}C and at pressures between 0.1 to 0.2 Torr using SiH_4 and GeH_4 as the precursor gas sources. The films were heavily doped with boron and phosphorus by ion implantation at a dose of 4×10^{15} cm^{-2} with energy 20 keV for boron and 60 keV for phosphorus and annealed at 900 ^{o}C for 40 min in Ar [166]. The resistivity of boron-doped poly-SiGe films was substantially less than that of poly-Si films with the same doping level, the resistivity decreasing with Ge mole fraction up to $x = 0.6$. On the other hand, the resistivity of phosphorus doped films decreased slightly up to a Ge mole fraction of $x = 0.45$ and increased considerably for higher Ge mole fractions.

The rapid thermal annealing led to a further reduction of resistivity [167] at a higher Ge fraction even at a low peak anneal temperature (500–700 ^{o}C). It was thus evident that the temperature required to activate boron decreases dramatically with Ge mole fraction. For example, as reported in References 166 and 168, the resistivity of $Si_{1-x}Ge_x$ film of thickness 25 nm annealed for 30 s at 500 ^{o}C is about 10 Ω cm which is half that of a Si film annealed for 30 s at 900–1000 ^{o}C. The mobility also increases significantly with Ge content. Heavily boron doped $Si_{1-x}Ge_x$ films with $x = 0.5$ may be a superior gate electrode material for MOSFETs and as a good interconnect layer in VLSI. On the other hand, heavily phosphorus doped poly-$Si_{1-x}Ge_x$ films degrade with increasing Ge content above 35%. Since the electron affinity does not change appreciably with the mixing of Ge in Si, the work function of p-type $Si_{1-x}Ge_x$ can be more effectively controlled by changing the Ge content compared with the n-type material. This was experimentally confirmed by King *et al.* [166].

Bang *et al.* [169] have reported sheet resistance, Hall mobility and effective carrier concentration as a function of annealing parameters for boron and phosphorus ion implanted films of poly-Si, $Si_{0.75}Ge_{0.25}$ and $Si_{0.50}Ge_{0.50}$. The films were ion implanted with boron or phosphorus at dosages between 5×10^{14} and 4×10^{15} cm^{-2} and then thermally annealed between 550 and 650 ^{o}C from 0.25 to 120 min. Boron doped films showed decreasing minimum sheet resistance with increasing Ge mole fraction (see figure 2.21), while phosphorus doped films exhibited the reverse trend as discussed earlier.

2.11   a-SiGe:H Films

Hydrogenated amorphous silicon–germanium alloys are being developed for the narrow bandgap material in tandem solar cells, for optical detection and image sensing. The bandgap of a-SiGe:H can be varied from 1.75 to 1.0 eV by changing the Ge content to make the material suitable for detection of light

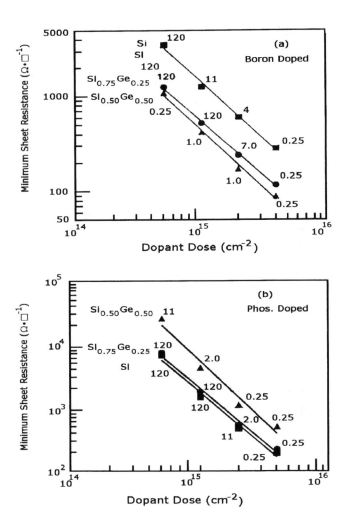

*Figure 2.21. Minimum sheet resistance of (a) boron and (b) phosphorus doped poly-$Si_{1-x}Ge_x$ thin films vs. dopant dose. Anneal temperature is 650 °C and anneal time (in min) required to reach minimum sheet resistances are indicated by each data point. After D. S. Bang et al., Appl. Phys. Lett., **66**, 195–197 (1995).*

emitted from commercial laser diodes or LEDs. The thin film can be deposited at a low temperature of about 250 °C on glass substrates.

Chemical vapor deposition methods, such as plasma enhanced CVD, microwave plasma CVD and photo CVD, have been used for growing a-SiGe:H thin films. The films are deposited by mixing H_2 with source gases: silane (SiH_4) and germane (GeH_4). The bandgap is primarily controlled by the germane fraction $f = GeH_4/(GeH_4+SiH_4)$. Typical conditions of rf plasma assisted deposition are: rf power 30–60 mW/cm^2, pressure 0.2 to 1.0 Torr, flow rates of germane–silane mixture 2–20 sccm, hydrogen flow rate 5–50 sccm. Dilution with hydrogen causes a small decrease of the bandgap and improves the structural and electronic properties. In the presence of dilute hydrogen gas, the dissociation of GeH_4 is enhanced relative to SiH_4 and the incorporation of Ge is thereby increased [170].

Evidence from small angle x-ray scattering confirms that the degradation of electronic properties of a-SiGe:H with Ge content greater than 0.2 is due to microstructural growth. Microstructural defects can be removed to some extent by proper use of hydrogen dilution of source gases and by controlling the rf power density.

2.12   Etching of Alloy Layers

At present, various processes, i.e., the non-passivated or passivated self-aligned double mesa process [171] and a passivated self-aligned single mesa process [172], are used for the fabrication of SiGe devices. The thickness of the $Si_{1-x}Ge_x$ base in HBTs is typically 500 Å or less, and for the fabrication of a device a process is required which can etch Si but stop on the $Si_{1-x}Ge_x$ base. Modulation doped field effect transistors grown on relaxed $Si_{1-x}Ge_x$ layers with a strained Si channel require etches which can etch $Si_{1-x}Ge_x$ selectively over Si.

Several studies using continuous wave rf plasma etch using both chlorine- and fluorine-based chemistries to etch $Si_{1-x}Ge_x$ have been reported [173–178]. Halogen-based plasma etch studies of $Si_{1-x}Ge_x$ alloys have shown that fluorine-based plasmas produce a significant increase in the $Si_{1-x}Ge_x$ etch rate as the Ge content increases. The $Si/Si_{1-x}Ge_x$ etch selectivity is controlled in a way similar to that of bulk Si and Ge and rf power is also a significant factor. Selective reactive ion etching of Si over $Si_{1-x}Ge_x$ and $Si_{1-x}Ge_x$ over Si has been performed using a modulation-frequency plasma-etch technique employing CHF_3 and H_2 as the etch precursor gases [179].

Etching was performed [179] for typical processing conditions: 3.5 sccm of CHF_3, 38 sccm of H_2, 50 mTorr chamber pressure, 0.5 W/cm^2 applied peak rf power and –360 V self-bias. The experimental results are shown in figure 2.22. The etch rates of both Si and $Si_{1-x}Ge_x$ were found to increase linearly with increasing power. Both chamber pressure and gas flow rate had little effect on the Si etch rate. The mean etch rate of strained $Si_{1-x}Ge_x$ alloy was investigated as a function of Ge mole fraction from $x = 0.0$ to $x = 0.20$, along with a relaxed $Si_{0.75}Ge_{0.25}$ film grown on a virtual substrate. The etch rate of the resist was approximately 30 Å/min in all cases.

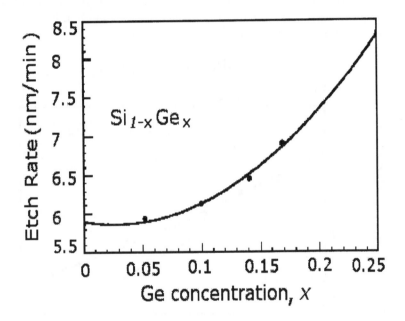

*Figure 2.22. Mean $Si_{1-x}Ge_x$ etch rate as a function of Ge mole fraction. After D. J. Paul et al., J. Vac. Sci. Technol., B, **13**, 2234–2237 (1995).*

Under these conditions the $Si_{1-x}Ge_x$ etch rate is found to increase geometrically as the $Si_{1-x}Ge_x$ alloy becomes more metallic in nature with increasing Ge mole fraction. An etch rate increase of 40% was observed for $Si_{0.75}Ge_{0.25}$ compared to Si (59 Å/min for bulk Si to 83 Å/min for $Si_{0.75}Ge_{0.25}$). The enhanced $Si_{1-x}Ge_x$ etch rate is consistent with the SiGe heterobond (Si–Si = 3.25 eV, Si–Ge = 3.12 eV) and chemical reactivity of the Ge outer electronic shell (Si = $3s^23p^2$, Ge = $4s^24p^2$). The results also indicate, within experimental error, that the $Si_{1-x}Ge_x$ etch rate is not significantly affected by strain within the $Si_{1-x}Ge_x$ lattice. A monotonic increase in the $Si_{1-x}Ge_x$ etch rate with

Ge content, however, has been observed for the chlorine-based etch precursor gases CF_2Cl_2 [177].

In figure 2.23, the etch rates of Si and $Si_{0.75}Ge_{0.25}$ are shown as a function of plasma modulation frequency. It can be seen that both films exhibit a strong exponential decay dependence with increasing modulation frequency. The Si etch rate dependence falls from 44 to 31 Å/min, while the $Si_{0.75}Ge_{0.25}$ etch rate dependence, in comparison, falls from 46 to 26 Å/min. For modulation frequencies less than 2 Hz, selective etching of $Si_{0.75}Ge_{0.25}$ over Si was achieved. For modulation frequencies greater than 3 Hz, Si over $Si_{0.75}Ge_{0.25}$ etch selectivity becomes increasingly strong.

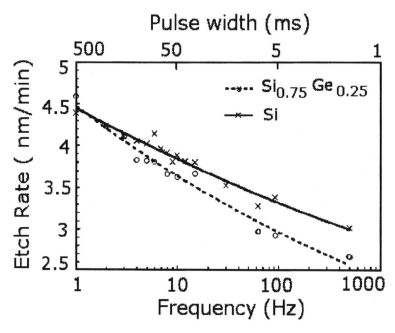

*Figure 2.23. Mean Si and $Si_{0.75}Ge_{0.25}$ etch rates as a function of modulation frequency. After D. J. Paul et al., J. Vac. Sci. Technol., B, **13**, 2234–2237 (1995).*

2.13 Summary

Growth of various alloy layers has been covered in this chapter. It is seen that each of the growth techniques presently in use has limitations and advantages.

MBE offers perhaps the broadest range of growth conditions and is mostly used as a research tool because of its low throughput. UHVCVD offers the possibility of multiwafer growth, which is certainly attractive for volume production. Atmospheric pressure CVD, using the dichlorosilane chemistry, represents an important development. It can be stated that insight into the detailed growth mechanism is rapidly progressing now based on investigations of fundamental surface science.

There now exists a detailed and mostly quantitative understanding of the epitaxial growth and cleaning procedures. However, growth rate data in the low temperature regime as a function of pressure over a broader pressure range (10^{-3}–100 Torr) are needed. Additionally, fundamental surface studies using modern analytical techniques, preferably in real CVD conditions in view of the very important role played by H, are necessary to remove much of the uncertainty from the development of epitaxial growth processes.

A variety of methods exists to deposit high quality alloy layers. In addition to depositing layers with germanium concentrations of at least 15%, control of the germanium profile to within 1% is desirable for bandgap grading. The use of $Si/Si_{1-x}Ge_x$ heteroepitaxial structures for heterojunction devices is hindered by the lattice mismatch between the two materials. However, strained $Si_{1-x}Ge_x$ layers can be deposited on silicon at or above the Matthews-Blakeslee critical thickness curve without interfacial dislocations. Typical bandgap engineering applications may require up to 150 meV bandgap difference. Therefore, the deposition technique must be able to deposit $Si_{1-x}Ge_x$ layers with germanium concentrations of at least 20%. Layers deposited above the Matthews-Blakeslee curve must contend with thermal relaxation during thermal processing. Unfortunately, the Matthew-Blakeslee critical thickness at 20% germanium is only about 20 nm, and is a limitation for applications requiring higher Ge mole fractions. Partially strain-compensated or fully strain-compensated SiGeC films may extend the application areas.

Since standard Si processing steps, such as implantation annealing, typically exceed the strained layer deposition temperature, thermal stability of strained layers is of utmost importance. The Matthews–Blakeslee curve imposes severe limitations on stable strained layer thickness and Ge concentration. Understanding the relaxation processes of metastable group-IV alloy layers is imperative if thicknesses and Ge concentrations greater than the equilibrium curve are needed.

2.14 References

1 B. S. Meyerson, "UHV/CVD Growth of Si and Si-Ge Alloys: Chemistry,

Physics, and Device Applications," *Proc. IEEE*, vol. 80, pp. 1592–1608, 1992.

2 J. C. Bean, "Silicon-Based Semiconductor Heterostructures: Column IV Bandgap Engineering," *Proc. IEEE*, vol. 80, pp. 571–587, 1992.

3 S. C. Jain, *Germanium–Silicon Strained Layers and Heterostructures*. Academic Press Inc., New York, 1994.

4 E. Kasper, Editor, *Properties of Strained and Relaxed Silicon Germanium*. INSPEC, Institute of Electrical Engineers, London, 1995.

5 C. A. King, J. L. Hoyt, C. M. Gronet, J. F. Gibbons, M. P. Scott and J. Turner, "Si/Si$_{1-x}$Ge$_x$ heterojunction bipolar transistors produced by limited reaction processing," *IEEE Electron Dev. Lett.*, vol. EDL-10, pp. 52–54, 1989.

6 B. S. Meyerson, "Low temperature silicon epitaxy by ultrahigh vacuum/chemical vapor deposition," *Appl. Phys. Lett.*, vol. 48, pp. 797–799, 1986.

7 P. Agnello, T. O. Sedgwick, M. S. Goorsky, J. Ott, T. S. Kuan and G. Scilla, "Selective growth of silicon–germanium alloys by atmospheric-pressure chemical vapor deposition at low temperatures," *Appl. Phys. Lett.*, vol. 59, pp. 1479–1481, 1991.

8 T. O. Sedgwick, M. Berkenblit and T. S. Kuan, "Low-temperature selective epitaxial growth of silicon at atmospheric pressure," *Appl. Phys. Lett.*, vol. 54, pp. 2689–2691, 1989.

9 T. O. Sedgwick, V. P. Kesan, P. D. Agnello, D. A. Grutzmacher, D. Nguyen-Ngoc, S. S. Iyer, D. J. Meyer and A. P. Ferro, "Characterization of Devices Fabricated in Films Grown at Low Temperature by Atmospheric Pressure CVD," in *IEEE IEDM Tech. Dig.*, pp. 451–454, 1991.

10 C. C. Cheng and J. T. Yates, Jr., "H-induced surface restructuring on Si(100): formation of higher hydrides," *Phys. Rev.*, vol. B43, pp. 4041–4045, 1991.

11 B. S. Meyerson, K. J. Uram and F. K. LeGoues, "Cooperative phenomena in silicon/germanium low temperature epitaxy," *Appl. Phys. Lett.*, vol. 53, pp. 2555–2557, 1988.

12 P. Garone, J. C. Sturm, P. V. Schartz, S. A. Schwartz and B. J. Wilkens, "Silicon vapor phase epitaxial growth catalysis by the presence of germane," *Appl. Phys. Lett.*, vol. 56, pp. 1275–1277, 1990.

13 S. M. Gates, C. M. Greenlief and D. B. Beach, "Decomposition mechanisms of SiH_x species on $Si(100)$–(2×1) for $x = 2$, 3 and 4," *J. Chem. Phys.*, vol. 93, pp. 7493–7503, 1990.

14 D. W. Greve, "Growth of epitaxial germanium–silicon heterostructures by chemical vapor deposition," *Mater. Sci. Eng.*, vol. B18, pp. 22–51, 1993.

15 F. Hirose, M. Suemitsu and N. Miyamoto, "Silane adsorption on $Si(001)2\times1$," *J. Appl. Phys.*, vol. 70, pp. 5380–5384, 1991.

16 C. M. Greenlief, P. C. Wankum, D. A. Klug and L. A. Keeling, "Surface reactions of Ge containing organometallics on $Si(100)$," *J. Vac. Sci. Technol. A*, vol. 10, pp. 2465–2469, 1992.

17 D. J. Robbins, J. L. Glasper, A. G. Cullis and W. Y. Leong, "A model of heterogeneous growth of $Si_{1-x}Ge_x$ films from hydrides," *J. Appl. Phys.*, vol. 69, pp. 3729–3732, 1991.

18 M. Racanelli and D. Greve, "Temperature dependence of growth of Ge_xSi_{1-x} by ultrahigh vacuum chemical vapor deposition," *Appl. Phys. Lett.*, vol. 56, pp. 2524–2526, 1990.

19 D. J. Robbins, P. Calcott and W. Y. Leong, "Electroluminescence from a pseudomorphic $Si_{0.8}Ge_{0.2}$ alloy," *Appl. Phys. Lett.*, vol. 59, pp. 1350–1352, 1991.

20 M. Racanelli, D. Greve, M. K. Hatalis and L. J. van Yzendoorn, "Alternate surface cleaning approaches for ultrahigh vacuum chemical vapor deposition epitaxy of Si and Ge_xSi_{1-x}," *J. Electrochem. Soc.*, vol. 138, pp. 3783–3789, 1991.

21 P. D. Angello, T. O. Sedgwick, M. S. Goorsky and J. Coue, "Heavy arsenic doping of silicon grown by atmospheric pressure chemical vapor deposition at low temperatures," *Appl. Phys. Lett.*, vol. 60, pp. 454–456, 1992.

22 P. D. Agnello, T. O. Sedgwick, K. C. Bretz and T. S. Kuan, "Silicon epitaxy from silane by atmospheric pressure chemical vapor deposition at low temperatures," *Appl. Phys. Lett.*, vol. 61, pp. 1298–1300, 1992.

23 P. A. Coon, P. Gupta, M. L. Wise and S. M. George, "Adsorption and desorption kinetics for SiH_2Cl_2 on $Si(111)7\times7$," *J. Vac. Sci. Technol. A*, vol. 10, pp. 324–333, 1992.

24 M. Kato, T. Sato, J. Murota and N. Mikoshiba, "Nucleation control of silicon on silicon oxide for low temperature CVD and silicon selective epitaxy," *J. Cryst. Growth*, vol. 99, pp. 240–244, 1990.

25 A. Zaslavsky, D. A. Grutzmacher, Y. H. Lee, W. Ziegler and T. O. Sedgwick, "Selective growth of Si/SiGe resonant tunnelling diodes by atmospheric pressure chemical vapor deposition," *Appl. Phys. Lett.*, vol. 61, pp. 2872–2874, 1992.

26 S. Fukatsu, K. Fujita, H. Yaguchi, Y. Shiraki and R. Ito, "Atomistic picture of interfacial mixing in the Si/Ge heterostructures," *Surf. Sci. Lett.*, vol. 267, pp. 79–82, 1992.

27 N. Usami, S. Fukatsu and Y. Shiraki, "Abrupt compositional transition in luminescent $Si_{1-x}Ge_x$/Si quantum well structures fabricated by segregant assisted growth using Sb adlayer," *Appl. Phys. Lett.*, vol. 63, pp. 388–390, 1993.

28 W. C. Tsai, C. Y. Chang, T. G. Jung, T. S. Liou, G. W. Huang, T. C. Chang, L. P. Chen and H. C. Lin, "Abruptness of Ge composition at the Si/SiGe interface grown by ultrahigh vacuum chemical vapor deposition," *Appl. Phys. Lett.*, vol. 67, pp. 1092–1094, 1995.

29 J. Mi, P. Warren, P. Letourneau, M. Judelewicz, M. Gailhanou, M. Dutoit, C. Dubois and J. C. Dupuy, "High quality $Si_{1-x-y}Ge_xC_y$ epitaxial layers grown on (100) Si by rapid thermal chemical vapor deposition using methylsilane," *Appl. Phys. Lett.*, vol. 67, pp. 259–261, 1995.

30 Z. Atzmon, A. E. Bair, E. J. Jaquez, J. W. Mayer, D. Chandrasekhar, D. J. Smith, R. L. Hervig and M. D. Robinson, "Chemical vapor deposition of heteroepitaxial $Si_{1-x-y}Ge_xC_y$ films on (001)Si Substrates," *Appl. Phys. Lett.*, vol. 65, pp. 2559–2561, 1994.

31 J. L. Regolini, F. Gisbert, G. Dolino and P. Boucaud, "Growth and characterization of strain compensated $Si_{1-x-y}Ge_xC_y$ epitaxial layers," *Mater. Lett.*, vol. 18, pp. 57–60, 1993.

32 J. Kouvetakis, M. Todd, D. Chandrasekhar and D. J. Smith, "Novel chemical routes to silicon–germanium–carbon materials," *Appl. Phys. Lett.*, vol. 65, pp. 2960–2962, 1994.

33 L. D. Lanzerotti, A. St. Amour, C. W. Liu and J. C. Sturm, "$Si_{1-x-y}Ge_xC_y$/Si heterojunction bipolar transistors," in *IEEE IEDM Tech. Dig.*, pp. 930–932, 1994.

34 P. Boucaud, C. Francis, A. Larre, F.H. Julien, J.-M. Lourtioz, D. Bouchier, S. Bodnar and J. L. Regolini, "Photoluminescence of strained $Si_{1-y}C_y$ alloys grown at low temperature," *Appl. Phys. Lett.*, vol. 66, pp. 70–72, 1995.

35 J. Mi, P. Warren, M. Gailhanou, J. D. Ganiere, M. Dutoit, P. H. Jouneau and R. Houriet, "Epitaxial growth of $Si_{1-x-y}Ge_xC_y$ alloy layers on (100) Si by rapid thermal chemical vapor deposition using methylsilane," *J. Vac. Sci. Technol. B*, vol. 14, pp. 1660–1669, 1996.

36 K. Eberl, K. Brunner and W. Winter, "Pseudomorphic $Si_{1-y}C_y$ and $Si_{1-x-y}Ge_xC_y$ alloy layers on Si," *Thin Solid Films*, vol. 294, pp. 98–104, 1997.

37 J. C. Bean, T. T. Sheng, L. C. Feldman, A. T. Fiory and R. T. Lynch, "Pseudomorphic growth of Ge_xSi_{1-x} on silicon by molecular beam epitaxy," *Appl. Phys. Lett.*, vol. 44, pp. 102–104, 1984.

38 G. S. Higashi, J. C. Bean, C. Buescher, R. Yadvish and H. Temkin, "Improved minority carrier lifetime in Si/SiGe heterojunction bipolar transistors grown by molecular beam epitaxy," *Appl. Phys. Lett.*, vol. 56, pp. 2560–2562, 1990.

39 H. Hirayama, M. Hiroi, K. Koyama and T. Tatsumi, "Selective heteroepitaxial growth of $Si_{1-x}Ge_x$ using gas-source molecular beam epitaxy," *Appl. Phys. Lett.*, vol. 56, pp. 1107–1109, 1990.

40 J. F. Gibbons, C. M. Gronet and K. E. Williams, "Limited reaction processing: Silicon epitaxy," *Appl. Phys. Lett.*, vol. 47, pp. 721–723, 1985.

41 T. Ghani, J. L. Hoyt, D. B. Noble and J. F. Gibbons, "Effect of oxygen on minority carrier lifetime and recombination currents in $Si_{1-x}Ge_x$ heterostructure devices," *Appl. Phys. Lett.*, vol. 58, pp. 1317–1319, 1991.

42 M. L. Green, D. Brasen and H. Luftman, "High-Quality Homoepitaxial Silicon Films Deposited by Rapid Thermal Chemical Vapor Deposition," *J. Appl. Phys.*, vol. 65, pp. 2558–2560, 1989.

43 D. W. McNeill, B. M. Armstrong and H. S. Gamble, "Low Temperature Epitaxial Silicon Growth in a Rapid Thermal Processor," *Mat. Res. Soc. Symp. Proc.*, vol. 224, pp. 235–240, 1991.

44 D. Noble, J. L. Hoyt, C. A. King and J. F. Gibbons, "Reduction in misfit dislocation density by the selective growth of $Si_{1-x}Ge_x$/Si in small areas," *Appl. Phys. Lett.*, vol. 56, pp. 51–53, 1990.

45 D. J. Meyer and T. I. Kamins, "The deposition of Si-Ge strained layers from GeH_4, SiH_2Cl_2, SiH_4 and Si_2H_6," *Thin Solid Films*, vol. 222, pp. 30–33, 1992.

46 T. I. Kamins and D. J. Meyer, "Kinetics of silicon–germanium deposition by atmospheric-pressure chemical vapor deposition," *Appl. Phys. Lett.*, vol. 59, pp. 178–180, 1991.

47 W. de Boer and D. Meyer, "Low-temperature chemical vapor deposition of epitaxial Si and SiGe at atmospheric pressure," *Appl. Phys. Lett.*, vol. 58, pp. 1286–1288, 1991.

48 H. Kuhne, Th. Morgenstern, P. Zaumseil, D. Kruger, E. Bugiel and G. Ritter, "Chemical vapor deposition of epitaxial SiGe thin films from SiH_4–GeH_4–HCl–H_2 gas mixtures in an atmospheric pressure process," *Thin Solid Films*, vol. 222, pp. 34–37, 1992.

49 T. I. Kamins, "Pattern sensitivity of selective $Si_{1-x}Ge_x$ chemical vapor deposition: Pressure dependence," *J. Appl. Phys.*, vol. 74, pp. 5799–5802, 1993.

50 T. I. Kamins, D. W. Vook, P. K. Yu and J. E. Turner, "Kinetics of selective epitaxial deposition of $Si_{1-x}Ge_x$," *Appl. Phys. Lett.*, vol. 61, pp. 669–671, 1992.

51 J. H. Comfort, L. M. Garverick and R. Reif, "Silicon surface cleaning by low dose argon ion bombardment for low-temperature (750 °C) epitaxial silicon deposition. I. Process considerations," *J. Appl. Phys.*, vol. 62, pp. 3388–3390, 1987.

52 L. M. Garverick, J. H. Comfort, T. R. Yew, R. Reif, F. A. Baiocchi and H. S. Luftman, "Silicon surface cleaning by low dose argon ion bombardment for low-temperature (750 °C) epitaxial silicon deposition. II. Epitaxial quality," *J. Appl. Phys.*, vol. 62, pp. 3398–3404, 1987.

53 T.-R. Yew and R. Reif, "Low temperature *in-situ* surface cleaning of oxide patterned wafers by Ar/H_2 plasma sputter," *J. Appl. Phys.*, vol. 68, pp. 4681–4693, 1990.

54 J. H. Comfort and R. Reif, "*In-situ* arsenic doping of epitaxial silicon at 800 °C by plasma enhanced chemical vapor deposition," *Appl. Phys. Lett.*, vol. 51, pp. 1536–1538, 1987.

55 T.-R. Yew and R. Reif, "Silicon selective epitaxial growth at 800 °C using SiH_4/H_2 assisted by H_2/Ar plasma sputter," *Appl. Phys. Lett.*, vol. 55, pp. 1014–1016, 1989.

56 T.-R. Yew and R. Reif, "Silicon selective epitaxial growth at 800 °C by ultralow-pressure chemical vapor deposition using SiH_4 and SiH_4/H_2," *J. Appl. Phys.*, vol. 65, pp. 2500–2507, 1989.

57 O. W. Holland, C. W. White and D. Fathy, "Novel oxidation process in Ge^+-implanted Si and its effect on oxidation kinetics," *Appl. Phys. Lett.*, vol. 51, pp. 520–522, 1987.

58 D. Fathy, O. W. Holland and C. W. White, "Formation of epitaxial layers of Ge on Si substrates by Ge implantation and oxidation," *Appl. Phys. Lett.*, vol. 51, pp. 1337–1339, 1987.

59 D. Srivastava and B. J. Garrison, "Growth mechanisms of Si and Ge epitaxial films on the dimer reconstructed Si100 surface via molecular dynamics," *J. Vac. Sci. Technol. A*, vol. 8, pp. 3506–3511, 1990.

60 A. Fukami, K. Shoji and T. Nagano, "Silicon heterostructure by germanium ion implantation," in *Extended Abstract SSDM*, pp. 337–340, 1990.

61 Ken-ichi Shoji, A. Fukami and T. Nagona, "Improved crystalline quality of $Si_{1-x}Ge_x$ formed by low temperature germanium ion implantation," *Appl. Phys. Lett.*, vol. 60, pp. 451–453, 1992.

62 F. Corni, S. Frabboni, G. Ottaviani, G. Queirolo, D.Bisero, C. Bresolin, R. Fabbri and M. Servidori, "Solid-phase epitaxial growth of Ge–Si alloys made by ion implantation," *J. Appl. Phys.*, vol. 71, pp. 2644–2649, 1992.

63 A. Fukami, Ken-ichi Shoji, T. Nagano and C. Y. Yang, "Characterization of SiGe/Si heterostructures formed by Ge and C implantation," *Appl. Phys. Lett.*, vol. 57, pp. 2345–2347, 1990.

64 G.-Q. Lu, E. Nygren and M. J. Aziz, "Pressure-enhanced crystallization kinetics of amorphous Si and Ge: Implications for point-defect mechanisms," *J. Appl. Phys.*, vol. 70, pp. 5323–5345, 1991.

65 T. L. Lin, T. George, E. W. Jones, A. Ksendzov and M. L. Huberman, "Elemental boron-doped p^+-SiGe layers grown by molecular beam epitaxy for infrared detector applications," *Appl. Phys. Lett.*, vol. 60, pp. 380–382, 1992.

66 H. Jorke and H. Kibbel, "Boron delta doping in Si and $Si_{0.8}Ge_{0.2}$ layers," *Appl. Phys. Lett.*, vol. 57, pp. 1763–1765, 1990.

67 P. Kuo, J. L. Hoyt, J. F. Gibbons, J. E. Turner and D. Lefforge, "Effects of strain on boron diffusion in Si and $Si_{1-x}Ge_x$," *Appl. Phys. Lett.*, vol. 66, pp. 580–582, 1995.

68 R. A. Metzger and F. G. Allen, "Evaporative antimony doping of silicon during molecular beam epitaxial growth," *J. Appl. Phys.*, vol. 55, pp. 931–940, 1984.

69 L. C. Markert, J. E. Greene, W.-X. Ni, G. V. Hansson and J.-E. Sundgren, "Concentration transient analysis of antimony surface segregation during Si(001) molecular beam epitaxy," *Thin Solid Films*, vol. 206, pp. 59–63, 1991.

70 Y. Koide, S. Zaima, N. Ohshima and Y. Yasuda, "Initial Stage of Growth of Ge on (100)Si by Gas Source Molecular Beam Epitaxy Using GeH_4," *Jap. J. Appl. Phys.*, vol. 28, pp. L690–L693, 1989.

71 H. Hirayama, T. Tatsumi and M. Aizaki, "Selective growth condition in disilane gas-source silicon molecular beam epitaxy," *Appl. Phys. Lett.*, vol. 52, pp. 2242–2244, 1988.

72 A. Yamada, M. Tanda, F. Kato, M. Konagai and K. Takahashi, "Gas-source molecular-beam epitaxy of Si and SiGe using Si_2H_6 and GeH_4," *J. Appl. Phys.*, vol. 69, pp. 1008–1012, 1991.

73 H. Hirayama and T. Tatsumi, "Phosphorus gas doping in gas-source silicon-MBE," *Thin Solid Films*, vol. 184, pp. 125–130, 1990.

74 M. L. Yu, D. J. Vitkavage and B. S. Meyerson, "Doping reaction of PH_3 and B_2H_6 with Si(100)," *J. Appl. Phys.*, vol. 59, pp. 4032–4037, 1986.

75 D. Greve and M. Racanelli, "Incorporation of Boron into UHV/CVD-Grown Germanium–Silicon Epitaxial Layers," *J. Electron. Mater.*, vol. 21, pp. 593–597, 1992.

76 B. S. Meyerson, F. K. LeGoues, T. N. Nguyen and D. L. Harame, "Nonequilibrium boron doping effects in low-temperature epitaxial silicon films," *Appl. Phys. Lett.*, vol. 50, pp. 113–115, 1987.

77 V. Venkataraman and J. C. Sturm, "Single and symmetric double two-dimensional hole gases at Si/SiGe heterojunctions grown by rapid thermal chemical vapor deposition," in *Mat. Res. Soc. Symp. Proc.*, vol. 220, pp. 391–396, 1991.

78 P. J. Wang, F. F. Fang, B. S. Meyerson, J. Nocera and B. Parker, "Two-dimensional hole gas in $Si/Si_{0.85}Ge_{0.15}/Si$ modulation-doped double heterostructures," *Appl. Phys. Lett.*, vol. 54, pp. 2701–2703, 1989.

79 B. S. Meyerson and M. L. Yu, "Phosphorus doped polycrystalline silicon via LPCVD, II: Surface interactions of the silane/phosphine/silicon system," *J. Electrochem. Soc.*, vol. 31, pp. 2366–2368, 1984.

80 N. Maity, L.-Q. Xia and J. R. Engstrom, "Effect of PH_3 on the associative chemisorption of SiH_4 and Si_2H_2 on Si(100): Implications on the growth of *in-situ* doped Si thin films," *Appl. Phys. Lett.*, vol. 66, pp. 1909–1911, 1995.

81 S. M. Jang, L. Liao and R. Reif, "Chemical-vapor-deposition of epitaxial silicon–germanium from silane and germane I – Kinetics," *J. Electrochem. Soc.*, vol. 142, pp. 3513–3520, 1996.

82 E. Kasper and C. M. Falco, "Molecular Beam Epitaxy of Silicon, Silicon Alloys and Metals," in *Advanced Silicon and Semiconducting Silicon-Alloy Based Materials and Devices* (J. F. A. Nijs, ed.), pp. 103–140, Institute of Physics Publishing, Bristol, 1995.

83 M. R. Caymax and W. Y. Leong, "Low Thermal Budget Chemical Vapor Deposition Techniques for Si and SiGe," in *Advanced Silicon and Semiconducting Silicon-Alloy Based Materials and Devices* (J. F. A. Nijs, ed.), pp. 141–184, Institute of Physics Publishing, Bristol, 1995.

84 E. Kasper and H. J. Herzog, "Elastic Strain and Misfit Dislocation Density in $Si_{0.92}Ge_{0.08}$ Films on Silicon Substrates," *Thin Solid Films*, vol. 44, pp. 357–370, 1977.

85 G. C. Osbourn, "Strained layer superlattices from lattice mismatched materials," *J. Appl. Phys.*, vol. 53, pp. 1586–1589, 1982.

86 J. C. Bean, L. C. Feldman, A. T. Fiory, S. Nakahara and I.K. Robinson, "Ge_xSi_{1-x}/Si strained layer superlattice growth by molecular beam epitaxy," *J. Vac. Sci. Technol. A*, vol. 2, pp. 436–440, 1984.

87 J. W. Matthews and A. E. Blakeslee, "Defects in epitaxial multilayers - I. Misfit dislocations," *J. Crystal Growth*, vol. 27, pp. 118–125, 1974.

88 F. C. Zhang, J. Singh and P. K. Bhattacharya, "Kinetics of $Si_{1-x}Ge_x$/Si $(0 < x < 1)$ growth by molecular beam epitaxy using disilane and germanium," *Appl. Phys. Lett.*, vol. 67, pp. 85–87, 1995.

89 S. M. Mokler, N. Ohtani, Y.-H. Xie, J. Zhang and B. A. Joyce, "Surface studies during growth of $Si_{1-x}Ge_x$/Si from gaseous Si and Ge hydrides," *J. Vac. Sci. Technol. B*, vol. 11, pp. 1073–1076, 1993.

90 J. Murota, N. Nakamura, M. Kato and N. Mikoshiba, "Low-temperature silicon selective deposition and epitaxy on silicon using the thermal decomposition of silane under ultraclean environment," *Appl. Phys. Lett.*, vol. 54, pp. 1007–1009, 1989.

91 M. Racanelli and D. Greve, "Low-temperature selective epitaxy by ultrahigh-vacuum chemical vapor deposition from SiH_4 and GeH_4," *Appl. Phys. Lett.*, vol. 58, pp. 2096–2098, 1991.

92 D. W. Greve and M. Racanelli, "Construction and operation of an ultrahigh vacuum chemical vapor deposition epitaxial reactor for growth of Ge_xSi_{1-x}," *J. Vac. Sci. Technol. B*, vol. 8, pp. 511–515, 1990.

93 M. L. Green, D. Brasen, H. Temkin, R. D. Yadvish, T. Boone, L. C. Feldman, M. Geva and B. E. Spear, "High Gain Si-Ge Heterojunction Bipolar Transistors Grown by Rapid Thermal Chemical Vapor Deposition," *Thin Solid Films*, vol. 184, pp. 107–115, 1990.

94 M. L. Green, B. E. Weir, D Brasen, W. F. Hsieh, G. Higashi, A. Feygenson, L. C. Feldman and R. L. Headrick, "Mechanically and Thermally Stable Si-Ge Films and Heterojunction Bipolar Transistors

Grown by Rapid Thermal Chemical Vapor Deposition at 900 °C," *J. Appl. Phys.*, vol. 69, pp. 745–751, 1991.

95 J. Hoyt, C. A King, D. B. Noble, C. M. Gronet, J. F. Gibbons, M. P. Scott, S. S. Laderman, S. J. Rosner, K. Nauka, J. Turner and T. I. Kamins, "Limited reaction processing: Growth of $Si_{1-x}Ge_x$/Si for heterojunction bipolar transistor applications," *Thin Solid Films*, vol. 184, pp. 93–106, 1990.

96 D. Dutartre, P. Warren, I. Berbezier and P. Perret, "Low temperature silicon and $Si_{1-x}Ge_x$ epitaxy by rapid thermal chemical vapor deposition using hydrides," *Thin Solid Films*, vol. 222, pp. 52–56, 1992.

97 T. Hsu, B. Anthony, R. Qian, J. Irby, D. Kinosky, A. Mahajan, S. Banerjee, C. Magee and A. Tasch, "Advance in Remote Plasma-enhanced Chemical Vapor Deposition for Low Temperature In Situ Hydrogen Plasma Clean and Si and $Si_{1-x}Ge_x$ Epitaxy," *J. Electron. Mater.*, vol. 21, pp. 65–74, 1992.

98 J. B. Posthill, R. A. Rudder, S. V. Hattangady, G. G. Fountain and R. J. Markunas, "On the feasibility of growing dilute C_xSi_{1-x} epitaxial alloys," *Appl. Phys. Lett.*, vol. 56, pp. 734–736, 1990.

99 S. S. Iyer, K. Eberl, M. S. Goorsky, F. K. LeGoues and J. C. Tsang, "Synthesis of $Si_{1-y}C_y$ alloys by molecular beam epitaxy," *Appl. Phys. Lett.*, vol. 60, pp. 356–358, 1992.

100 S. S. Iyer, K. Eberl, A. R. Powell and B. R. Ek, "$Si_{1-x-y}Ge_xC_y$ ternary alloys - extending Si-based heterostructures," *Microelectron. Engineering*, vol. 19, pp. 351–356, 1992.

101 K. Eberl, S. S. Iyer, J. C. Tsang, M. S. Goorsky and F. K. LeGoues, "The growth and characterization of $Si_{1-y}C_y$ alloys on Si(001) substrate," *J. Vac. Sci. Technol. B*, vol. 10, pp. 934–936, 1992.

102 M. S. Goorsky, S. S. Iyer, K. Eberl, F. LeGoues, J. Angiletto and F. Cardone, "Thermal stability of $Si_{1-x}C_x$/Si strained layer superlattices," *Appl. Phys. Lett.*, vol. 60, pp. 2758–2760, 1992.

103 K. Eberl, S. S. Iyer and F. K. LeGoues, "Strain symmetrization effects in pseudomorphic $Si_{1-y}C_y$/$Si_{1-x}Ge_x$ superlattices," *Appl. Phys. Lett.*, vol. 64, pp. 739–741, 1994.

104 W. Faschinger, S. Zerlauth, G. Bauer and L. Palmetshofer, "Electrical properties of $Si_{1-x}C_x$ alloys and modulation doped $Si/Si_{1-x}Ge_x/Si$ structures," *Appl. Phys. Lett.*, vol. 67, pp. 3933–3935, 1995.

105 J. Mi, P. Letourneau, J. D. Ganiere, M. Gailhanou, M. Dutoit, C. Dubois and J. C. Dupuy, "Silicon–Carbon Random Alloy Epitaxy on Silicon by Rapid Thermal Chemical Vapor Deposition," *Mat. Res. Soc. Symp. Proc.*, vol. 342, pp. 255–259, 1994.

106 S. K. Ray, D. W. McNeill, D. L. Gay, C. K. Maiti, G. A. Armstrong, B. M. Armstrong and H. S. Gamble, "Comparison of $Si_{1-y}C_y$ films produced by solid-phase epitaxy and rapid thermal chemical vapor deposition," *Thin Solid Films*, vol. 294, pp. 149–152, 1997.

107 J. W. Strane, H. J. Stein, S. R. Lee, S. T. Picraux, J. K. Watanabe and J. W. Mayer, "Precipitation and relaxation in strained $Si_{1-y}C_y/Si$ heterostructures," *J. Appl. Phys.*, vol. 76, pp. 3656–3668, 1994.

108 K. Eberl, S. S. Iyer, S. Zollner, J. C. Tsang and F. K. LeGoues, "Growth and strain compensation effects in the ternary $Si_{1-x-y}Ge_xC_y$ alloy system," *Appl. Phys. Lett.*, vol. 60, pp. 3033–3035, 1992.

109 S. John, E. J. Quinones, B. Ferguson, S. K. Ray, C. B. Mullins and S. K. Banerjee, "Surface morphology of $Si_{1-x-y}Ge_xC_y$ epitaxial films deposited by low temperature UHVCVD," in *Mat. Res. Symp. Proc.*, pp. 275-279, 1996.

110 H. J. Osten, E. Bugiel and P. Zaumseil, "Growth of an inverse tetragonal distorted SiGe layer on Si(001) by adding small amounts of carbon," *Appl. Phys. Lett.*, vol. 64, pp. 3440–3442, 1994.

111 H. J. Osten, W. Kissinger, M. Weidner and M. Eichler, "Optical Transition in Strained $Si_{1-y}C_y$ and $Si_{1-x-y}Ge_xC_y$ Layers on Si(001)," in *Mat. Res. Soc. Symp. Proc.*, vol. 379, pp. 199–204, 1995.

112 A. R. Powell and S. S. Iyer, "Silicon–Germanium–Carbon Alloys Extending Si Based Heterostructure Engineering," *Jap. J. Appl. Phys.*, vol. 33, pp. 2388–2391, 1994.

113 B. A. Orner, J. Olowolafe, K. Roe, J. Kolodzey, T. Laursen, J. W. Mayer and J. Spear, "Band gap of Ge rich $Si_{1-x-y}Ge_xC_y$ alloys," *Appl. Phys. Lett.*, vol. 69, pp. 2557–2559, 1996.

114 B. Dietrich, H. J. Osten, H, H. Rucker, M. Methfessel and P. Zaumseil, "Lattice distortion in a strain-compensated $Si_{1-x-y}Ge_xC_y$ layers on silicon," *Phys. Rev. B*, vol. 49, pp. 185–190, 1994.

115 S. Im, J. Washburn, R. Gronsky, N. W. Cheung, K. M. Yu and J. W. Ager, "Optimization of Ge/C ratio for compensation of misfit strain in solid phase epitaxial growth of SiGe layers," *Appl. Phys. Lett.*, vol. 63, pp. 2682–2684, 1993.

116 J. W. Strane, H. J. Stein, S. R. Lee, B. L. Doyle, S. T. Picraux and J. W. Mayer, "Metastable SiGeC formation by solid phase epitaxy," *Appl. Phys. Lett.*, vol. 63, pp. 2786–2788, 1993.

117 X. Lu and N. W. Cheung, "Synthesis of SiGe and SiGeC alloys formed by Ge and C implantation," *Appl. Phys. Lett.*, vol. 69, pp. 1915–1917, 1996.

118 J. Menendez, P. Gopalan, G. S. Spencer, N. Cave, J. W. Strane, "Raman spectroscopy study of microscopic strain in epitaxial $Si_{1-x-y}Ge_xC_y$ alloys," *Appl. Phys. Lett.*, vol. 66, pp. 1160–1162, 1995.

119 C. K. Maiti, "Technical Report, Contract No. CI1*-CT94-0051," 1996.

120 K. Rim, S. Takagi, J. L. Hoyt and J. F. Gibbons, "Capacitance–Voltage Characteristics of p-Si/SiGeC MOS Capacitors," in *Mat. Res. Soc. Symp. Proc.*, vol. 379, pp. 327–332, 1995.

121 P. Boucaud, C. Francis, F. H. Julien, J. M. Lourtioz, D. Bouchier, D. Bodnar, B. Lambert and J. L. Regolini, "Band-edge and deep level photoluminescence of pseudomorphic $Si_{1-x-y}Ge_xC_y$ alloys," *Appl. Phys. Lett.*, vol. 64, pp. 875–877, 1994.

122 A. St. Amour, L. D. Lanzerottu, C. L. Chang and J. C. Sturm, "Optical and electrical properties of $Si_{1-x-y}Ge_xC_y$ thin films and devices," *Thin Solid Films*, vol. 294, pp. 112–117, 1997.

123 A. St. Amour, C. W. Lice, J. C. Sturm, Y. Lacroix and M. L. W. Thewalt, "Defect-free band-edge photo-luminescence and bandgap measurement of pseudomorphic $Si_{1-x-y}Ge_xC_y$ alloy layers on Si(100)," *Appl. Phys. Lett.*, vol. 67, pp. 3915–3917, 1995.

124 J. L. Regolini, S. Bodnar, J. C. Obertin, F. Ferrieu, M. Gauneau, B. Lambert and P. Boucaud, "Strain compensated heterostructures in

the $Si_{1-x-y}Ge_xC_y$ ternary system," *J. Vac. Sci. Technol. A*, vol. 12, pp. 1015–1019, 1994.

125 C. L. Chang, A. St. Amour and J. C. Sturm, "The effect of carbon on the valence band offset of compressively strained $Si_{1-x-y}Ge_xC_y/(100)$ Si heterostructures," *Appl. Phys. Lett.*, vol. 70, pp. 1557–1559, 1997.

126 C. W. Liu, A. St. Amour, J. C. Sturm, Y. R. J. Lacroix, M. L. Thewalt, C. W. Magee and D. Eaglesham, "Growth and photoluminescence of high quality SiGeC random alloys on silicon substrates," *J. Appl. Phys.*, vol. 80, pp. 3043–3047, 1996.

127 C. Guedj, X. Portier, A. Hairie, D. Bouchier, G. Calvarin and B. Piriou, "Carbon and germanium distributions in $Si_{1-x-y}Ge_xC_y$ layers epitaxially grown on Si(001) by RTCVD," *Thin Solid Films*, vol. 294, pp. 129–132, 1997.

128 J. D. Lorentzen, G. H. Loecheit, M. Melendez-Lira, J. Menendez, S. Sego, R. J. Culbertson, W. Windl, O. F. Sankey, A. E. Blair and T. L. Alford, "Photoluminescence in $Si_{1-x-y}Ge_xC_y$ alloys," *Appl. Phys. Lett.*, vol. 70, pp. 2353–2355, 1997.

129 R. Garcia, K. E. Daley, S. Sego, R. J. Culbertson and D. B. Poker, "Sb ion implantation and annealing of SiGeC heteroepitaxial layers on Si(001)," *J. Vac. Sci. Technol. A*, vol. 13, pp. 662–665, 1995.

130 M. Todd, P. Matsunaga and J. Kouvetakis, "Growth of heteroepitaxial $Si_{1-x-y}Ge_xC_y$ alloys on silicon using novel deposition chemistry," *Appl. Phys. Lett.*, vol. 67, pp. 1247–1249, 1995.

131 H.-T. Chiu, C.-S. Shie and S.-H. Chuang, "Growth of ternary $Si_{1-x-y}Ge_xC_y$ thin films from a single source precursor, $Ge(SiMe_3)_4$," *J. Mater. Res.*, vol. 10, pp. 2257–2259, 1995.

132 T. Akane, H. Okumura, J. Tanaka and S. Matsumoto, "New Ge substrate cleaning method for $Si_{1-x-y}Ge_xC_y$ MOMBE growth," *Thin Solid Films*, vol. 294, pp. 153–156, 1997.

133 H. J. Osten, E. Bugiel and P. Zaumseil, "Antimony mediated growth of epitaxial $Ge_{1-y}C_y$ layers on Si(001)," *J. Crystal Growth*, vol. 142, pp. 322–326, 1994.

134 H. J. Osten and J. Klatt, "*In situ* monitoring of strain relaxation during antimony-mediated growth of Ge and $Ge_{1-y}C_y$ layers on Si(001) using reflection high energy electron diffraction," *Appl. Phys. Lett.*, vol. 65, pp. 630–632, 1994.

135 J. Kolodzey, P. A. O'Neil, S. Zhang, B. A. Orner, K. Roe, K. M. Unruh, C. P. Swann, M. M. White and A. I. Shah, "Growth of germanium-carbon alloys on silicon substrates by molecular beam epitaxy," *Appl. Phys. Lett.*, vol. 67, pp. 1865–1867, 1995.

136 F. Chen, R. T. Troger, K. Roe, M. D. Dashell, R. Jonczyk, D. S. Holmes, R. G. Wilson and J. Kolodzey, "Electrical Properties of $Si_{1-x-y}Ge_xC_y$ and $Ge_{1-y}C_y$," *J. Electron. Mater.*, vol. 26, pp. 1371–1375, 1997.

137 H. P. Zeindl, T. Wegehaupt, I. Risele, H. Oppolzer, H. Reissinger, G. Tempel and F. Koch, "Growth and characterization of delta-function doping layer in Si," *Appl. Phys. Lett.*, vol. 50, pp. 1164–1166, 1987.

138 K. S. Min and H. A. Atwater, "Ultrathin pseudomorphic Sn/Si and Sn_xSi_{1-x}/Si heterostructures," *Appl. Phys. Lett.*, vol. 72, pp. 1884–1886, 1998.

139 G. He, M. D. Savellano and H. A. Atwater, "Synthesis of dislocation free $Si_y(Sn_xC_{1-x})_{1-y}$ alloys by molecular beam deposition and solid phase epitaxy," *Appl. Phys. Lett.*, vol. 65, pp. 1159–1161, 1994.

140 E. A. Fitzgerald, Y.-H. Xie, M. L. Green, D. Brasen, A. R. Kortan, J. Michel, Y.-J. Mii and B. E. Weir, "Totally relaxed Ge_xSi_{1-x} layers with low threading dislocation densities grown on Si substrates," *Appl. Phys. Lett.*, vol. 59, pp. 811–813, 1991.

141 F. K. LeGoues, B. S. Meyerson and J. F. Morar, "Anomalous Strain Relaxation in SiGe Films and Superlattices," *Phys. Rev. B*, vol. 66, pp. 2903–2906, 1991.

142 F. Schaffler, D. Tobben, H.-J. Herzog, G. Abstreiter and B. Hollander, "High-electron-mobility Si/SiGe heterostructures: influence of the relaxed SiGe buffer layer," *Semicond. Sci. Technol.*, vol. 7, pp. 260–266, 1992.

143 Y. H. Xie, E. A. Fitzgerald, P. J. Silverman, A. R. Kortan and B. E. Weir, "Fabrication of relaxed GeSi buffer layers on Si(100) with low threading dislocation density," *Mater. Sci. Eng.*, vol. B14, pp. 332–335, 1992.

144 R. Hull, J. C. Bean, G. S. Higashi, M. L. Green, L. Peticolas, D. Bahnck and D. Brasen, "Improvement in heteroepitaxial film quality by a novel substrate patterning geometry," *Appl. Phys. Lett.*, vol. 60, pp. 1468–1470, 1992.

145 A. R. Powell, S. S. Iyer and F. K. LeGoues, "New approach to the growth of low dislocation relaxed SiGe material," *Appl. Phys. Lett.*, vol. 64, pp. 1856–1858, 1994.

146 F. K. LeGoues, M. M. Mooney and J. O. Chu, "Crystallographic tilting resulting from nucleation limited relaxation," *Appl. Phys. Lett.*, vol. 62, pp. 140–142, 1993.

147 F. K. LeGoues, B. S. Meyerson, J. F. Morar and P. D. Kirchner, "Mechanism and conditions for anomalous strain relaxation in graded thin films and superlattices," *J. Appl. Phys.*, vol. 71, pp. 4230–4243, 1992.

148 P. M. Mooney, J. L. Jordan-sweet, K. Ismail, J. O. Chu, R. M. Feenstra and F. K. LeGoues, "Relaxed $Si_{0.7}Ge_{0.3}$ buffer layers for high-mobility devices," *Appl. Phys. Lett.*, vol. 67, pp. 2373–2375, 1995.

149 J. W. P. Hsu, E. A. Fitzgerald, Y. H. Xie, P. J. Silverman and M. J. Cardillo, "Surface morphology of relaxed Ge_xSi_{1-x} films," *Appl. Phys. Lett.*, vol. 61, pp. 1293–1295, 1992.

150 M. A. Lutz, R. M. Feenstra, F. K. LeGoues, P. M. Mooney and J. O. Chu, "Influence of misfit dislocations on the surface morphology of $Si_{1-x}Ge_x$ films," *Appl. Phys. Lett.*, vol. 66, pp. 724–726, 1995.

151 J. H. Li, E. Koppensteiner, G. Bauer, M. Hohnisch, H. J. Herzog and F. Schaffler, "Evolution of strain relaxation in compositionally graded $Si_{1-x}Ge_x$ films on Si(001)," *Appl. Phys. Lett.*, vol. 67, pp. 223–225, 1995.

152 P. M. Mooney, J. L. Jordan-Sweet, J. O. Chu and F. K. LeGoues, "Evolution of strain relaxation in step-graded SiGe/Si structures," *Appl. Phys. Lett.*, vol. 66, pp. 3642–3644, 1995.

153 F. Schaffler, "High-mobility Si and Ge structures," *Semicond. Sci. Technol.*, vol. 12, pp. 1515–1549, 1997.

154 K. H. Jung, Y. M. Kim and D. L. Kwong, "Relaxed Ge_xSi_{1-x} films grown by rapid thermal processing chemical vapor deposition," *Appl. Phys. Lett.*, vol. 56, pp. 1775–1777, 1990.

155 G. Kissinger, T. Morgenstern, G. Morgenstern and H. Richter, "Stepwise equilibrated graded Ge_xSi_{1-x} buffer with very low threading dislocation density on Si(001)," *Appl. Phys. Lett.*, vol. 66, pp. 2083–2085, 1995.

156 Y.-H. Xie, E. A. Fitzgerald, D. Monroe, G. P. Watson and P. J. Silverman, "From Relaxed GeSi Buffers to Field Effect Transistors: Current Status and Future Prospects," *Jap. J. Appl. Phys.*, vol. 33, pp. 2372–2377, 1994.

157 E. A. Fitzgerald, Y. H. Xie, D. Monroe, P. J. Silverman, J. M. Kuo and A. R. Kortan, "Relaxed Ge_xSi_{1-x} structures for III–V integration with Si and high mobility two-dimensional electron gases in Si," *J. Vac. Sci. Technol. B*, vol. 10, pp. 1807–1819, 1992.

158 D. K. Nayak, N. Usami, H. Sunamura, S. Fukatsu and Y. Shiraki, "Band-edge photoluminescence of SiGe/strained-Si/SiGe type-II quantum wells on Si(100)," *Jap. J. Appl. Phys.*, vol. 32, pp. L1391–L1393, 1993.

159 D. K. Nayak, N. Usami, S. Fukatsu and Y. Shiraki, "Band-edge photoluminescence of SiGe/strained-Si/SiGe type-II quantum wells on Si(100)," *Appl. Phys. Lett.*, vol. 63, pp. 3509–3511, 1993.

160 S. Fukatsu, H. Yoshida, A. Fujiwara, Y. Takahashi and Y. Shiraki, "Spectral blue shift of photoluminescence in strained layer $Si_{1-x}Ge_x$/Si quantum well structures grown by gas-source Si MBE," *Appl. Phys. Lett.*, vol. 61, pp. 804–806, 1992.

161 Y. Kato, S. Fukatsu and Y. Shiraki, "Postgrowth of a Si contact layer on air-exposed $Si_{1-x}Ge_x$/Si single quantum well grown by gas-source molecular beam epitaxy, for use in an electroluminescent device," *J. Vac. Sci. Technol,* vol. B13, pp. 111–117, 1995.

162 M. T. Currie, S. B. Samavedam, T. A. Langdo, C. W. Leitz and E. A. Fitzgerald, "Controlling threading dislocation densities in Ge on Si using graded SiGe layers and chemical-mechanical polishing," *Appl. Phys. Lett.*, vol. 72, pp. 1718–1720, 1998.

163 H. J. Osten and E. Bugiel, "Relaxed $Si_{1-x}Ge_x$/$Si_{1-x-y}Ge_xC_y$ buffer structures with low threading dislocation density," *Appl. Phys. Lett.*, vol. 70, pp. 2813–2815, 1997.

164 E. A. Fitzgerald, Y. H. Xie, D. Brasen, M. L. Green, J. Michel, P. E. Freeland and B. E. Weir, "Elimination of dislocations in heteroepitaxial

MBE and RTCVD Ge_xSi_{1-x} grown on patterned Si substrates," *J. Electron. Mat.*, vol. 19, pp. 949–955, 1990.

165 T. J. King and K. C. Saraswat, "Deposition and properties of low-pressure chemical-vapor deposited polycrystalline silicon–germanium films," *J. Electrochem. Soc.*, vol. 141, pp. 2235–2241, 1994.

166 T. J. King, J. P. McVittie, K. C. Saraswat and J. R. Pfiester, "Electrical properties of heavily doped polycrystalline silicon–germanium films," *IEEE Trans. Electron Dev.*, vol. 41, pp. 228–232, 1994.

167 T.-J. King and K. C. Saraswat, "A Low-Temperature ($\leq 550\ ^oC$) Silicon–Germanium MOS Thin-Film Transistor Technology for Large-Area Electronics," in *IEEE IEDM Tech. Dig.*, pp. 567–570, 1991.

168 T. J. King, J. R. Pfriester, J. D. Scott, J. P. McVittie and K. C. Saraswat, "A polycrystalline SiGe gate CMOS technology," in *IEEE IEDM Tech. Dig.*, pp. 253–256, 1990.

169 D. S. Bang, M. Cao, A. Wang, K. C. Saraswat and T.-J. King, "Resistivity of boron and phosphorus doped polycrystalline $Si_{1-x}Ge_x$ films," *Appl. Phys. Lett.*, vol. 66, pp. 195–197, 1995.

170 A. R. Middya, S. C. De and S. Ray, "Improvement in the properties of a-SiGe:H films: Roles of deposition rate and hydrogen dilution," *J. Appl. Phys.*, vol. 73, pp. 4622–4630, 1993.

171 A. Gruhle, "SiGe-HBTs," in *Silicon-based Millimetre Wave Devices*, pp. 149–192, Springer-Verlag, Berlin, 1994.

172 A. Schuppen, A. Gruhle, H. Kibbel and U. Konig, "Mesa and planar SiGe-HBTs on MBE-wafers," *J. Mater. Sci.: Mater. Electron.*, vol. 6, pp. 298–305, 1995.

173 A. A. Bright, S. S. Lyer, S. W. Robey and S. L. Delage, "Technique for selective etching of Si with respect to Ge," *Appl. Phys. Lett.*, vol. 53, pp. 2328–2329, 1988.

174 G. S. Oehrlein, T. D. Bestwick, P. L. Jones and J. W. Corbett, "Selective dry etching of silicon with respect to germanium," *Appl. Phys. Lett.*, vol. 56, pp. 1436–1438, 1990.

175 G. S. Oehrlein, G. M. W. Kroeson, E. de Fresart, Y. Zhang and T. D. Bestwick, "Studies of the reactive ion etching of SiGe alloys," *J. Vac. Sci. Technol. A*, vol. 9, pp. 768–774, 1991.

176 G. S. Oehrlein, Y. Zhang, G. M. W. Kroeson, E. de Fresart and T. D. Bestwick, "Interactive effects in the reactive ion etching of SiGe alloys," *Appl. Phys. Lett.*, vol. 58, pp. 2252–2254, 1991.

177 Y. Zhang, G. S. Oehrlein and E. de Fresart, "Reactive ion etching of SiGe alloys using CF_2/Cl_2," *J. Appl. Phys.*, vol. 71, pp. 1936–1942, 1992.

178 J. G. Couillard and H. G. Craighead, "High-resolution reactive ion etching of SiGe alloys," *J. Vac. Sci. Technol. B*, vol. 11, pp. 717–719, 1993.

179 D. J. Paul, V. J. Law and G. A. C. Jones, "$Si_{1-x}Ge_x$ pulsed plasma etching using CHF_3 and H_2," *J. Vac. Sci. Technol. B*, vol. 13, pp. 2234–2237, 1995.

Chapter 3

Electronic Properties of Alloy Layers

The active region of a Si heterostructure consists of strained layers of different composition, thickness and doping. The composition of a layer determines the bandgap and band offset with the adjacent layers. The polarity and magnitude of the band offset define the depth of a well or the height of a barrier, and thus the channel region. One can assign an average lattice constant to a strained layer. Elastic strains develop when there is a large misfit between adjacent layers. The thickness of a strained layer has a critical upper value. Some properties of binary alloys follow Vegard's law permitting interpolation between the properties of the constituents. The interpolation is not always valid however. Current state of knowledge of critical thickness is reviewed in this chapter.

The intrinsic optical and electronic properties of a semiconducting region are determined by the band structure which depends on the composition, orientation and strain. The first parameter of concern in optical transitions and many electronic processes is the bandgap. Strain in the alloy layer lifts the degeneracy of the conduction and valence bands. An important associated property is the band alignment at a heterojunction which determines whether a layer will act as a well for a channel or provide a barrier defining the channel boundary. The bandgap of a bulk material or thin film is commonly measured by optical means. Band offset can be measured by photoelectron spectroscopy and also from capacitance–voltage (C-V) characteristics.

A basic transport property of concern is the velocity–electric field characteristics of the carriers. This is described at low fields by mobility. For strained alloys mobility depends significantly on the orientation of the field with respect to the growth direction of the alloy. Extensive theoretical studies have been carried out for elucidating the mobility behavior of silicon–germanium alloys. The scattering processes included are intrinsic lattice scattering (acoustic and optical), impurity scattering and alloy scattering. At low fields, intervalley scattering is ignored, but it becomes important as the field and consequently the carrier energy increase.

In a physical device the mobility of prime concern is drift mobility which for the bulk is measured typically by a time-of-flight method. Hall mobility provides an indirect estimate of the mobility behavior and is often a good indicator. Mobility in an actual MOS device is inferred indirectly from the current–voltage relation. Experimental results of electron and hole mobility in strained Si, SiGe and SiGeC are discussed and compared with predictions from simulation.

3.1   Critical Layer Thickness

When an epitaxial Si film is grown on a Si substrate, there is a natural matching of the crystal lattice, and a high quality single crystal layer results. However, it is often desirable to obtain epitaxial layers of materials that differ somewhat from the substrate. This can be accomplished easily if the lattice structure and lattice constant of the two materials match. For example, GaAs and AlAs both have the zinc blende structure, with a lattice constant of about 5.65 Å. As a result, epitaxial layers of the ternary alloy AlGaAs can be grown on GaAs substrates with little lattice mismatch. Similarly, GaAs can be grown on Ge substrates.

Figure 3.1 shows the energy bandgap as a function of lattice constant for elemental semiconductors, Si and Ge, and several III–V and II–VI compound semiconductors. Some of the relevant structural and electronic properties of group-IV materials, i.e., Si, Ge, C and α-Sn are listed in table 3.1. The ratio of the electron to hole mobility (μ–ratio) may be used as a measure of the feasibility of complementary structures for semiconductors. A μ–ratio closer to unity is a good indicator for symmetric complementary structures. Table 3.2 gives the low field mobility and the μ-ratio for common semiconductors. It may be noted from table 3.2 that all III–V compound semiconductors have μ–ratios higher than 8, reaching 21 and 36 for GaAs and InP, respectively. While Ge has a more favorable μ–ratio than Si, some technological problems have excluded Ge from complementary circuit applications.

It will be noticed that group-IV elements have large differences in lattice constants. Due to the large lattice mismatch between Ge and Si ($\approx$ 4.17%), it was not possible until 1985 to grow commensurate layers of $Si_{1-x}Ge_x$ on Si. The development of MBE has made possible the growth of very thin layers (in atomic level) of useful semiconductor materials with precise control of layer thickness and composition. Using MBE, the concept of strained layer

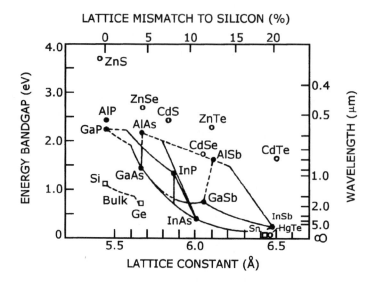

*Figure 3.1. Plot of semiconductor lattice constant a_0, vs. minimum bandgap E_g, for column IV, II–VI, and III–V semiconductors. Lines indicate the bandgap for alloys of semiconductors at the end points. Solid lines denote a resulting direct minimum bandgap, dashed lines an indirect bandgap. Also shown are photon energies equivalent to the minimum bandgap and percentage lattice mismatch to silicon's 5.42 Å lattice constant. After J. C. Bean, Proc. IEEE, **80**, 571–587 (1992), copyright ©1992 IEEE.*

epitaxy [1, 2] was applied to the SiGe/Si material system by Bean *et al.* [3] in 1984. In the case of strained layer epitaxy, lattice mismatched strained layers can be grown without misfit dislocations if the thickness is kept below a critical thickness. In the last decade, increasing focus has been placed upon lattice mismatched combination of materials as these material systems also offer unique new prospects, such as the use of strain to modify electronic and optical properties. Recently, alloys of group-IV elements including C and Sn are being actively pursued for use in heterojunction devices compatible with conventional Si processing technology.

In the plane of the interface between a film and the substrate on which it is grown, the film is forced to adopt the lattice structure of its host substrate. Under the stress imposed by the substrate, the film deforms. A deformed crystalline film is referred to as pseudomorphic since, in planes parallel to

Table 3.1. Room temperature materials data of selected group-IV elements.

Element	C	Si	Ge	α-Sn
lattice	diamond	diamond	diamond	diamond
lattice constant, a_0 (Å)	3.5668	5.431	5.657	6.489
density (g/cm^3)	3.515	2.329	5.323	7.285
TCE α (10^{-6} K^{-1})	1.0	2.56	5.9	4.7
bandgap E_g (eV)	5.48	1.11	0.664	-
dielectric constant ϵ	5.7	11.9	16.2	24
electron mobility μ_e (cm^2/Vs)	1800	1450	3900	1400
hole mobility μ_h (cm^2/Vs)	1600	450	1900	1200
effective mass m^*				
electron $m_e^*(\perp)$	-	0.19	0.08	0.024
electron $m_e^*(\parallel)$	-	0.92	0.64	0.2–0.45
light-hole, $m_h^*(l)$	0.7	0.15	0.043	-
heavy-hole, $m_h^*(h)$	2.18	0.54	0.28–0.38	-

Table 3.2. Bulk undoped low field mobility for electrons and holes for common semiconductors. The ratio of electron and hole mobilities is also indicated.

Semiconductor	μ_n [cm^2/Vs]	μ_p [cm^2/Vs]	μ_n/μ_p
Si	1450	450	3.2
Ge	3900	1900	2.0
GaAs	8500	400	21.2
InP	5400	150	36.0
InAs	33000	450	73.3
GaSb	7700	850	9.0
InSb	70000	8500	8.2

the substrate interface, it acquires the lattice structure of the host substrate, although in the perpendicular direction it may be quite different. The Si$_{1-x}$Ge$_x$ alloy is a substitutional alloy and may be viewed as a crystal in which a fraction x of the Si ions have been randomly substituted by Ge ions. It is recognized that there will be compositional randomness from lattice site to lattice site. Yet the lattice can be approximately treated as though it has a single, average lattice constant given by Vegard's rule.

Due to lattice mismatch in strained layers, strain may develop between the film and the substrate. The strain energy in the epilayer grows in proportion to the pseudomorphic epilayer thickness. When this strain energy exceeds a certain threshold energy, the film strain tends to give way to misfit dislocations. The maximum thickness of the pseudomorphic epilayer, which is thermodynamically stable, is referred to as the "critical layer thickness" and is one of the most important parameters in the pseudomorphic growth of heteroepitaxial films.

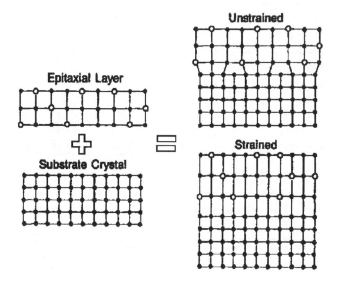

Figure 3.2. Schematic representation of strained layer growth.

Because a significant lattice mismatch exists, the epitaxial layer of $Si_{1-x}Ge_x$ on Si can range from a fully strained to a fully relaxed state as shown in the two-dimensional cross-sectional schematic of figure 3.2. In the fully strained case, the larger $Si_{1-x}Ge_x$ horizontal lattice compresses to match the Si substrate and the $Si_{1-x}Ge_x$ vertical lattice constant expands to accommodate the horizontal compression. The structure stores a high amount of elastic strain energy, because interatomic bond lengths in the epilayer are significantly stretched or compressed with respect to their equilibrium value. At some epilayer thickness, generally called the critical layer thickness h_c, it becomes energetically favorable to reduce this elastic strain energy by introduction of misfit dislocations, as illustrated in figure 3.2, which allow the epilayer to relax towards its bulk lattice parameter.

The stability of strained layers was first studied by Frank and van der

Merwe [4]. The term critical thickness was initially defined to denote the transition from a strained to a relaxed layer. Using a continuum model, they calculated the critical layer thickness, below which the strained layer is expected to be in the thermodynamically stable state. Above h_c, it is energetically favorable to relieve the strain via dislocations. Since then, there have been a number of reports on calculations of h_c [5–9], most of which differ from each other in the assumed energy stored in a dislocation. Van der Merwe [10, 11] was the first to calculate the critical thickness as a function of increased lattice mismatch by minimizing the sum of the interfacial and strain energy. However, most of the published literature favor the mechanical equilibrium theory of Matthews and Blakeslee [6, 12] as defining the transition from the stable to metastable regimes.

Mechanical equilibrium theory assumes the existence of a threading dislocation. The energy required to glide a threading dislocation into a misfit dislocation is balanced with the strain energy from the lattice mismatch to define the critical thickness as a function of lattice mismatch. When the strain energy exceeds a critical value, a misfit dislocation forms to relieve the strain energy. However, the transition from the strained to relaxed case is not abrupt and is not clearly defined as different degrees of strain relaxation can exist [13].

Figure 3.3 shows three regimes in the plot of $Si_{1-x}Ge_x$ layer thickness on Si vs. Ge concentration. The Ge concentration is directly related to the lattice mismatch (the upper x-axis label) according to Vegard's law. Although the Matthews–Blakeslee equilibrium theory is widely cited, strained $Si_{1-x}Ge_x$ layers have been deposited much thicker than the theory predicts. Bean and his coworkers [3] deposited strained layers by MBE at 550 °C with a thickness an order of magnitude or more above the Matthews–Blakeslee curve as shown by the solid data points in figure 3.3.

The authors measured the axial strain as a function of alloy composition using Rutherford backscattering, x-ray diffraction and resonant Raman scattering. Their data were fitted with a theoretical model due to Hull *et al.* [14] and the critical layer thickness was found to vary as

$$h_c = \frac{1-\gamma}{1+\gamma} \frac{b^2}{8\pi w} \epsilon_\parallel^{-2} \ln\left(\frac{h_c}{b}\right) \tag{3.1}$$

where w is the width of an isolated dislocation, taken as 20 Å, γ is the Poisson ratio (≈ 0.3), b is the dislocation Burger's vector (≈ 4 Å) and $\epsilon_\parallel$ is the in-plane strain. The above equation indicates that the critical thickness varies roughly as the reciprocal of the square of the lattice misfit. The critical layer thickness for 1% net strain ($x \sim 0.25$) is about 1000 Å and decreases to 100 Å

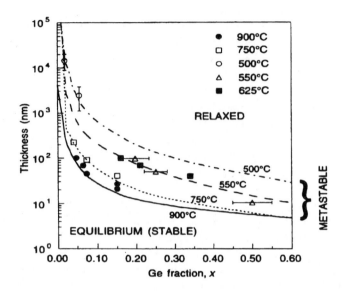

Figure 3.3. Critical thickness of $Si_{1-x}Ge_x$ layers as a function of Ge mole fraction. Lines show theoretical kinetic model for various growth temperatures. After J. J. Welser, The application of strained-silicon/relaxed-silicon germanium heterostructures to metal–oxide–semiconductor field–effect transistors, *PhD Thesis, Stanford University, 1994.*

for 2% strain ($x \sim 0.50$). Under normal growth conditions, there is a thickness dependent energy barrier for the introduction of dislocations [15].

The measured critical thickness depends on growth temperature and growth rate. This was found experimentally: h_c increases with decreasing substrate temperature [9, 16]. The dotted line joining the data points demarcates the metastable and dislocation regimes. Above the dotted line, Bean and his coworkers found strained $Si_{1-x}Ge_x$ layers were impossible to deposit. People and Bean [9] sought to reconcile these differences by including the kinetics of relaxation in their calculation and the critical thickness prediction fits their experimental data. Between the solid mechanical equilibrium line and the dotted MBE data points, layers are labeled metastable. Layers in the metastable regime are strained even though the layers are above the Matthews–Blakeslee critical thickness. Many other researchers have contributed with critical thickness theories based on energy, mechanical equilibrium and kinetics of dislocations [17–19].

Instead of using thin layers of $Si_{1-x}Ge_x$ on a Si substrate, thin layers of Si can be grown epitaxially on higher lattice constant $Si_{1-x}Ge_x$ substrates. The Si film will be strained in biaxial tension. However, bulk $Si_{1-x}Ge_x$ crystalline substrates are not available. The simplest method of strain adjustment in a buffer layer is to grow a constant-Ge SiGe layer well beyond the critical thickness when a high degree of strain relaxation takes place. Initially, $Si_{1-x}Ge_x$ buffer layers with constant Ge content were grown to a thickness somewhat above the critical thickness for strain relaxation by the formation of misfit dislocations [20].

The alternative is to grow Si on a thick relaxed $Si_{1-x}Ge_x$ buffer layer that achieves the lattice constant of the bulk. To reduce the propagation of threading dislocations in the strained Si layer, it is necessary to grow a graded $Si_{1-y}Ge_y$ layer several microns thick in which most of the dislocations are trapped, and then grow a SiGe buffer layer of constant Ge concentration. Experimental studies in the last few years on strained SiGe materials have resulted in significant progress in the understanding of strain relaxation kinetics and optimization of graded buffer layers with respect to relaxation and surface morphology [21–25]. These parameters are of crucial importance as they are interdependent and are affected by growth temperature, grading rate and composition. Recently, a review on the present status of the growth of strained Si on relaxed SiGe buffer layers on Si using various techniques and applications of strained Si films in SiGe-based CMOS technology has been made (Reference 26 and references therein).

A major drawback of growing SiGe on Si is the relatively large lattice mismatch between Si and Ge ($\approx$ 4.17%). There is an equilibrium critical thickness for pseudomorphic growth of $Si_{1-x}Ge_x$ on Si and thermal stability of the film is a concern for device fabrication, especially if the film thickness exceeds this equilibrium critical thickness, which is less than 10 nm for $Si_{0.75}Ge_{0.25}$. If the layer thickness exceeds this critical value, mismatch leads to strain-relieving misfit dislocations. Consequently, the design flexibility is restricted especially for applications which involve high Ge concentrations.

In an effort to extend the $Si_{1-x}Ge_x$ strained layer technology and to search for new materials, experimental work was started in the 1990s on $Si_{1-x}C_x$ and $Si_{1-x-y}Ge_xC_y$ alloys and recently on $Ge_{1-y}C_y$ alloys. Strain compensation effects were observed in $Si_{1-x}Ge_x$ alloys doped with boron by Herzog *et al.* [27] and have been also been observed by several other workers [28, 29]. Recent studies on the incorporation of a small amount of C atoms in the Si/SiGe system to develop new types of buffer layers with reduced misfit dislocations have been reviewed by several workers [30, 31].

A different concept for strain adjustment has been suggested by adding carbon into the Si/SiGe material system [32]. Recently, growth of a relaxed 1 μm thick stepwise graded buffer, based on a combination of $Si_{1-x}Ge_x$ and $Si_{1-x-y}Ge_xC_y$ layers, has been demonstrated [33] indicating that the addition of carbon is a promising way for new relaxed buffer concepts with low threading dislocation densities. The addition of carbon to a SiGe layer not only reduces the strain, but also stabilizes the layer [32]. Attempts have been made to form strained alloys on Si or Ge substrates containing Sn as a constituent. Synthesis of dislocation-free $Si_y(Sn_xC_{1-x})_{1-y}$ [34] and growth of quaternary alloys of $Si_{1-x-y-z}Ge_xC_ySn_z$ have also been announced [35].

As the lattice parameter of carbon (3.546 Å) is much smaller than that of Si (5.431 Å) or that of Ge (5.658 Å), C may be used as a substitutional impurity in the SiGe to decrease the lattice mismatch of the SiGe system. In case of a ternary alloy like $Si_{1-x-y}Ge_xC_y$, assuming Vegard's law and for a fully relaxed film, the lattice parameter can be written as

$$a_{SiGeC} = a_{Si} + x(a_{Ge} - a_{Si}) + y(a_C - a_{Si}) \qquad (3.2)$$

where a_i is the lattice parameter of the ith component. As the third term is negative, it is possible to adjust the composition of the alloy to cancel the second and third terms leading to an alloy with exactly the Si lattice parameter (i.e. zero net strain).

The lattice constant mismatch is +4.18% for Ge on Si and –50% for carbon on Si. According to Vegard's law, for about 12% Ge in Si and 1% C in Si, the mismatch is equal and opposite and a strain symmetrized structure with zero average strain may be obtained as shown in the two-dimensional cross-sectional schematic of figure 3.4. It is therefore expected that the ternary alloys will be more stable. Also, in principle it should be possible to incorporate larger C concentrations enabling the growth of unstrained films on Si(100) with a commensurate film/substrate interface which are not limited by a critical thickness [36, 37]. Addition of substitutional carbon to the $Si_{1-x}Ge_x$ material system can provide an additional design parameter in band structure engineering on Si substrates. Since large bandgap variation from 5.5 eV (diamond) to 0.66 eV (Ge) exists, the $Si_{1-x-y}Ge_xC_y$ system may result in an increase in the bandgap to values greater than those of SiGe and Si, in addition to other interesting properties such as the highest known thermal conductivity (diamond), high hole mobility (Ge) and matured processing technology (Si). The incorporation of C, however, presents difficult challenges due to large lattice mismatch between C and Si, low solubility of C in Si and silicon carbide (SiC) precipitation.

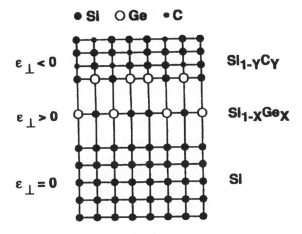

Figure 3.4. Schematic representation of strain symmetrized structure with zero average strain.

3.2   Energy Gap and Band Structure

The band structure determines the electronic and optical properties of semiconductors. Transport properties, the bandgap and band shapes control optical absorption and photoluminescence. The knowledge of band alignment is essential to determine the band offsets which in turn play an important role in the design of heterostructure devices. The high lattice mismatch in SiGe heterostructures significantly modifies the band structures giving rise to many complexities in band structure calculations. The effective mass approximation model, valid for lattice matched heterostructures such as GaAs–AlGaAs, does not hold good for a strained layer system which requires a microscopic formalism taking into account the precise atomic positions and potential variations in the crystal. In the empirical pseudopotential (EPP) approach used for bulk semiconductors [38], the empirically known band structure properties are reproduced by choosing a suitable pseudopotential. A perturbation potential is then constructed for each atom in the unit cell to account for the lattice strain introduced into the bulk crystal. The consideration of microscopic variation of potential due to individual atoms gives rise to a clear description of bulk momentum mixing resulting from the scattering of electrons from the misplaced atoms. The scheme accurately describes the states at and around the conduction and valence band edges.

Therefore, the pseudopotential method has been found to be successful in predicting various optical and transport properties of strained SiGe layers. The approach has been found to have limitations in the presence of impurity atoms when the deviations from the bulk-like structure are too great due to structural relaxation.

On the other hand, the local density approximation (LDA) [39, 40] and *ab-initio* pseudopotential calculations provide a microscopic description of SiGe alloys without any reference to the bulk crystal properties and describe the properties of a particular atomic species regardless of its local environment. The LDA schemes, which include the microscopic potential variation in the interface region, give a very accurate description of the valence band through minimization of ground-state energy and predict structural relaxation very precisely at the cost of an increased computational effort. However, the value of the bandgap is vastly underestimated due to an inadequate description of the excited states of the system in the LDA approach. Both the EPP and LDA methods have contributed significantly to an understanding of the band structure and transport properties of strained SiGe layer heterostructures. The suitability of a particular method depends upon the particular structure, the nature of the system and specific properties that need to be studied. In the following, we present the band structure and transport properties of group-IV alloy semiconductor heterostructures and strained layer superlattices.

3.2.1 Band structure: Bulk Si and Ge

Silicon, the most important material in the semiconductor industry, is far from ideal as far as band structure is concerned. Figure 3.5 shows the calculated band structure [38] of bulk Si. The indirect bandgap in Si greatly limits the application of Si in optical devices, particularly for light emitting devices. The conduction band minimum in Si is located at $(2\pi/a)(0.85,0,0)$ near the X-point of the Brillouin zone. There are six degenerate X-points due to the symmetry of the fcc lattice leading to 6-fold degenerate conduction band valleys. The constant energy surfaces of Si in the conduction band shown in figure 3.6 are ellipsoids with longitudinal (m_l^*) and transverse (m_t^*) electron effective masses of $0.98m_0$ and $0.19m_0$, respectively. Taking account of the anisotropic conduction band property, the density of state mass for a single valley in Si is given by

$$m_{dos}^* = (m_l^* \cdot m_t^{*2})^{1/3} \tag{3.3}$$

For six conduction band edge valleys in Si, the total density of states is obtained by multiplying the density of states calculated for one valley. The very large density of states ($\sim 7.73 \times 10^{21} E^{1/2}$/eV cm^{-3}) near the band edge

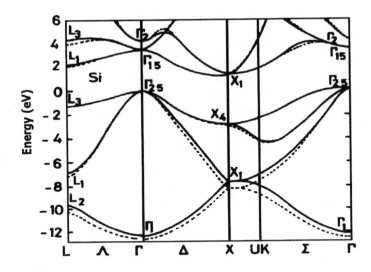

*Figure 3.5. Band structure of Si showing the top of the valence band and bottom of the conduction band at different k-points. After J. R. Chelikowsky and M. L. Cohen, Phys. Rev. B, **14**, 556–582 (1976).*

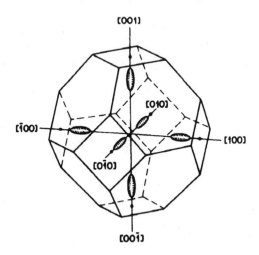

Figure 3.6. Constant energy ellipsoids for Si conduction band showing six equivalent valleys in Si at the band edge.

due to six equivalent valleys results in a high scattering rate responsible for the poor electron transport properties in Si. For the conduction band edge near $(2\pi/a)[100]$, the longitudinal direction is the x-axis and the transverse plane is the y–z plane. But for the valley at $(2\pi/a)[001]$, the longitudinal direction is the z-axis while the transverse plane is the x–z plane. So there is strong anisotropy in each valley though the average electron transport involves the electrons moving in all of the valleys, with a characteristic conductivity mass given by

$$m_\sigma^* = 3(2/m_l^* + 1/m_t^*)^{-1} \qquad (3.4)$$

While the density of state mass represents the properties of the electrons at a constant energy surface in the conduction band, the conductivity effective mass affects the response of electrons to an external potential.

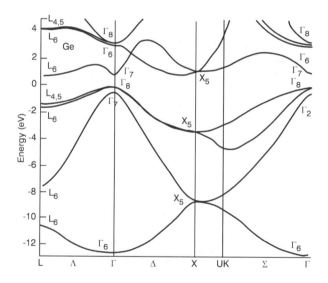

*Figure 3.7. Band structure of Ge. After J. R. Chelikowsky and M. L. Cohen, Phys. Rev. B, **14**, 556–582 (1976).*

The calculated band structure [38] for bulk Ge is shown in figure 3.7. The bottom of the conduction band in Ge, which is also an indirect bandgap semiconductor, occurs at the L-point, $(2\pi/a)$ (1/2,1/2,1/2), of the Brillouin

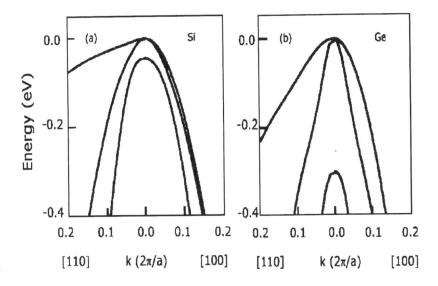

Figure 3.8. Band structure of unstrained $Si_{0.5}Ge_{0.5}$ alloy along the symmetry lines L–Γ–X, where Γ is the Brillouin zone center and L and X are the wave vectors $2\pi/a(1/2,1/2,1/2)$ and $2\pi/a(1,0,0)$, respectively.

zone. There are eight L-points, but only four of them are independent since the others are connected by a reciprocal lattice vector. Comparatively lower values of the density of states and effective mass give rise to higher electron mobility in Ge compared to Si.

In contrast to the conduction band, spin–orbit coupling has a significant effect on the near band edge structure of the valence band. Figure 3.8 shows the top of the valance band in Si and Ge with degeneracy of the heavy-hole (HH) and light-hole (LH) bands at the Γ-point ($k = 0$). The split-off band (SO) in Si due to spin–orbit coupling is separated by only 44 meV from the top of the band. This is one of the smallest split-off energies of any semiconductor. The large hole effective mass and the close proximity of the SO band are the reasons for poor hole transport properties in Si. In contrast, Ge has lower hole masses and has very large spin–orbit splitting (figure 3.8b) leading to perhaps the best hole properties of any semiconductor. However, several important material and processing properties like the existence of stable SiO_2, ease of processing and nearly equal saturation velocity at high fields make Si much more useful for commercial applications.

3.2.2   Bulk SiGe Alloy

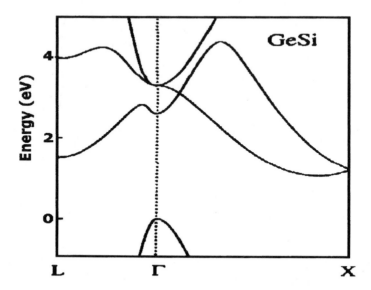

Figure 3.9.   The valence band structure of Si and Ge showing lower hole effective masses and very large spin–orbit splitting in Ge as compared to Si.

In theoretical calculations of the band structure, it is difficult to define a unit cell of the randomly distributed SiGe alloy due to breaking of translational symmetry of the crystal. Though this can be defined by considering a very large unit cell for the empirical pseudopotential formalism, the treatment becomes computationally intensive. The virtual crystal approximation (VCA) [41] assuming linear interpolation of various physical properties of Si and Ge restores the translational symmetry of the alloy system. Therefore the empirical pseudopotential formalism can be used by treating the SiGe alloy as a "virtual crystal" having the interpolated lattice constant, potential and so on. The band structure of unstrained $Si_{0.5}Ge_{0.5}$ alloy using the VCA is shown in figure 3.9 along the symmetry lines L–Γ–X. The bandgap of unstrained bulk SiGe alloy was measured at 4.2 K [42]. The excitonic bandgap varies smoothly with Ge content from the Si gap at 1.155 eV to the Ge gap at 0.740 eV. For the $Si_{1-x}Ge_x$ alloy, the conduction band structure is Si-like up to 85% Ge with minima at the X-point. However, it exhibits Ge-like character with minima at the L-point for Ge fractions above 85%. The experimentally measured values could be fitted well by the quadratic equation [43]

$$E_{\mathrm{g}}(x) = (1.155 - 0.43x + 0.0206x^2) \ \text{eV for } 0 < x < 0.85 \qquad (3.5)$$

and

$$E_{\mathrm{g}}(x) = (2.010 - 1.27x) \ \text{eV for } 0.85 < x < 1 \qquad (3.6)$$

3.2.3 Strained SiGe Alloy

As discussed earlier, the pseudomorphic growth of strained SiGe on Si results in a tetragonal distortion of the lattice lowering the symmetry of the lattice from cubic to tetragonal. The broken symmetry of the Brillouin zone causes the splitting of both the valence and conduction band edges and therefore modifies the energy gap of a strained material compared to an unstrained lattice. In addition, the exact nature of the splitting and deformation will determine both the magnitude of the band offset occurring in the valence and conduction bands and any change in the shape of the bands in k-space.

Figure 3.10 shows the indirect energy gap of strained $Si_{1-x}Ge_x$ alloys for different substrates. The closed symbols represent the data of Lang *et al.* [44] from photocurrent measurements (at 90 K) on p-i-n diode structures for Ge fraction varying from 0 to 70%. The corresponding value measured using photoluminescence data [45] for low Ge content (up to 24%) is shown as open symbols in figure 3.10. The bandgap obtained from the low temperature photoluminescence measurements ($T = 6$ K) are fitted to the following quadratic equation [45]

$$E_{\mathrm{g}} = (1.171 - 1.01x - 0.835x^2) \ \text{eV for } x < 0.25 \qquad (3.7)$$

The bandgap E_{g} is determined by adding the binding energy of the free exciton, which is estimated by linear extrapolation between the values for Si (14.7 meV) and Ge (4.15 meV), to the experimentally determined no-phonon free-exciton luminescence line.

There is a large body of experimental data on the effects of compressive strain on the bandgap, but comparatively less for the effects of tensile strain, i.e., for $Si_{1-x}Ge_x$ grown on a Ge substrate or Si grown on a relaxed $Si_{1-x}Ge_x$ substrate. However, theoretical calculations of the bandgap of arbitrary strained alloy layers have been reported by several authors [46–48].

The phenomenological deformation potential can be used as one of the methods to calculate the effects of strain on the indirect bandgap from the following:

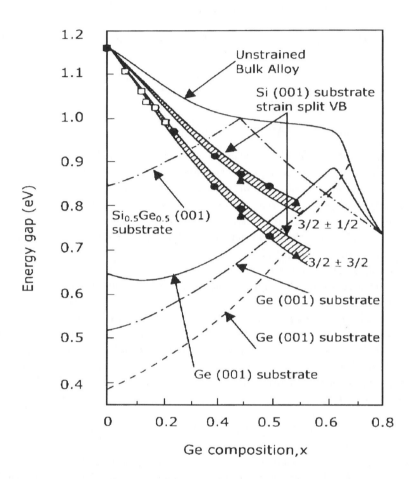

Figure 3.10. *Fundamental indirect bandgap of strained* $Si_{1-x}Ge_x$ *alloys in comparison with bulk alloy. The full (open) symbols are the results of photocurrent measurements at* $T = 90$ K *for pseudomorphic* $Si_{1-x}Ge_x$ *films on Si(001) substrates. The hatched areas between the full lines show the theoretical results of People et al. [46, 47] for transitions involving heavy holes (±3/2) and light holes (±1/2). Also shown are the calculated data for pseudomorphic films on either* $Si_{0.5}Ge_{0.5}$ *or Ge substrates [46–48]. After T. Fromherz and G. Bauer, Properties of Strained and Relaxed Silicon Germanium, Edited by E. Kasper, IEE, 87–93 (1995).*

(i) the elements of the strain tensor calculated,

(ii) the known indirect bandgap of the unstrained bulk alloy,

(iii) the deformation potential which describes the uniaxial splitting of the valence band edge, and

(iv) the hydrostatic deformation potential, which is required for finite change in the volume of the unit cell.

For compressively strained $Si_{1-x}Ge_x$ on Si substrate, the theoretical results of People [46, 47] using deformation potential theory, indicated by the hatched lines in figure 3.10, are in good agreement with the experimental data.

The bandgap of compressively strained $Si_{1-x}Ge_x$ layer can be approximated by the relationship [49]

$$E_g(x, T) = E_0(T) - 0.96x + 0.43x^2 - 0.13x^3 \qquad (3.8)$$

where E_g is the difference between the lowest lying conduction band and the highest lying valence band edge for a $Si_{1-x}Ge_x$ strained layer grown on Si and $E_0(T)$ is the bandgap of bulk Si.

As seen from figure 3.10, the theoretically reported values for the bandgap of strained SiGe alloys on a Ge or a $Si_{0.5}Ge_{0.5}$ substrate differ quite substantially. The solid line was reported by People [46, 47] by considering the strain induced shift and splitting of conduction and valence bands using the deformation potential parameters of bulk Si for $x < 0.85$ and those for bulk Ge for $x > 0.85$. However, the use of deformation potential parameters by linearly extrapolating the respective values of Si and Ge yielded a different result, as shown by the dashed-dotted lines in figure 3.10. On the other hand, the calculated values of bandgap based on local density fluctuations and *ab initio* pseudopotentials [48] are shown by the dashed line. Therefore, experimental values of bandgap are necessary for tensile strained SiGe alloys grown on higher lattice constant substrates.

3.2.4   Effect of Strain

Lattice strain alters the band structure of a semiconductor by either shifting it in energy, distorting it, removing degeneracy effects, or any combination of the three. Hydrostatic strain shifts the energy level of a band. On the other hand, a uniaxial or biaxial strain splits the degenerate bands. Uniaxial strain splits the band into twofold degenerate bands at lower energy plus a nondegenerate band at higher energy while keeping the average energy over the three bands equal. The energy shift due to strain does not alter the shape of the conduction band and keeps the effective mass unaltered. However, for multi-valleyed semiconductors like Si, Ge and $Si_{1-x}Ge_x$ alloys the carrier population in individual bands may be altered. On the other hand, the strain has a significant effect on the shape of the valence band structure, resulting in a change in effective mass and mobility. So the piezo resistance effect is much more predominant in p-type semiconductors than in n-type semiconductors. Additionally, reduced symmetry of the band structure leads to a shift in energy of the valence band.

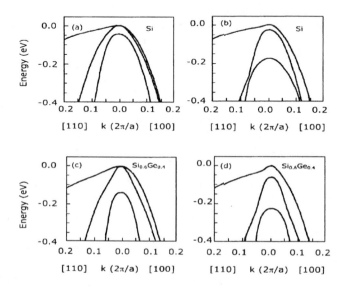

*Figure 3.11. Effect of strain and alloying on valence band structure: (a) bulk Si; (b) biaxially compressed Si, $\epsilon_{xx} = \epsilon_{yy} = -0.0156$, $\epsilon_{zz} = 0.0119$, which is strain-equivalent to $Si_{0.6}Ge_{0.4}/Si$; (c) bulk (unstrained) $Si_{0.6}Ge_{0.4}$; (d) pseudomorphic $Si_{0.6}Ge_{0.4}/Si(001)$, $\epsilon_{xx} = \epsilon_{yy} = -0.0156$, $\epsilon_{zz} = 0.0119$. After J. M. Hinckley and J. Singh, Phys. Rev. B, **41**, 2912–2926 (1990).*

Hinckley and Singh [50] calculated the valence band structure of strained SiGe for the heavy-hole, light-hole and split-off-hole bands by solving the Shockley **kp** matrix, in which spin–orbit coupling has been included. The effect of strain due to lattice mismatch was included in the band structure calculation with the use of valence band deformation potential theory. The calculated valence band structures of bulk Si and strained $Si_{0.6}Ge_{0.4}$, showing the effects of both strain and alloying, are presented in figure 3.11. Figure 3.11a shows the bulk Si band structure, while figure 3.11b shows the band structure of 1.56% compressively strained Si (without any Ge) which corresponds to strain in pseudomorphic $Si_{0.6}Ge_{0.4}$ on Si. Figure 3.11c shows the band structure of unstrained or bulk $Si_{0.6}Ge_{0.4}$, while figure 3.11d shows the band structure of strained $Si_{0.6}Ge_{0.4}/Si$ incorporating the effects of both strain and alloying.

The topmost valence band edge in Si and Ge is threefold degenerate in the absence of spin. The spin–orbit interaction results in a fourfold ($J = 3/2$) upper set of states having symmetry Γ_8^+ and a split-off doublet ($J = 1/2$) having symmetry Γ_7^+ [5, 51]. At the center of the Brillouin zone ($k = 0$), the uniaxial component of the lattice strain splits the fourfold ($J = 3/2$) upper set of valence band states into a set of doublets denoted by $(3/2, \pm 3/2)$ and $(3/2, \pm 1/2)$ [5]. The top of the valence band (lowest hole energy) will be in the $(3/2, \pm 3/2)$ state in the case of a compressively strained $Si_{1-x}Ge_x$ film on Si. The upward shift of the top of the valence band as a function of Ge fraction contributes to the lowering of the bandgap in the case of SiGe alloy. The state of the topmost valence band will be reversed, i.e., $(3/2, \pm 1/2)$, in the case of tensile strained Si grown on a larger lattice constant substrate.

Therefore, in unstrained material, the valence band maximum is composed of three bands: the degenerate heavy-hole (HH) and light-hole (LH) bands at $k = 0$, and the split-off (SO) band which is slightly lower in energy. Very close proximity (44 meV) of the SO band to the top of the valence band is one of the reasons for poor hole transport properties in Si. It should be noted that the "heavy" and "light" refer to the masses of the holes in each band relative to the other band in the unstrained case only (figure 3.12a). However, for convenience, the distinction is made between the two bands by assigning HH and LH labels so that they can easily be tracked as they move under the influence of strain.

The biaxial stress can be resolved into a hydrostatic and a uniaxial stress component. The hydrostatic stress equally shifts all three valence bands, whereas the uniaxial stress lifts the degeneracy between LH and HH bands. In addition, the spin–orbit band is lowered in energy with respect to the other two bands (figures 3.12b and 3.12c). The effect of both kinds of strain (biaxially compressive and tensile) on the valence bands is shown in figures 3.12b and 3.12c. Strain of either type deforms both bands as it removes the degeneracy of the valence bands, changing the relative masses in the bands considerably.

The conduction band structure of $Si_{1-x}Ge_x$ is silicon-like below $x = 0.85$. The conduction band minimum in Si occurs at the Δ-point between the Γ- and X-valleys and is strongly affected by a tetragonal strain. The L conduction band minimum of Ge has a fourfold spatial degeneracy, which is not lifted by a tetragonal strain. Since the L-valley for Ge becomes important only for very high Ge concentration, the sixfold degenerate bands at the Γ-point are

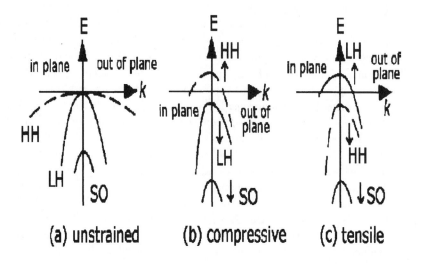

Figure 3.12. E–k diagram for the Si valence band under three strain conditions. The shifts and deformations are schematic only and are exaggerated somewhat for emphasis.

considered here for $Si_{1-x}Ge_x$ alloys of high practical interest. The six ellipsoidal energy shells located at equivalent positions in k-space in $< 001 >$ directions are shown in figure 3.13a. The strain has no apparent effect on the ellipsoidal shape of the energy surface but causes a shift in energy [52, 53]. So unlike the valence band cases, the effective mass of each band remains unchanged [48]. The relative shift in energy in the case of a compressively strained film is shown in figure 3.13b where [001] conduction bands (twofold degenerate Δ_2 bands) move up in energy, while the [010] and [100] bands (fourfold degenerate Δ_4 bands) are shifted downwards in energy. Figure 3.13 shows the constant energy surface diagram indicating the size change of each valley to a smaller one for an upward energy shift and to a larger one for a downward shift. In the case of tensile strained film (film grown pseudomorphically on a substrate with larger lattice constant), the direction of motion of the split levels is reversed. Here, the Δ_2 band moves lower in energy relative to the Δ_4 band as shown in figure 3.13c.

The splitting of the conduction band minima for compressively strained $Si_{1-x}Ge_x$ films as a function of Ge concentration is shown in figure 3.14. The energy difference between the separated bands due to splitting depends on the

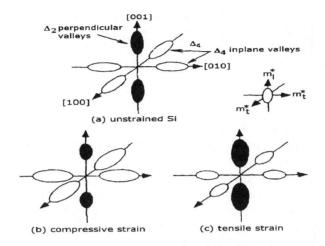

Figure 3.13. Constant energy surfaces for the Si conduction band. The size change of each valley in the constant energy surface diagram indicates a shift up (smaller) or down (larger) in energy. After T. Manku and A. Nathan, IEEE Trans. Electron. Dev., 39, 2082–2089 (1992), copyright ©1992 IEEE.

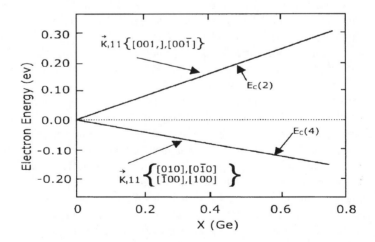

Figure 3.14. Δ_1 conduction band splitting (uniaxial components only) for bulk $Si_{1-x}Ge_x$ alloys grown on Si(001) substrates. After R. People, Phys. Rev. B, 32, 1405–1408 (1985).

magnitude of the strain and can be calculated by the deformation potential theory [5, 51]. Deformation potential theory has also been used by van de Walle and Martin [48] to give expressions for the perturbed conduction band levels of Si and Ge.

3.3   Band Alignment

Band alignments in heterostructures giving rise to band offsets or band discontinuities are very important for junction device applications. For bipolar heterostructure devices, the band offset determines the forward injection efficiency. On the other hand, it controls the fraction of charge carriers confined in the high mobility channel in MOSFETs. For an atomically sharp interface, the band discontinuities are sharp on an atomic length scale in contrast to the much larger length in band bending associated with depletion layers. Therefore, the band discontinuities are more fundamental in nature and are used as boundary conditions in the solution of Poisson's equation to yield the band bending.

In case of an unstrained system, i.e., for a lattice matched heterojunction, the band offsets can be simply derived from the band line-up at the interface. The calculation is more complicated in the case of strained heterostructures with group-IV alloys. Besides the normal band offsets, the splitting of bands due to uniaxial strain and additional shifts due to hydrostatic strain need to be computed. It should be noted that the band discontinuities in strained heterostructures are well defined only if the interface is pseudomorphic so that the in-plane lattice constant is continuous across the interface. The best theoretical study on the estimation of heterojunction discontinuities has been carried out by van de Walle and Martin [48].

In order to determine the band alignment for a given $Si_{1-x}Ge_x/Si$ heterointerface, the bandgap of the constituent layers as well as the band discontinuity, ΔE_c or ΔE_v, must be known. Strain induced modification of Si/SiGe films is found to have a significant impact on the band structure and carrier transport. When a thin film with a larger lattice constant (e.g., $Si_{1-x}Ge_x$) is grown on a smaller lattice constant substrate (e.g., Si), the film maintains the in-plane lattice constant of the substrate and is under a biaxially compressive strain. Figure 3.15 shows the band offset between a strained $Si_{0.7}Ge_{0.3}$ film grown on Si. This is known as the type-I band alignment where almost the entire band offset occurs in the valence band (figure 3.15a) while the band offset in the conduction band is very small. This type of structure is favorable for hole confinement and has been exploited in novel heterostructure devices, namely p-MOSFETs, p-MODFETs and HBTs [54].

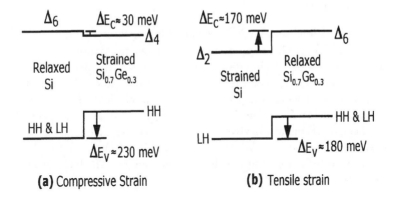

Figure 3.15. Band alignments between Si and $Si_{0.70}Ge_{0.30}$ on two substrates: (a) Si and (b) $Si_{0.70}Ge_{0.30}$.

Similarly, a smaller lattice constant Si epilayer will be under biaxial tension when grown on a larger lattice constant relaxed $Si_{1-x}Ge_x$ substrate. Figure 3.15b shows the band offset for a strained Si epilayer grown on relaxed $Si_{0.70}Ge_{0.30}$. In this case, type-II band offset occurs and the structure has several advantages over the more common type-I band alignment, as a large band offset is obtained in both the conduction and valence bands, relative to the relaxed $Si_{1-x}Ge_x$ layer [5]. This allows both electron and hole confinements making it useful for both n- and p-type devices for strained Si/SiGe-based CMOS technology. Since strained Si provides larger conduction and valence band offsets and does not suffer from alloy scattering (hence mobility degradation) [55], a significant improvement in carrier mobility can be achieved. Strained Si is more difficult to grow than strained $Si_{1-x}Ge_x$, since bulk $Si_{1-x}Ge_x$ substrate is currently not available and, until recently, growth of relaxed $Si_{1-x}Ge_x$ without forming a large concentration of defects due to dislocations was difficult. However, the ability to achieve both n-MOS and p-MOS devices using strained Si provides a promising alternative for the next generation high performance SiGe-CMOS technology (see, for example, reviews by Schäffler [56] and Maiti *et al.* [26] and references therein).

The bandgap difference ΔE_g between the emitter and the base of a transistor consists of the valence band discontinuity ΔE_v and the conduction band discontinuity ΔE_c. In a SiGe-HBT, most of the bandgap reduction results from a shift in the valence band edge. The conduction band discontinuity is usually a small fraction of the total bandgap difference.

3.4   Bandgap Narrowing

In many applications such as SiGe-HBTs, it is necessary to dope the SiGe alloy to a very high level. This causes bandgap narrowing. Bandgap narrowing due to a heavy doping effect changes the physical constants used in the minority carrier transport equation and is crucial for accurate determination of collector current. Jain and Roulston [57] reported general closed-form equations for bandgap narrowing for n- and p-type Si, Ge, GaAs and $Si_{1-x}Ge_x$ strained layers. The equations are derived by identifying the exchange energy shift of the majority band edge, correlation energy shift of the minority band edge, and impurity interaction shifts of the two band edges. The bandgap narrowing (for $N > 10^{18}$ cm^{-3}) is given by

$$\Delta E_g = A_1 \left(\frac{N}{10^{18}} \right)^{1/3} + A_2 \left(\frac{N}{10^{18}} \right)^{1/4} + A_3 \left(\frac{N}{10^{18}} \right)^{1/2} \tag{3.9}$$

where A_1, A_2 and A_3 for n- and p-type Si, Ge and GaAs are given in Reference 58. For p-type $Si_{1-x}Ge_x$ alloy with a Ge content of less than 0.3, the bandgap narrowing is given by

$$\Delta E_g = 11.07(1 - 0.35x)\left(\frac{N}{10^{18}} \right)^{1/3} + 15.17(1 - 0.54x)\left(\frac{N}{10^{18}} \right)^{1/4} + \\ 5.07(1 + 0.18x)\left(\frac{N}{10^{18}} \right)^{1/2} \tag{3.10}$$

Matutinovic-Krstelj *et al.* [59] have extracted the effective bandgap narrowing as a function of Ge concentration for different doping levels as shown in figure 3.16. For devices with similar doping, the linear dependence on Ge concentration is obvious. Fitting the data at the same doping level gives an effective bandgap reduction with respect to Si of about 7 meV for 1% Ge. The authors have also extracted the $\Delta E_{g,dop}$ for several reported SiGe-HBTs, as shown in figure 3.17.

Assuming that the linear dependence on Ge concentration is independent of doping, the effects of bandgap reduction due to Ge concentration and bandgap narrowing due to heavy doping are separated

$$\Delta E_{g,eff} = \Delta E_{g,dop} + \Delta E_{g,Ge} \tag{3.11}$$

Assuming a $\Delta E_{g,dop}$ of the form

$$\Delta E_{g,dop} = A + B\log\left(\frac{N_A}{10^{18}} \right) \tag{3.12}$$

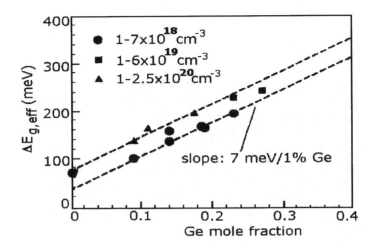

Figure 3.16. Effective bandgap reduction with respect to intrinsic Si vs. Ge concentration. After Z. Matutinovic-Krstelj et al., IEEE Trans. Electron Dev., **43**, *457–466 (1996), copyright ©1996 IEEE.*

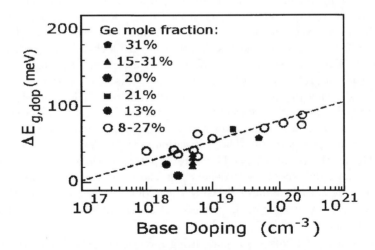

Figure 3.17. Heavy doping contribution to the effective bandgap reduction in SiGe-HBTs. The broken line is a fit of equation 3.13 with x = 0.0. After Z. Matutinovic-Krstelj et al., IEEE Trans. Electron Dev., **43**, *457–466 (1996), copyright ©1996 IEEE.*

and a linear dependence of $\Delta E_{g,Ge}$ on x, a three-parameter best fit was found to be

$$\Delta E_{g,eff} = 28.6 + 27.4\log_{10}\left(\frac{N_A}{10^{18}}\right) + 688x \tag{3.13}$$

where N_A is the base doping and x is the Ge mole fraction and equation 3.13 is valid only for doping levels beyond 10^{18} cm^{-3}. The first two terms represent bandgap narrowing due to heavy doping and the last term is the bandgap reduction due to the Ge contribution. $\Delta E_{g,eff}$ is not a measure of the actual bandgap reduction but the effective (apparent) bandgap reduction relevant for minority carrier concentration and thus electron transport across the $Si_{1-x}Ge_x$ base. The apparent bandgap is larger than the true bandgap due to valence band filling in the degenerately doped semiconductor. The effective bandgap reduction is a useful parameter to model the collector current of the $Si_{1-x}Ge_x$ bipolar transistor.

3.5   Scattering Mechanisms

The drift mobility and velocity field characteristics of a bulk semiconductor are limited by intrinsic lattice scattering through interaction between the phonons and the carriers, impurity scattering in doped semiconductors and alloy scattering in semiconductor alloys. Phonon scattering is contributed by acoustic phonons and optical phonons each with its characteristic dispersion relations. These are specified by the relevant deformation potentials, which determine the rate of scattering from an initial momentum to a final momentum, with the states in the same or a different valley, i.e., intravalley scattering or intervalley scattering. The actual scattering rate depends on the band structure details, the number of equivalent and non-equivalent valleys and their energy separation. The interaction potential for impurity scattering is related directly to the screened Coulombic potential while for alloy scattering an alloy potential whose effect depends on the mole fractions of the mixture is used. The actual mobility measured in a device is smaller than the bulk mobility however because of the presence of interfaces invariably present with their defects and charges.

The general formalism for computing momentum relaxation time in the case of acoustic phonon scattering τ_{ac} and zero-order optical phonon scattering τ_{op} is given by [60]

$$\tau_{ac} = \frac{\pi\hbar^4 c_L}{2^{1/2}\Xi^2 m^{*3/2}k_B T E_k^{1/2}} \tag{3.14}$$

and

$$\tau_{\mathrm{op}} = \frac{2^{1/2}\pi\hbar^3\omega_0\rho}{D_0^2 m^{*3/2}} \left\{ n(\omega_0)(E_k + \hbar\omega_0)^{1/2} + [n(\omega_0) + 1](E_k - \hbar\omega_0)^{1/2} \right\}^{-1} \quad (3.15)$$

where Ξ is the deformation potential, D_0 is the interaction constant, $n(\omega_0)$ is the Bose–Einstein occupancy function with optical phonon frequency ω_0, E_k is the carrier energy, ρ is the mass density, $\hbar$ is the reduced Planck's constant and c_{L} is the elastic constant of the lattice in the longitudinal mode. The expressions show that the acoustic phonon limited mobility increases as T^{-1}. The variation of optical phonon limited mobility arising from the Bose–Einstein occupancy function $n(\omega_0)$ shows a faster increase with inverse temperature.

In devices impurity scattering plays an important role in limiting the achievable mobility. The Brooks–Herring and Cornwell–Weisskopf approaches [60] are used in modeling three-dimensional charged impurity scattering including the effect of screening by carriers. The expression for momentum relaxation time of charged impurity scattering in the Brooks–Herring model is given by [60]

$$\tau_{\mathrm{imp}} = \frac{2^{9/2}\pi\epsilon^2 m^{*1/2} E_k^{3/2}}{Z^2 e^4 N_{\mathrm{I}}} \left[\log\left(1 + \frac{8m^* E_k}{\hbar^2 q_0^2}\right) - \frac{1}{1 + \hbar^2 q_0^2/8m^* E_k} \right]^{-1} \quad (3.16)$$

where q_0 is the reciprocal screening length, N_{I} is the number of impurity centers per unit volume and Z is the atomic number. Mobility limited by impurity scattering decreases with temperature, varies inversely with doping concentration and increases with carrier concentration. The scattering rate depends on the separation between the carrier supply layer and the channel and is therefore high for accumulation mode devices when the dopants are separated from the channel. This effect is exploited in modulation doped structures and may be computed in a manner similar to that employed for finding the effect of oxide charge.

The band structure of an alloy is described using the virtual crystal approximation assuming a uniform distribution of the constituents. Fluctuations from uniformity give rise to local changes in the potential experienced by the carriers and hence scattering. The scattering rate is proportional to the magnitude of the scattering potential V_{a} and the product $x(1-x)$ where x is the Ge mole fraction in the binary alloy. The alloy scattering limited mobility in a $\mathrm{Si}_{1-x}\mathrm{Ge}_x$ alloy is given by [61]

$$\mu_{\mathrm{alloy}} = \frac{e\hbar^3}{m_{\mathrm{t}}^{*2}\Omega_0 V_{\mathrm{a}}^2 x(1-x)}\left(\frac{16}{3b}\right) \quad (3.17)$$

where m_t^* is the transverse effective mass, Ω_0 is the atomic volume, V_a is the alloy disorder potential and b is the variational parameter given by

$$b = \left(\frac{33 m_l^* e^2 n}{8 \epsilon \hbar^2} \right)^{1/3} \tag{3.18}$$

where m_l^* is the longitudinal effective mass and n is the carrier concentration. Thus, the alloy scattering mobility has dependence on the masses and the carrier concentration. It is presumed that alloy scattering limited mobility in relaxed and strained SiGe materials would be similar. An estimation of the alloy disorder potential V_a is important as it is squared in the formula.

Computations of band properties in relation to determination of scattering processes and bulk mobility of carriers have been made by Hinckley and Singh [50]. Acoustic phonon and optical phonon scattering rates and their dependence on energy at room temperature have also been computed. At very low energies, the optical phonon emission process is absent and the acoustic phonon scattering rate (W_{ac}) and non-polar optical phonon absorption rate (NPOA) are comparable. It can be concluded that (i) the non-polar optical (NPO) scattering rate (W_{np}) is greater than the acoustic phonon scattering rate, (ii) at low energies the impurity scattering rate may exceed W_{np}, and (iii) W_{alloy} for an alloy potential of 0.8 eV exceeds W_{np} and phonon scattering rates for Ge are smaller typically by a factor of about two. This combined with a smaller effective mass for holes in Ge gives higher mobility.

Physically based models have been developed for the transport behavior of Si materials and devices based on extensive measurements on current–voltage relations in bulk Si devices. The models provide estimates of mobilities limited by scattering due to acoustical phonons and optical phonons, impurities, interface charges and the surface roughness of active surfaces, and also carrier–carrier scattering. The dependence of scattering rates on temperature and carrier density has also been studied. Unfortunately alloy scattering has not received the attention due. It has been conjectured that strained SiGe alloys would have smaller effective hole mass than that in Si and the scattering rates would also be smaller. Computations by Fischetti and Laux [62] show that this would have been the case were it not for the high alloy scattering rates. Early workers assumed an alloy scattering potential of about 0.2 eV for holes in SiGe. Experimental results indicate that the alloy scattering potential for holes is close to 0.6 eV. Scattering rates of SiGeC alloys do not appear to have been studied in detail.

3.6   SiGe: Hole Mobility

While a large amount of experimental data is available for accurate mobility models of Si and Ge, the data on bulk and relaxed $Si_{1-x}Ge_x$ alloys are still scattered and limited. This is mainly due to technological problems in growing $Si_{1-x}Ge_x$ bulk crystals with homogeneous Ge content and doping concentration over the whole alloy composition range. Electron and hole mobilities in bulk $Si_{1-x}Ge_x$ alloys reported by several authors up to 1988 have been reviewed by Jain and Hayes [63]. With the availability of advanced low temperature epitaxy techniques, a lot of experimental data are now available on carrier mobilities in pseudomorphic alloys of $Si_{1-x}Ge_x$.

Calculation of transport properties of holes in strained $Si_{1-x}Ge_x$ layers grown on a Si(001) substrate was attempted by Hinckley *et al.* [64] and Hinckley and Singh [50]. They include the changes in valence band structure via a **kp** calculation (including heavy-hole (hh), light-hole (lh) and spin–orbit (so) band) and valence band deformation potential theory. The scattering mechanisms included are deformation potential acoustical phonon scattering, deformation potential optical phonon scattering and random alloy disorder scattering. Acoustic phonon and alloy scattering were treated as being elastic, while optical phonon emission and absorption were treated as distinct inelastic scattering processes. The high field carrier transport properties for the undoped material were obtained using Monte Carlo calculation.

Theoretical calculations of bulk, strained $Si_{1-x}Ge_x$ grown over Si have demonstrated that the light-hole and heavy-hole degeneracy is lifted as a direct effect of the biaxial compressive strain [46, 65]. These calculations suggest that the hh band is strongly affected by the strain and moves toward higher energies. The density of state and the carrier concentration effective masses are both reduced compared to bulk Si as a result of the strain and a reduction in the effective mass (see figure 3.18), from $1.1m_0$ for bulk Si to $0.35m_0$ for $Si_{0.7}Ge_{0.3}$, is assumed [66].

The calculated in-plane carrier velocity and mobility are shown in figure 3.19. Both low field and high field velocities of the strained $Si_{1-x}Ge_x$ layer are enhanced with an increase in x. The transport properties of a $Si_{0.6}Ge_{0.4}$ layer are found to be similar to those of bulk Ge. For a strained $Si_{0.7}Ge_{0.3}$ layer the low field mobility is about 3–3.5 times larger than that of bulk Si. At a high field (20 kV/cm), the carrier saturation velocity of a strained $Si_{0.7}Ge_{0.3}$ layer is about 20%–25% higher than that of bulk Si. It should be noted that the above results are obtained without knowing exactly the value of the alloy scattering potential. The estimated value of 0.2 eV (i.e., half of the bandgap difference between Si and Ge) is used in the above calculation [50, 64].

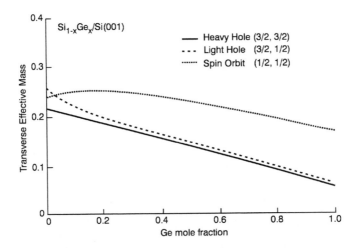

*Figure 3.18. In-plane effective hole masses of strained $Si_{1-x}Ge_x$ on Si(001) calculated using an alloy scattering potential of 0.2 eV. After S. K. Chun and K. L. Wang, IEEE Trans. Electron Dev., **39**, 2153–2164 (1992), copyright ©1992 IEEE.*

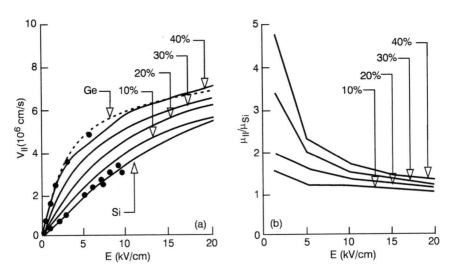

*Figure 3.19. In-plane hole (a) velocity and (b) mobility of strained $Si_{1-x}Ge_x$. After J. M. Hinckley and J. Singh, Phys. Rev. B, **41**, 2912–2926 (1990).*

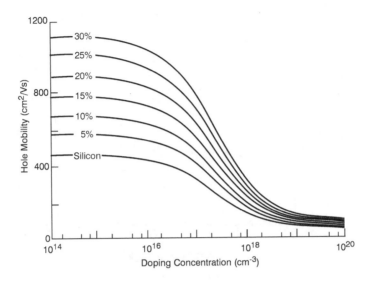

Figure 3.20. The in-plane hole drift mobility for strained SiGe as a function of doping concentration and Ge fraction. After T. Manku et al., IEEE Trans. Electron Dev., **40**, *1990–1996 (1993), copyright ©1993 IEEE.*

The hole mobility of doped SiGe has been reported [67] using the relaxation times associated with acoustical, optical, alloy and ionized impurity scattering. The mobility was found to increase with increasing Ge fraction in both the unstrained and strained alloys. The in-plane hole drift mobility for strained SiGe as a function of doping concentration and Ge fraction is shown in figure 3.20. A good agreement between experimental and theoretical values of mobility was found for a boron doping concentration of 2×10^{19} cm^{-3}. For doping less than 5×10^{15} cm^{-3}, the mobility of Si$_{0.7}$Ge$_{0.3}$ (1120 cm^2/Vs) is about 2.5 times higher than that of bulk Si. At a 5×10^{16} cm^{-3} doping, which is typical for a Si device, the mobility (900 cm^2/Vs) of the Si$_{0.7}$Ge$_{0.3}$ is about 2.3 times higher than that of bulk Si.

An improvement by at least a factor of four in the low field mobility of holes has been achieved in modulation-doped Si/SiGe heterostructures compared to Si p-MOS, as shown in figure 3.21. A 1000 cm^2/Vs low field mobility at room temperature has been demonstrated [68]. This value is more than four times higher than that in state-of-the-art Si p-MOS. From figure 3.21, it is also clear that a very good confinement is obtained, indicated by the almost constant

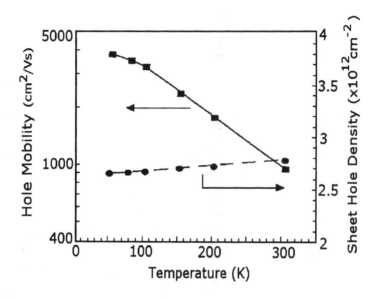

*Figure 3.21.   Dependence of hole mobility (squares) and hole sheet concentration (circles) on temperature for high mobility modulation doped heterostructures. After K. Ismail and B. S. Meyerson, J. Mat. Sci.: Mater. Electron. **6**, 306–310 (1995).*

carrier concentration in the channel as a function of temperature. Also, the monotonic increase of the mobility as the temperature is lowered demonstrates the excellent quality of the alloy layers. Figure 3.22 shows the velocity–field characteristics of strained SiGe compared to those of Si. It is seen that at each value of the field a factor of 5 improvement in the hole velocity over Si p-MOS is obtained. At higher electric fields, signs of velocity saturation start to occur at room temperature and become more evident at 77 K for strained SiGe.

Carns *et al.* [69] obtained the apparent drift mobility of the compressively strained $Si_{1-x}Ge_x$ samples by measuring both the boron concentration using secondary ion mass spectrometry (SIMS) and the resistivity using van der Pauw's method. The apparent drift mobility, however, is increased by 50% over the Ge content range studied.

Figure 3.23 shows the average apparent drift mobilities vs. the Ge content. The error bars are used to reflect the total uncertainty of the measurements from finite contact sizes (±5%), contact resistance (±1%), SIMS (±10%), ionization (< 10%), Rutherford backscattering spectrometry

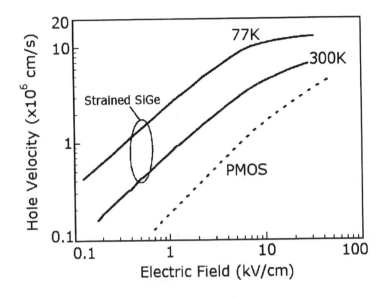

Figure 3.22. Velocity–field characteristics of strained SiGe channels (solid lines) at room temperature and 77 K compared to room temperature p-MOS (dashed line). After M. A. Arafa, Silicon/Silicon–Germanium Modulation Doped Field-Effect Transistors for Complementary Circuit Applications, *PhD Thesis, Univ. of Illinois at Urbana-Champaign, 1997.*

measurements of thickness (±5%) and Ge content (±5%), and changes in thickness due to surface and junction depletion (±1%). The results in figure 3.23 do provide an indication that increasing Ge content may enhance the drift mobility in $Si_{1-x}Ge_x$ compared to bulk Si. This is in good agreement with the mobility measurements in strained SiGe and strained SiGe channel p-MOSFETs reported by Nayak *et al.* [70].

For device modeling purposes, the data for the SiGe-HBT, shown in figure 3.23, can be fitted to the following empirical equation [58]

$$\mu_p = \mu_{min} + \frac{\mu_{po}}{1 + (N/2.35 \times 10^{17})^{0.9}} \tag{3.19}$$

where $\mu_{p,min} = 44 - 20x + 850x^2$ and $\mu_{po} = 400 + 29x + 4737x^2$.

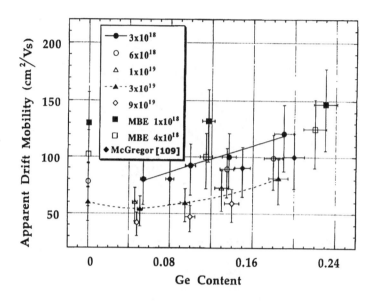

*Figure 3.23. Apparent room temperature drift mobility as a function of Ge content obtained from SIMS and resistivity measurements for CVD-grown and MBE-grown samples with various doping concentrations. After T. K. Carns et al., IEEE Trans. Electron Dev., **41**, 1273–1281 (1994), copyright ©1994 IEEE.*

3.6.1   Doping Dependence

Recently, ohmic majority and minority drift mobilities have been calculated for holes in unstrained and strained SiGe up to Ge contents of 30% and doping concentrations of 10^{20} cm^{-3} by Bufler *et al.* [71]. The transport model considered was based on three analytical hole bands. The phonon scattering mechanisms included are optical phonons and acoustic phonons in the isotropic and elastic equipartition approximation. In SiGe both Si-type and Ge-type phonons are considered. Alloy scattering, intraband impurity scattering based on the Brooks–Herring formulation and the doping-dependent screening length, adjusted to experimental drift mobility data for majority and minority holes in unstrained Si at room temperature, are shown in figure 3.24.

In figure 3.25 the theoretical in-plane component of the drift mobility in strained and heavily doped SiGe is compared [71] to experimentally measured mobility data. Experimental data [67] from the literature are also shown. It can be seen in figure 3.25 that the scattering of the available experimental data

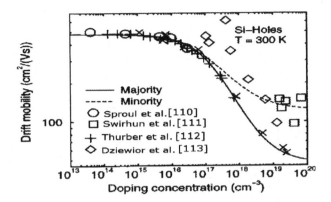

Figure 3.24. Doping-dependent adjustment of the impurity scattering model to measurements of majority and minority mobility in Si at room temperature. After F. M. Bufler et al., J. Vac. Sci. Technol. B, 16, 1667–1669 (1998).

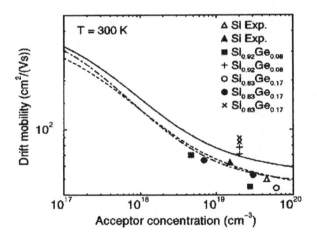

Figure 3.25. In-plane component of majority drift mobility in strained SiGe: open symbols, molecular beam epitaxy, antimony doped samples; filled symbols, low pressure chemical vapor deposition, phosphorus doped samples; ×,+, T. Manku et al.; lines, theory; dot-dashed, Si; dashed, SiGe with 8% Ge; solid, SiGe with 17% Ge. After F. M. Bufler et al., J. Vac. Sci. Technol. B, 16, 1667–1669 (1998).

is rather large and does not provide a clear picture. It should be remarked that there are also noticeable differences in the heavy doping regime between the theoretical mobilities based on analytical band models as in figure 3.25. Further experimental and theoretical work will be necessary for an accurate determination of the hole mobilities in strained and heavily doped SiGe.

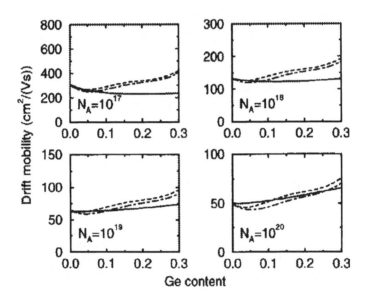

*Figure 3.26. Majority hole mobilities of SiGe for four doping levels (in cm^{-3}) at 300 K: dot-dashed line, $\mu_\perp$ of strained SiGe; solid line, μ of unstrained SiGe; dashed line, $\mu_\parallel$ of strained SiGe. After F. M. Bufler et al., J. Vac. Sci. Technol. B, **16**, 1667–1669 (1998).*

In figures 3.26 and 3.27 the ohmic drift mobilities in unstrained and strained SiGe are shown for majority and minority holes, respectively. The general behavior of the mobility as a function of the Ge content is determined by two competing trends. The effective mass decreases for growing Ge content while the alloy scattering becomes stronger as x increases. The final result is an appreciable mobility enhancement in strained SiGe for high Ge contents. The stronger reduction of the mobility for lower Ge contents in strained SiGe compared to unstrained SiGe is due to the fact that the impurity scattering rate decreases for increasing effective mass. Note also that the anisotropy of the hole mobility in strained SiGe is opposite to that of the electron mobility, i.e., the in-plane mobility is larger than the out-of-plane mobility.

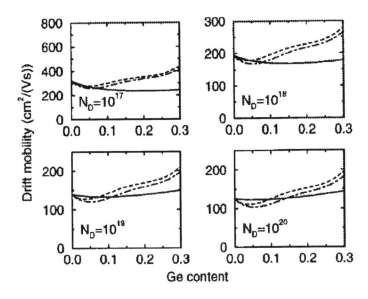

*Figure 3.27. Minority hole mobilities of SiGe for four doping levels (in cm^{-3}) at 300 K: dot-dashed line, $\mu_\perp$ of strained SiGe; solid line, μ of unstrained SiGe; dashed line, $\mu_\parallel$ of strained SiGe. After F. M. Bufler et al., J. Vac. Sci. Technol. B, **16**, 1667–1669 (1998).*

3.7   Strained Si: Electron Mobility

When a thin epitaxial Si layer is grown on top of a relaxed SiGe buffer, the biaxial tensile strain splits the sixfold degenerate conduction bands. The twofold degenerate bands perpendicular to the (100) direction shift toward lower energies. As a result, the population of these bands increases. The twofold degenerate bands are the contributors to transport at low field. The in-plane effective mass of the electrons occupying these bands approximately equals the Si transverse effective mass ($m_t^* = 0.19m_0$). Assumption of no deformation of the band shape has resulted from strain, results in an increase in the low field mobility. On the other hand, the effective mass perpendicular to the transport plane is equal to the longitudinal effective mass ($m_l^* = 0.92m_0$). An enhancement factor between 2 and 5 in the low field mobility compared to bulk Si and Si/SiO$_2$ inversion layers has been demonstrated [72]. The large improvement in the low field mobility has been obtained by ensuring a low density of interface charges in Si/SiGe [73].

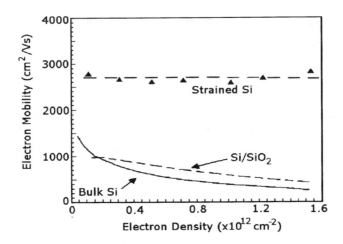

*Figure 3.28. Room temperature low field mobility dependence on the sheet density for strained Si, bulk Si and Si/SiO$_2$. The figure shows the advantage that strained Si offers, namely, the possibility of having an arbitrary carrier concentration while maintaining a mobility at least 2.5 times higher that of Si/SiO$_2$. After K. Ismail and B. S. Meyerson, J. Mat. Sci.: Mater. Electron., **6**, 306–310 (1995).*

Measurements from modulation doped SiGe over relaxed Si$_{0.7}$Ge$_{0.3}$ have shown that excellent transport properties for electrons and holes can be achieved. A factor of 2.5 improvement in the low field bulk mobility is achieved for electrons in strained Si n-MOS compared to Si n-MOS [74], as demonstrated in figure 3.28. Of greater interest is the fact that the carrier concentration in the channel can be increased without affecting the mobility. A mobility as high as 2700 cm^2/Vs is maintained for electron densities ranging from 1×10^{11} cm^{-2} to 1.5×10^{12} cm^{-2}. On the other hand, for bulk Si and for Si n-MOSFET the higher the carrier concentration, the lower the mobility because of higher impurity scattering in Si and increased roughness at Si/SiO$_2$ interface.

For indirect semiconductors such as Si and Ge, the drift velocity is proportional to the electric field at low electric field. When the fields are sufficiently large, nonlinearity in mobility and, in some cases, saturation of drift velocity are observed. For devices with submicron channel length, higher fields are present, necessitating the investigation of the velocity–field characteristics. Figure 3.29 displays the velocity–field characteristics of a

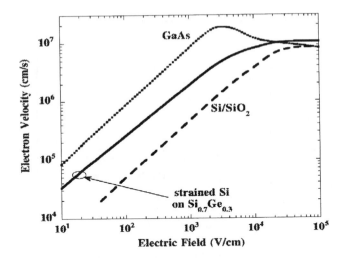

Figure 3.29. The electron velocity as a function of electric field for Si/SiO₂ MOSFETs, GaAs and strained Si on relaxed Si₀.₇Ge₀.₃. After D. J. Paul, Advanced Materials, 11, 191–204 (1999).

strained Si channel and a bulk Si MOS structure. While it is true that the saturation velocities in both structures are relatively close at about 9×10^6 cm/s, a considerable improvement of the velocity at lower field (about 3 times) is observed for the strained Si case. Consequently, the strained Si layer has a significantly lower critical field resulting in a lower drain bias operation for transistors fabricated using these layers.

In-plane drift velocities determined by full band Monte Carlo simulations and drift mobility calculations have been reported for electrons in strained Si grown on $Si_{1-x}Ge_x$ substrates up to $x = 0.4$ at 77 and 300 K by Bufler *et al.* [75]. In figure 3.30 the in-plane velocity–field characteristics of unstrained and strained Si at 77 and 300 K computed using full band Monte Carlo simulations for electrons are shown. The velocity enhancement in strained Si at 300 K can be understood analogously to the case of the ohmic mobilities while the behavior at 77 K is strongly influenced by valley repopulation.

Figure 3.31 shows the in-plane drift mobility of electrons in unstrained and strained Si as a function of lattice temperature along with experimental data for unstrained Si. The general feature of the mobility enhancement in strained Si as a result of the strain-induced valley shifts can be seen which is due to the reduced intervalley scattering and the small in-plane effective mass.

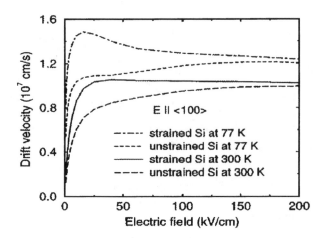

*Figure 3.30. Full band Monte Carlo results for in-plane electron drift velocity as a function of electric field strength in strained Si grown on $Si_{0.7}Ge_{0.3}$ and in unstrained Si at 77 and 300 K. After F. M. Bufler et al., Appl. Phys. Lett., **70**, 2144–2146 (1997).*

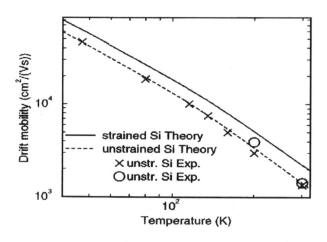

*Figure 3.31. Calculated low field in-plane electron drift mobility in strained Si grown on $Si_{0.7}Ge_{0.3}$ and in unstrained Si in comparison with experimental data for unstrained Si. After F. M. Bufler et al., Appl. Phys. Lett., **70**, 2144–2146 (1997).*

3.8 Strained Si: Hole Mobility

Nayak and Chun [55] have used strain Hamiltonian and **kp** theory to calculate low field hole mobility of strained Si layers on (100) $Si_{1-x}Ge_x$ substrates. Non-parabolicity and the warped nature of the valence bands are included. At room temperature, in-plane hole mobilities of strained Si were reported to be 1103 and 2747 cm^2/Vs for $x = 0.1$ and 0.2, respectively. These hole mobilities are, respectively, 2.4 and 6 times higher than that of bulk Si. This improvement in the mobility results is due to the large splitting energy between the occupied light-hole band and the empty heavy-hole band and smaller hole effective mass. The effect of p-type doping on mobility was also presented.

Monte Carlo simulation of hole transport in uniaxially strained Si has been reported by Dijkstra and Wenckebach [76]. They calculated the drift mobility $\mu_E = v_d(E)/E$ in the $< 100 >$ direction in strained Si for several strain levels as a function of the electric field up to 10 kV/cm at 300 K. The computation included the spin–orbit, light-hole and heavy-hole subbands, anisotropy, non-parabolicity and strain. Optical phonon and acoustical phonon scattering were also implemented. In weak electric fields, the drift mobility is shown to increase with strain and the increase depends on the strain being tensile or compressive. The difference found in the two cases can be explained by the larger scattering rate in the tensile case which is caused by the greater density of states in the heavy-hole band. At higher fields the drift mobility drops rapidly in strained Si as a function of the electric field.

The drift mobility in unstrained Si was shown to be almost independent of the electric field with a value of about 500 cm^2/Vs. The independence of μ_E on electric field means that for Si the linear transport regime extends up to a few kilovolts per centimeter. The drift mobility in tensilely strained Si at 300 K as a function of the electric field and strain (figure 3.32) was calculated using the diffusion coefficient and the Einstein relation at $E = 0$ kV/cm. The parameters used for the simulation in Si were taken from Hinckley and Singh [77]. From figure 3.32 one observes that for tensilely strained Si the drift mobility increases compared to unstrained Si. The effect is highest at low electric field and diminishes as the electric field strength increases.

The drift mobility in compressive strained Si at 300 K as a function of the electric field and strain is shown in figure 3.33. The increase is much stronger for compressive strain. In this case the strong decrease of the drift mobility at higher electric fields indicates the saturation of the drift velocity already at a few kilovolts per centimeter. In all cases most of the holes ($> 85\%$) are in the heavy-hole band, less than 14% in the light-hole band and a very small percentage ($< 1\%$) in the spin–orbit band.

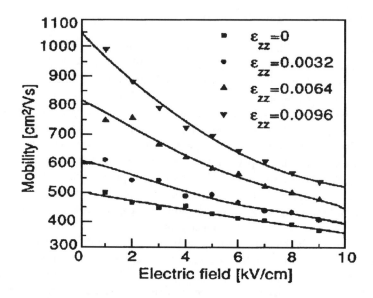

Figure 3.32. The drift mobility in tensile strained Si at 300 K as a function of the electric field and strain. The symbols are calculated points, the lines are guides for the eyes. After J. E. Dijkstra and W. Th. Wenckebach, J. Appl. Phys., **81**, *1259–1263 (1997).*

3.9 Mobility: SiC and SiGeC

As the $Si_{1-x}Ge_x$ on Si(001) system exhibits severe limitations such as a critical thickness for perfect pseudomorphic growth, attention is now drawn to other material systems such as $Si_{1-y}C_y$ and $Si_{1-x}Ge_xC_y$. An important question concerns the relation between substitutional and interstitial carbon incorporation, which can have a large impact on the electrical and optical properties of these layers. Osten and Gaworzewski [78] have investigated the temperature dependence of Hall mobilities in tensilely strained $Si_{1-y}C_y$ and in compressively strained $Si_{1-x}Ge_xC_y$ layers. The measured Hall mobilities are found to monotonically decrease with increasing carbon content for electrons in $Si_{1-y}C_y$ and for holes in $Si_{1-x}Ge_xC_y$. The reason for this is that electrically active defects are formed with the addition of carbon. These defects are presumably connected with carbon/Si interstitials or other C-related complexes.

Hall mobilities of electrons as a function of temperature for Sb-doped Si

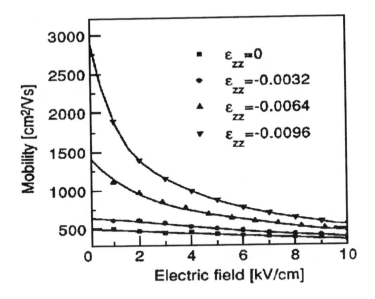

*Figure 3.33. The drift mobility in compressively strained Si at 300 K as a function of the electric field and strain. The symbols are calculated points, the lines are guides for the eyes. After J. E. Dijkstra and W. Th. Wenckebach, J. Appl. Phys., **81**, 1259–1263 (1997).*

and strained $Si_{1-y}C_y$ layers on Si(001) are shown in figure 3.34. The electron mobility in the bulk Si layer at 300 K is about 650 cm^2/Vs, close to the known value of $\approx$ 700 cm^2/Vs for bulk Si doped with Sb to the 10^{17} cm^{-3} level. The commonly observed $T^{-3/2}$ and $T^{3/2}$ dependences for the high and the low temperature regime, respectively, are indicated in the figure. Within the investigated temperature range, the electron mobilities of $Si_{1-y}C_y$ alloys are always smaller than that of the Si reference sample.

Contrary to Brunner *et al.* [79], results reported by Osten and Gaworzewski [78] on electron transport in $Si_{1-y}C_y$ alloys indicate that the effect of alloy, ionized, and neutral impurity scattering dominates over the expected gain due to strain. Adding C to Si reduces the electron mobility, and the effect becomes more pronounced at low temperatures (figure 3.35).

The influence of carbon on Hall mobilities in $Si_{1-x-y}Ge_xC_y$ layers has been studied and compared to homogeneously 1×10^{17} cm^{-3} boron-doped, pseudomorphically grown strained $Si_{1-x}Ge_x$ layers with 18% Ge. The measured temperature dependences of the Hall mobilities are shown in figure

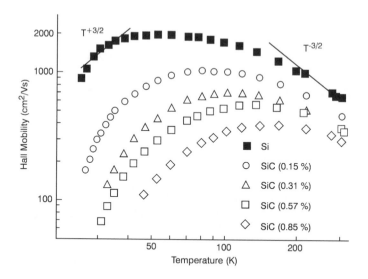

*Figure 3.34. Hall mobility of electrons as a function of temperature for Sb-doped Si and strained $Si_{1-y}C_y$ layers on Si(001). After H. J. Osten and P. Gaworzewski, J. Appl. Phys., **82**, 4977–4980 (1997).*

3.36. The $T^{-3/2}$ and $T^{3/2}$ dependences for the high and the low temperature regime, respectively, are indicated in the figure. The hole mobility of the strained $Si_{1-x}Ge_x$ layer at 300 K is 97 cm^2/Vs and is even larger than those found by Manku *et al.* [80] for their highest Ge content ($x = 0.1$) and the same doping level. Over the investigated temperature range, the hole mobilities of $Si_{1-x-y}Ge_xC_y$ alloys are always smaller than that of the $Si_{1-x}Ge_x$ sample. In analogy with the $Si_{1-y}C_y$ case, the increase in carbon content results in remarkable changes of the $\mu(T)$ dependence at lower temperatures. Thus, the addition of carbon is accompanied by the formation of electrically active defects, whose density increases with y, resulting in a substantial decrease in mobility at a low temperature.

Figure 3.37 summarizes the hole mobility values vs. carbon content at room temperature and at 77 K. The introduction of carbon leads to enhanced scattering and reduced mobility, with the effect being much larger at lower temperatures. As in the $Si_{1-y}C_y$ case, the increase in the density of ionized impurity centers with increasing C content is the main reason for the observed temperature dependence of mobility, especially at low temperatures.

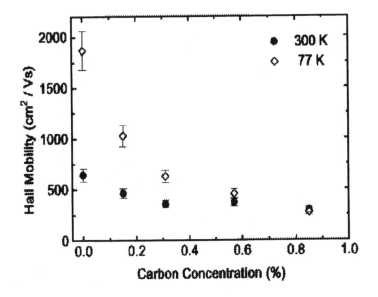

*Figure 3.35. Experimentally obtained electron mobilities at 300 and 77 K as a function of carbon content in homogeneously Sb-doped $Si_{1-y}C_y$ layers on Si(001). After H. J. Osten and P. Gaworzewski, J. Appl. Phys., **82**, 4977–4980 (1997).*

3.10 Determination of Band Offset

Accurate knowledge of band alignments is essential for heterojunction device design and performance analysis as the conduction and valence band discontinuities determine the electronic properties of heterojunctions. Doping profile, Si cap thickness and δE_v for a $Si/Si_{1-x}Ge_x$ heterostructure extracted from the characteristics of MOS capacitors have been reported by several workers [81–83]. MOS capacitance–voltage $(C - V)$ characteristics of Si/SiGe heterostructures may be utilized for extracting the apparent doping profile. Figure 3.38 shows a typical high frequency (HF, 1 MHz) $C - V$ characteristics [83]. The plateau in accumulation capacitance shows the hole confinement at the Si/SiGe interface. From high frequency capacitance–voltage $(C_{HF}\text{-}V_G)$ characteristics, the depletion depth (X_{dHF}) and apparent doping (N_{appHF}) as a function of applied gate potential (V_G) are obtained as [81]

$$X_{dHF}(V_G) = \epsilon_{Si}\left[\frac{1}{C_{HF}(V_G)} - \frac{1}{C_{ox}}\right] \tag{3.20}$$

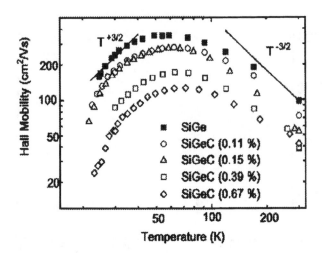

*Figure 3.36. Hall mobility of holes as a function of temperature for B-doped strained $Si_{1-x}Ge_x$ and $Si_{1-x-y}Ge_xC_y$ layers with 18% Ge on Si(001). After H. J. Osten and P. Gaworzewski, J. Appl. Phys., **82**, 4977–4980 (1997).*

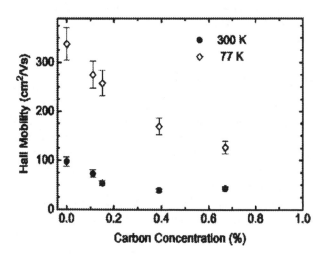

*Figure 3.37. Experimentally obtained hole mobilities at 300 and 77 K as a function of carbon content in homogeneously B-doped $Si_{1-x-y}Ge_xC_y$ layers on Si(001). After H. J. Osten and P. Gaworzewski, J. Appl. Phys., **82**, 4977–4980 (1997).*

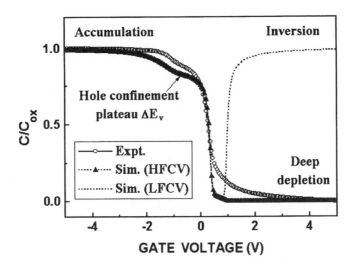

Figure 3.38. High frequency (HF, 1 MHz) $C - V$ characteristics of a MOS capacitor. Simulated HF and low frequency (LF) $C - V$ characteristics are also shown.

$$\frac{1}{N_{\text{appHF}}(V_{\text{G}})} = \frac{q\epsilon_{\text{Si}}}{2}\frac{\delta[1/C_{\text{HF}}^2(V_{\text{G}})]}{\delta V_{\text{G}}} \tag{3.21}$$

where ϵ_{Si} is the Si permittivity, C_{ox} is the gate oxide capacitance per unit area, and q is the electron charge.

The relationship between the threshold voltages and valence band offset (ΔE_{v}) is given by [84, 85]

$$\Delta E_{\text{v}} = \phi_{\text{H}} - 2\phi_{\text{F}} + \frac{kT}{q}\ln\left[\frac{[1 + C_{\text{ox}}\frac{t_{\text{cap}}}{\epsilon_{\text{Si}}} + \frac{C_{\text{ox}}(\Delta V_{\text{T}} - \Delta E_{\text{v}})}{qNX_{\text{dm}}}]^2 - 1}{2\epsilon_{\text{Si}}NkT/(qNX_{\text{dm}})^2}\right] \tag{3.22}$$

$$\phi_{\text{H}} = \phi_{\text{TH}} - \frac{kT}{q}\ln\left[\frac{[\frac{\epsilon_{\text{Si}}(\phi_{\text{H}} - 2\phi_{\text{F}})}{qNX_{\text{dm}}t_{\text{cap}}}]^2 - 1}{2\epsilon_{\text{Si}}NkT/(qNX_{\text{dm}})^2}\right] \tag{3.23}$$

where ϕ_{F} is the Fermi potential, ϕ_{H} is the potential at the top Si/Si$_{1-x}$Ge$_x$ heterointerface, ϕ_{TH} is the potential at threshold at the top Si/Si$_{1-x}$Ge$_x$ interface, q is the electronic charge, N is the effective doping concentration in the bulk of the semiconductor, X_{dm} is the maximum depletion layer width in strong inversion, t_{ox} is the oxide thickness, k is the Boltzmann constant, T is the temperature, t_{cap} is the Si cap layer thickness, and $\Delta V_{\text{T}} = V_{\text{TH}} - V_{\text{TS}}$.

3.10.1   SiGe

For a Si/SiGe heterostructure an experimental and simulated valence band offset (ΔE_{v}) values of 157 meV and 140 meV, respectively, are obtained by iterating equations 3.22 and 3.23 using the values of doping concentration and threshold voltages obtained from the experimental high frequency apparent doping vs. gate voltage characteristics as shown in figures 3.39 and 3.40, respectively [83]. When compared with the reported experimental ΔE_{v} values [82, 86–91], it is seen that the agreement is fairly good as shown in figure 3.41, for Ge mole fraction in the range $0 < x < 0.25$.

Gan *et al.* [86] measured the valence band discontinuity of strained $Si_{1-x}Ge_x$ on unstrained Si using semiconductor insulator semiconductor structures with Ge compositions in the range 10 to 25%. The epitaxial heterostructures were grown by chemical vapor deposition. The experimental data indicate that the valence band discontinuity between Si and $Si_{1-x}Ge_x$ can be approximated by $\Delta E_{\mathrm{v}} = 7.4x$ meV for $0 < x < 17.5\%$.

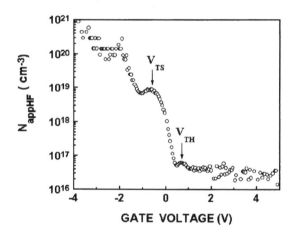

Figure 3.39. Apparent doping vs. distance from the Si/SiO_2 interface obtained from the high frequency $C - V$ characteristics.

3.10.2   SiGeC

$Si_{1-x-y}Ge_xC_y$ offers considerably greater flexibility, compared to that available in the $Si/Si_{1-x}Ge_x$ material system, to control strain and electronic properties

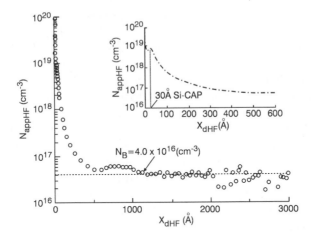

Figure 3.40. Experimental high frequency apparent doping vs. gate voltage characteristics for a Si/SiGe/Si MOS capacitor.

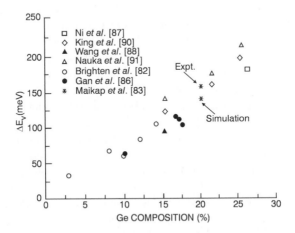

Figure 3.41. Summary of experimental measurement data of ΔE_v in strained $Si_{1-x}Ge_x$.

in group-IV heterostructures [26]. This leads to the possibility of fabricating group-IV heterostructure devices lattice matched to Si [92–96]. Novel heterojunction bipolar transistor [97], p-type metal oxide semiconductor field effect transistor (p-MOSFET) [98], and photodetector [99] devices using $Si_{1-x-y}Ge_xC_y$ layers have already been reported. For determining the valence band offset of Si/SiGeC heterostructures with different carbon concentrations, $C-V$ characteristics of MOS capacitors have been used.

Theoretical calculations for band offsets based on the electronic structures of the heterointerface, involving a variety of SiGeC layers on Si substrates, have been employed for the prediction of band offsets [100]. However, empirical rules, derived from experimental measurements, give better estimates of band offsets than theoretical calculations [101]. Experimental determination of the valence band offset between partially strained $Si_{1-x-y}Ge_xC_y$ and Si has been reported by several workers using different techniques such as admittance spectroscopy [102], photoluminescence measurements [96, 103] and capacitance–voltage characteristics [104].

Maikap *et al.* [105] have studied the effect of C on the valence band offsets in UHVCVD grown epitaxial layers of $Si_{1-x-y}Ge_xC_y$ on an epitaxial Si-buffer layer. The values of valence band offset (ΔE_v) were found to be 125 meV and 170 meV for $Si_{0.79}Ge_{0.2}C_{0.01}$ and $Si_{0.685}Ge_{0.3}C_{0.015}$ respectively, by iterating equations 3.22 and 3.23 and using the values of doping concentration and threshold voltages obtained from the experimental high frequency apparent doping vs. gate voltage characteristics. The reported experimental ΔE_v values obtained using different techniques by several authors are shown in figure 3.42. It is observed that irrespective of measurement techniques employed [102, 104], the valence band offset values agree well. The extracted values of ΔE_v for fully strained $Si_{1-x}Ge_x$ samples are plotted against Ge fraction along with the theoretical prediction. It is seen that as the carbon content increases, the band offset decreases. Measurements show a 23–35 meV reduction in valence band discontinuity per 1.0% C incorporation. Similar reduction has also been reported by Chang *et al.* [106].

Kolodzey *et al.* [107] have reported on conduction and valence band offsets in thick, relaxed Ge-rich $Si_{1-x-y}Ge_xC_y$ alloys grown by solid source molecular beam epitaxy on Si(100) substrates. X-ray photoemission spectroscopy was used to measure the valence band energies with respect to atomic core levels and showed that C increased the valence band maximum of SiGeC by +48 meV/%C. The bandgap energies were obtained from optical absorption and were combined with the valence band offsets to yield the conduction band offsets. For SiGeC/Si heterojunctions, the offsets were typically 0.6 eV for the valence band and 0.38 eV for the conduction band, with a staggered type-II

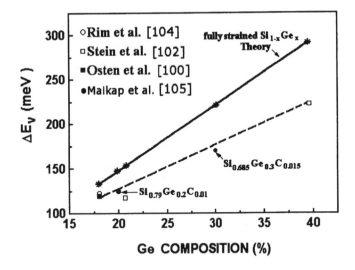

Figure 3.42. Summary of experimental measurement data of ΔE_v in a $Si/Si_{1-x-y}Ge_xC_y$ heterostructure.

alignment. These amounts of offset are sufficient to provide significant electron and hole confinement for device applications.

3.10.3   Strained Si

Heterojunction conduction and valence band offsets (ΔE_c, ΔE_v) of strained Si/SiGe heterostructures have also been determined from experimental threshold voltages of a strained Si channel p-MOSFET [85]. Threshold voltages at a strained-Si/relaxed-SiGe heterointerface (V_{TH}) and a strained-Si/SiO$_2$ interface (V_{TS}) have been extracted from I_D–V_G and $I_D/g_m^{1/2}$ characteristics of MOSFETs.

An experimental valence band offset (ΔE_v) value of 160 meV is obtained by iterating equations 3.22 and 3.23 using the values of doping concentration and threshold voltages obtained from the experimental high frequency apparent doping vs. gate voltage characteristics. Using the valance band offset data, conduction band offset (ΔE_c) is obtained from [108]

$$\Delta E_c = E_g(Si_{1-x}Ge_x) + \Delta E_v(Si_{1-x}Ge_x/Si) - E_g(\text{strained Si}) \qquad (3.24)$$

where E_g(strained Si) is given by [5, 20]

$$E_g = 1.11 - 0.4x \tag{3.25}$$

and x is the Ge concentration in the top part of a completely relaxed SiGe buffer. A value of 126 meV is obtained for ΔE_c. A comparison of extracted values of ΔE_v and ΔE_c with those reported is given in table 3.3.

Table 3.3. Band offsets of strained Si grown on relaxed $Si_{0.80}Ge_{0.20}$ buffer.

Reference	[85]	[108]	[5]
ΔE_c (meV)	126	120	125
ΔE_v (meV)	160	115	130

3.11 Summary

In this chapter material properties such as bandgap, lattice constants, velocity–field characteristics, conduction and valence band discontinuities, heavy doping effect, and carrier mobilities of SiGe, SiGeC and strained Si layers have been presented. It is seen that strain in SiGe films may be utilized for control of bandgap alignment and improved mobility. Two effects, namely impurity scattering and alloy scattering, often thwart the efforts for mobility enhancement. While impurity scattering may be partly mitigated by modulation doping and choice of operating conditions, alloy scattering appears to be a deterrent. In brief, the low field in-plane hole mobility of a strained $Si_{1-x}Ge_x$ layer is significantly higher than that of bulk Si. The low field mobility of a strained $Si_{0.7}Ge_{0.3}$ epilayer on Si(001) is about 2–3 times higher than of bulk Si. The saturation carrier velocity of a strained $Si_{0.7}Ge_{0.3}$ epilayer on Si(001) can be 20%–25% higher than that of bulk Si. The temperature dependence of Hall mobilities in tensilely strained $Si_{1-y}C_y$ layers and in compressively strained $Si_{1-x-y}Ge_xC_y$ layers has been covered. Band alignment and the $C - V$ technique used for the determination of band discontinuities have been discussed.

3.12 References

1 E. Kasper and H. J. Herzog, "Elastic Strain and Misfit Dislocation Density in $Si_{0.92}Ge_{0.08}$ Films on Silicon Substrates," *Thin Solid Films*, vol. 44, pp. 357–370, 1977.

2 G. C. Osbourn, "Strained layer superlattices from lattice mismatched materials," *J. Appl. Phys.*, vol. 53, pp. 1586–1589, 1982.

3 J. C. Bean, L. C. Feldman, A. T. Fiory, S. Nakahara and I.K. Robinson, "Ge$_x$Si$_{1-x}$/Si strained layer superlattice growth by molecular beam epitaxy," *J. Vac. Sci. Technol. A*, vol. 2, pp. 436–440, 1984.

4 F. C. Frank and J. H. van der Merwe, "One-dimensional dislocations. II. Misfitting monolayers and oriented overgrowth," *Proc. Roy. Soc.*, vol. A198, pp. 216–225, 1949.

5 R. People, "Physics and Applications of Ge$_x$Si$_{1-x}$/Si Strained Layer Heterostructures," *IEEE J. Quantum Elec.*, vol. QE-22, pp. 1696–1710, 1986.

6 J. W. Matthews and A. E. Blakeslee, "Defects in epitaxial multilayers - I. Misfit dislocations," *J. Crystal Growth*, vol. 27, pp. 118–125, 1974.

7 J. W. Matthews and A. E. Blakeslee, "Defects in epitaxial multilayers - II. Dislocation pile-ups, threading dislocations, slip lines and cracks," *J. Crystal Growth*, vol. 29, pp. 273–280, 1975.

8 J. W. Matthews and A. E. Blakeslee, "Defects in epitaxial multilayers - III. Preparation of almost perfect multilayers," *J. Crystal Growth*, vol. 32, pp. 265–273, 1976.

9 R. People and J. C. Bean, "Calculation of critical layer thickness versus lattice mismatch for Ge$_x$Si$_{1-x}$/Si strained layer heterostructures," *Appl. Phys. Lett.*, vol. 47, pp. 322–324, 1985.

10 J. H. van der Merwe, "Crystal Interfaces. Part II. Finite Overgrowths (see also Erratum p. 3420 (1963))," *J. Appl. Phys.*, vol. 34, pp. 123–127, 1963.

11 J. H. van der Merwe, "Structure of epitaxial crystal interfaces," *Surf. Sci.*, vol. 31, pp. 198–228, 1972.

12 J. W. Matthews, "Defects Associated with the Accommodation of Misfit Between Crystals," *J. Vac. Sci. Technol.*, vol. 12, pp. 126–133, 1975.

13 D. Eaglesham, E. Kvam, D. Maher, C. Humphreys, G. Green, B. Tanner and J. Bean, "X-ray topography of the coherency breakdown in Ge$_x$Si$_{1-x}$/Si (100)," *Appl. Phys. Lett.*, vol. 53, pp. 2083–2085, 1988.

14 R. Hull, J. C. Bean, F. Cerdeira, A. T. Fiory and J. M. Gibson, "Stability of semiconductor strained-layer superlattices," *Appl. Phys. Lett.*, vol. 48, pp. 56–58, 1986.

15 P. Voisin, J. Bleuse, C. Bouche, S. Gillard, C. Alibert and A. Regreny, "Observation of the Wannier–Stark quantization in a semiconductor superlattice," *Phys. Rev. Lett.*, vol. 61, pp. 1639–1642, 1988.

16 E. Kasper, "Growth and Properties of Si/SiGe Superlattices," *Surface Sci.*, vol. 174, pp. 630–639, 1986.

17 R. van de Leur, A. Schellingerhout, F. Tuinstra and J. Mooij, "Critical thickness for pseudomorphic growth of Si/Ge alloys and superlattices," *J. Appl. Phys.*, vol. 64, pp. 3043–3050, 1988.

18 D. Chidambarrao, G. Srinivasan, B. Cunningham and C. S. Murthy, "Effects of Peierls barrier and epithreading dislocation orientation on the critical thickness in heteroepitaxial structures," *Appl. Phys. Lett.*, vol. 57, pp. 1001–1003, 1990.

19 Y. H. Lo, "New approach to grow pseudomorphic structures over the critical thickness," *Appl. Phys. Lett.*, vol. 59, pp. 2311–2313, 1991.

20 G. Abstreiter, H. Brugger, T. Wolf, H. Jorke and H. J. Herzog, "Strain-Induced Two-Dimensional Electron Gas in Selectively Doped Si/Si_xGe_{1-x} Superlattices," *Phys. Rev.*, vol. 54, pp. 2441–2444, 1985.

21 F. K. LeGoues, M. M. Mooney and J. O. Chu, "Crystallographic tilting resulting from nucleation limited relaxation," *Appl. Phys. Lett.*, vol. 62, pp. 140–142, 1993.

22 F. K. LeGoues, B. S. Meyerson, J. F. Morar and P. D. Kirchner, "Mechanism and conditions for anomalous strain relaxation in graded thin films and superlattices," *J. Appl. Phys.*, vol. 71, pp. 4230–4243, 1992.

23 P. M. Mooney, J. L. Jordan-sweet, K. Ismail, J. O. Chu, R. M. Feenstra and F. K. LeGoues, "Relaxed $Si_{0.7}Ge_{0.3}$ buffer layers for high-mobility devices," *Appl. Phys. Lett.*, vol. 67, pp. 2373–2375, 1995.

24 J. W. P. Hsu, E. A. Fitzgerald, Y. H. Xie, P. J. Silverman and M. J. Cardillo, "Surface morphology of relaxed Ge_xSi_{1-x} films," *Appl. Phys. Lett.*, vol. 61, pp. 1293–1295, 1992.

25 M. A. Lutz, R. M. Feenstra, F. K. LeGoues, P. M. Mooney and J. O. Chu, "Influence of misfit dislocations on the surface morphology of $Si_{1-x}Ge_x$ films," *Appl. Phys. Lett.*, vol. 66, pp. 724–726, 1995.

26 C. K. Maiti, L. K. Bera and S. Chattopadhyay, "Strained Si Heterostructure Field Effect Transistors," *Semicond. Sci. Technol.*, vol. 13, pp. 1225–1246, 1998.

27 H. J. Herzog, L. Csepregi and H. Seidel, "X-ray investigation of boron- and germanium-doped silicon epitaxial layers," *J. Electrochem. Soc.*, vol. 131, pp. 2969–2974, 1984.

28 W. P. Maszara and T. Thompson, "Strain compensation by Ge in B-doped silicon epitaxial films," *J. Appl. Phys.*, vol. 72, pp. 4477–4479, 1992.

29 B. Tillack, O. Zaumseil, G. Morgenstern, D. Kruger and G. Ritter, "Strain compensation in $Si_{1-x}Ge_x$ by heavy boron doping," *Appl. Phys. Lett.*, vol. 67, pp. 1143–1145, 1995.

30 S. C. Jain, H. J. Osten, B. Dietrich and H. Rucker, "Growth and properties of strained $Si_{1-x-y}Ge_xC_y$ layers," *Semicond. Sci. Technol.*, vol. 10, pp. 1289–1302, 1995.

31 H. J. Osten, *Carbon-Containing Layers on Silicon - Growth, Properties and Applications.* Trans-Tech Publications, Switzerland, 1999.

32 H. J. Osten, E. Bugiel and P. Zaumseil, "Growth of an inverse tetragonal distorted SiGe layer on Si(001) by adding small amounts of carbon," *Appl. Phys. Lett.*, vol. 64, pp. 3440–3442, 1994.

33 H. J. Osten and E. Bugiel, "Relaxed $Si_{1-x}Ge_x/Si_{1-x-y}Ge_xC_y$ buffer structures with low threading dislocation density," *Appl. Phys. Lett.*, vol. 70, pp. 2813–2815, 1997.

34 G. He, M. D. Savellano and H. A. Atwater, "Synthesis of dislocation free $Si_y(Sn_xC_{1-x})_{1-y}$ alloys by molecular beam deposition and solid phase epitaxy," *Appl. Phys. Lett.*, vol. 65, pp. 1159–1161, 1994.

35 A. T. Khan, P. R. Berger, F. J. Guarin and S. S. Iyer, "Near band edge photoluminescence from pseudomorphically tensilely strained $Si_{0.985}C_{0.015}$," *Thin Solid Films*, vol. 294, pp. 122–124, 1997.

36 S. Bodnar and J. L. Regolini, "Growth of ternary alloy $Si_{1-x-y}Ge_xC_y$ by rapid thermal chemical vapor deposition," *J. Vac. Sci. Technol. A*, vol. 13, pp. 2336–2340, 1995.

37 J. Xiang, N. Herbots, H. Jacobsson, O. Ye, S. Heame and S. Whaley, "Comparative study on dry oxidation of heteroepitaxial $Si_{1-x}Ge_x$ and $Si_{1-x-y}Ge_xC_y$ on Si(100)," *J. Appl. Phys.*, vol. 80, pp. 1857–1866, 1996.

38 J. R. Chelikowsky and M. L. Cohen, "Nonlocal pseudopotential calculations for the electronic structure of eleven diamond and zinc-blende semiconductors," *Phys. Rev. B*, vol. 14, pp. 556–582, 1976.

39 P. Hohenberg and W. Kohn, "Inhomogeneous Electron Gas," *Phys. Rev. B*, vol. 136, pp. 864–871, 1964.

40 S. Froyen, D. M. Wood and A. Zunger, "Structural and electronic properties of epitaxial thin layer superlattices," *Phys. Rev.B*, vol. 37, pp. 6893–6907, 1988.

41 M. Jaros, "Electronic properties of semiconductor alloy systems," *Reports on Progress in Physics*, vol. 48, pp. 1091–1154, 1985.

42 R. Braunstein, A. R. Moore and F. Herman, "Intrinsic optical absorption in germanium–silicon alloys," *Phys. Rev.*, vol. 109, pp. 695–710, 1958.

43 J. Weber and M. I. Alonso, "Near-band-gap photoluminescence of SiGe alloys," *Phys. Rev. B*, vol. 40, pp. 5683–5693, 1989.

44 D. V. Lang, R. People, J. C. Bean and A. M. Sergent, "Measurement of the band gap of Ge_xSi_{1-x}/Si strained-layer heterostructures," *Appl. Phys. Lett.*, vol. 47, pp. 1333–1335, 1985.

45 D. Dutartre, G. Bremond, A. Soufi and T. Benyattou, "Excitonic photoluminescence from Si capped strained $Si_{1-x}Ge_x$ layers," *Phys. Rev. B*, vol. 44, pp. 11525–11527, 1991.

46 R. People, "Indirect band gap of coherently strained Ge_xSi_{1-x} alloys on < 001 > silicon substrates," *Phys. Rev. B*, vol. 32, pp. 1405–1408, 1985.

47 R. People, "Indirect bandgap and band alignment of coherently strained Si_xGe_{1-x} bulk alloys on Ge(001) substrate," *Phys. Rev. B*, vol. 34, pp. 2508–2510, 1986.

48 C. G. van de Walle and R. M. Martin, "Theoretical calculations of heterojunction discontinuities in the Si/Ge system," *Phys. Rev. B*, vol. 34, pp. 5621–5634, 1986.

49 J. C. Bean, "Silicon-Based Semiconductor Heterostructures: Column IV Bandgap Engineering," *Proc. IEEE*, vol. 80, pp. 571–587, 1992.

50 J. M. Hinckley and J. Singh, "Hole transport theory in pseudomorphic $Si_{1-x}Ge_x$ alloys grown on Si(100) substrates," *Phys. Rev. B*, vol. 41, pp. 2912–2926, 1990.

51 R. People and J. C. Bean, "Band alignments of coherently strained Ge_xSi_{1-x}/Si heterostructures on (001) Ge_ySi_{1-y} substrate," *Appl. Phys. Lett.*, vol. 48, pp. 1338–1340, 1986.

52 T. Manku and A. Nathan, "Electron drift mobility model for devices based on unstrained and coherently strained $Si_{1-x}Ge_x$ grown on (001) silicon substrate," *IEEE Trans. Electron Dev.*, vol. 39, pp. 2082–2089, 1992.

53 C. Herring and E. Vogt, "Transport and deformation-potential theory for many-valley semiconductors with anisotropic scattering," *Phys. Rev.*, vol. 101, pp. 944–961, 1956.

54 U. Konig and H. Daembkes, "SiGe HBTs and HFETs," *Solid-State Electron.*, vol. 38, pp. 1595–1602, 1995.

55 D. K. Nayak and S. K. Chun, "Low field mobility of strained Si on (100) $Si_{1-x}Ge_x$ substrate," *Appl. Phys. Lett.*, vol. 64, pp. 2514–2516, 1994.

56 F. Schaffler, "High-mobility Si and Ge structures," *Semicond. Sci. Technol.*, vol. 12, pp. 1515–1549, 1997.

57 S. C. Jain and D. J. Roulston, "A Simple Expression for Band Gap Narrowing (BGN) in Heavily Doped Si, Ge, GaAs and Ge_xSi_{1-x} Strained Layers," *Solid-State Electron.*, vol. 34, pp. 453–465, 1991.

58 J. S. Yuan, *SiGe, GaAs, and InP Heterojunction Bipolar Transistors.* John Wiley & Sons, New York, 1999.

59 Z. Matutinovic-Krstelj, V. Venkataraman, E. J. Prinz, J. C. Sturm and C. W. Magee, "Base resistance and effective bandgap reduction in n-p-n Si/$Si_{1-x}Ge_x$/Si HBTs with heavy base doping," *IEEE Trans. Electron Dev.*, vol. 43, pp. 457–466, 1996.

60 B. K. Ridley, *Quantum Processes in Semiconductors*. Oxford Univ. Press, New York, 3rd Ed., 1993.

61 V. Venkataraman, C. W. Liu and J. C. Sturm, "Alloy scattering limited transport of two-dimensional carriers in strained $Si_{1-x}Ge_x$ quantum wells," *Appl. Phys. Lett.*, vol. 63, pp. 2795–2797, 1993.

62 M. V. Fischetti and S. E. Laux, "Band structure, deformation potentials and carrier mobility in strained Si, Ge, and SiGe alloys," *J. Appl. Phys.*, vol. 80, pp. 2234–2252, 1996.

63 S. C. Jain and W. Hayes, "Structure, properties and applications of Ge_xSi_{1-x} strained layers and superlattices," *Semicond. Sci. Technol.*, vol. 6, pp. 547–576, 1991.

64 J. M. Hinckley, V. Sankaran and J. Singh, "Charged carrier transport in $Si_{1-x}Ge_x$ pseudomorphic alloys matched to Si-strain-related transport improvements," *Appl. Phys. Lett.*, vol. 55, pp. 2008–2010, 1989.

65 T. Manku and A. Nathan, "Energy-Band Structure for Strained p-type $Si_{1-x}Ge_x$," *Phys. Rev. B*, vol. 43, pp. 12634–12637, 1991.

66 T. Manku and A. Nathan, "Effective mass for strained p-type $Si_{1-x}Ge_x$," *J. Appl. Phys.*, vol. 69, pp. 8414–8416, 1991.

67 T. Manku, J. M. McGregor, A. Nathan, D. J. Roulston, J.-P. Noel and D. C. Houghton, "Drift hole mobility in strained and unstrained doped $Si_{1-x}Ge_x$ alloys," *IEEE Trans. Electron Dev.*, vol. 40, pp. 1990–1996, 1993.

68 K. Ismail, J. O. Chu and B. S. Meyerson, "High hole mobility in SiGe alloys for device applications," *Appl. Phys. Lett.*, vol. 64, pp. 3124–3126, 1994.

69 T. K. Carns, S. K. Chun, M. O. Tanner, K. L. Wang, T. I. Kamins, J. E. Turner, D. Y. C. Lie, M.-A. Nicolet and R. G. Wilson, "Hole mobility measurements in heavily doped $Si_{1-x}Ge_x$ strained layers," *IEEE Trans. Electron Dev.*, vol. 41, pp. 1273–1281, 1994.

70 D. K. Nayak, J. C. S. Woo, J. S. Park, K. L. Wang and K. P. MacWilliams, "High-mobility p-channel metal–oxide–semiconductor field-effect transistor on strained Si," *Appl. Phys. Lett.*, vol. 62, pp. 2853–2855, 1993.

71 F. M. Bufler, P. Graf, B. Meinerzhagen, G. Fischer and H. Kibbel, "Hole transport investigation in unstrained and strained SiGe," *J. Vac. Sci. Technol. B*, vol. 16, pp. 1667–1669, 1998.

72 K. Ismail, S. F. Nelson, J. O. Chu and B. S. Meyerson, "Electron transport properties of Si/SiGe heterostructures: measurements and device implications," *Appl. Phys. Lett.*, vol. 63, pp. 660–662, 1993.

73 F. Stern and S. E. Laux, "Charge transfer and low-temperature electron mobility in a strained Si layer in relaxed $Si_{1-x}Ge_x$," *Appl. Phys. Lett.*, vol. 61, pp. 1110–1112, 1992.

74 K. Ismail and B. S. Meyerson, "Si/SiGe Quantum-Wells – Fundamentals to Technology," *J. Mater. Sci.: Mater. Electron.*, vol. 6, pp. 306–310, 1995.

75 F. M. Bufler, P. Graf, S. Keith and B. Meinerzhagen, "Full band Monte Carlo investigation of electron transport in strained Si grown on $Si_{1-x}Ge_x$ substrates," *Appl. Phys. Lett.*, vol. 70, pp. 2144–2146 (see also Erratum: *Appl. Phys. Lett.*, vol. 71, p. 3447, 1997), 1997.

76 J. E. Dijkstra and W. Th. Wenckebach, "Hole transport in strained Si," *J. Appl. Phys.*, vol. 81, pp. 1259–1263, 1997.

77 J. M. Hinckley and J. Singh, "Monte Carlo studies of ohmic hole mobility in silicon and germanium: Examination of the optical phonon deformation potential," *J. Appl. Phys.*, vol. 76, pp. 4192–4200, 1994.

78 H. J. Osten and P. Gaworzewski, "Charge transport in strained $Si_{1-y}C_y$ and $Si_{1-x-y}Ge_xC_y$ alloys on Si(001)," *J. Appl. Phys.*, vol. 82, pp. 4977–4981, 1997.

79 K. Brunner, W. Winter, K. Eberl, N. Y. Jin Phillipp and F. Phillipp, "Fabrication and band alignment of pseudomorphic $Si_{1-y}C_y$, $Si_{1-x-y}Ge_xC_y$ and coupled $Si_{1-y}C_y/Si_{1-x-y}Ge_xC_y$ quantum well structures on Si substrates," *J. Cryst. Growth*, vol. 175, pp. 451–458, 1997.

80 T. Manku, S. C. Jain and A. Nathan, "On the reduction of hole mobility in strained p-SiGe layers," *J. Appl. Phys.*, vol. 71, pp. 4618–4619, 1992.

81 S. P. Voinigescu, K. Iniewski, R. Lisak, C. A. T. Salama, J. P. Noel and D. C. Houghton, "New Technique for the Characterization of Si/SiGe

layers Using Heterostructure MOS Capacitors," *Solid-State Electron.*, vol. 37, pp. 1491–1501, 1994.

82 J. C. Brighten, I. D. Hawkins, A. R. Peaker, E. H. C. Parker and T. E. Whall, "The determination of valence band discontinuities in $Si/Si_{1-x}Ge_x/Si$ heterojunctions by capacitance–voltage techniques," *J. Appl. Phys.*, vol. 74, pp. 1894–1899, 1993.

83 S. Maikap, L. K. Bera, S. K. Ray and C. K. Maiti, "Electrical characterization of $Si/Si_{1-x}Ge_x/Si$ quantum well heterostructures using MOS capacitor," *Solid-State Electron.*, vol. 44, pp. 1029–1034, 2000.

84 K. Iniewski, S. Voinigescu, J. Atcha and C. A. T. Salama, "Analytical modeling of threshold voltages in p-channel Si/SiGe/Si MOS structures," *Solid-State Electron.*, vol. 36, pp. 775–783, 1993.

85 L. K. Bera, S. K. Ray, D. K. Nayak, N. Usami, Y. Shiraki and C. K. Maiti, "Gas Source Molecular Beam Epitaxy Grown Strained Si Films on Step-Graded Relaxed $Si_{1-x}Ge_x$ for MOS Applications," *J. Electron. Mater.*, vol. 28, pp. 98–104, 1999.

86 C. H. Gan, J. A. del Alamo, B. R. Bennett, B. S. Meyerson, E. F. Crabbe, C. G. Sodini, L. R. Reif, "$Si/Si_{1-x}Ge_x$ valence band discontinuity measurements using a semiconductor–insulator–semiconductor (SIS) heterostructure," *IEEE Trans. Electron Dev.*, vol. 41, pp. 2430–2439, 1994.

87 W.-X. Ni, J. Knall and G. V. Hansoon, "New Method to Study Band Offsets Applied to Strained $Si/Si_{1-x}Ge_x$ (100) Heterojunction Interfaces," *Phys. Rev. B*, vol. 36, pp. 7744–7747, 1987.

88 P. J. Wang, F. F. Fang, B. S. Meyerson, J. Nocera and B. Parker, "Two-dimensional hole gas in $Si/Si_{0.85}Ge_{0.15}/Si$ modulation-doped double heterostructures," *Appl. Phys. Lett.*, vol. 54, pp. 2701–2703, 1989.

89 S. S. Iyer, G. L. Patton, J. M. C. Stork, B. S. Meyerson and D. L. Harame, "Heterojunction bipolar transistors using Si–Ge alloys," *IEEE Trans. Electron Dev.*, vol. ED-36, pp. 2043–2064, 1989.

90 C. A. King, J. L. Hoyt and J. F. Gibbons, "Bandgap and Transport Properties of $Si_{1-x}Ge_x$ by Analysis of Nearly Ideal $Si/Si_{1-x}Ge_x/Si$ Heterojunction Bipolar Transistors," *IEEE Trans. Electron Dev.*, vol. 36, pp. 2093–2104, 1989.

91 K. Nauka, T. I. Kamins, J. E. Turner, C. A. King, J. L. Hoyt and J. F. Gibbons, "Admittance spectroscopy measurements of band offsets in Si/Si$_{1-x}$Ge$_x$/Si heterostructures," *Appl. Phys. Lett.*, vol. 60, pp. 195–197, 1992.

92 K. Eberl, S. S. Iyer, S. Zollner, J. C. Tsang and F. K. LeGoues, "Growth and strain compensation effects in the ternary Si$_{1-x-y}$Ge$_x$C$_y$ alloy system," *Appl. Phys. Lett.*, vol. 60, pp. 3033–3035, 1992.

93 J. Kolodzey, P. R. Berger, B. A. Orner, D. Hits, F. Chen, A. Khan, X. Shao, M. Waite, S. I. Shah, C. P. Swann and K. M. Unruh, "Optical and electronic properties of SiGeC alloys grown on Si substrates," *J. Cryst. Growth*, vol. 157, pp. 386–391, 1995.

94 C. W. Liu, A. St. Amour, J. C. Sturm, Y. R. J. Lacroix, M. L. Thewalt, C. W. Magee and D. Eaglesham, "Growth and photoluminescence of high quality SiGeC random alloys on silicon substrates," *J. Appl. Phys.*, vol. 80, pp. 3043–3047, 1996.

95 R. A. Soref, "Optical band gap of the ternary semiconductor Si$_{1-x-y}$Ge$_x$C$_y$," *J. Appl. Phys.*, vol. 70, pp. 2470–2472, 1991.

96 O. G. Schmidt and K Eberl, "Photoluminescence of tensile strained, exactly strain compensated, and compressively strained Si$_{1-x-y}$Ge$_x$C$_y$ layers on Si," *Phys. Rev. Lett.*, vol. 80, pp. 3396–3399, 1998.

97 L. D. Lanzerotti, A. St. Amour, C. W. Liu, J. C. Sturm, J. K. Watanabe and N. D. Theodore, "Si/Si$_{1-x-y}$Ge$_x$C$_y$/Si Heterojunction Bipolar Transistors," *IEEE Electron Dev. Lett.*, vol. 17, pp. 334–337, 1996.

98 S. K. Ray, S. John, S. Oswal and S. K. Banerjee, "Novel SiGeC Channel Heterojunction PMOSFET," in *IEEE IEDM Tech. Dig.*, pp. 261–264, 1996.

99 F. Y. Huang and K. L. Wang, "Normal-incidence epitaxial SiGeC photodetector near 1.3 μm wavelength grown on Si substrate," *Appl. Phys. Lett.*, vol. 69, pp. 2330–2332, 1996.

100 H. J. Osten, "Band-gap changes and band offsets for ternary Si$_{1-x-y}$Ge$_x$C$_y$ alloys," *J. Appl. Phys.*, vol. 84, pp. 2716–2721, 1998.

101 A. Ichii, Y. Tsou and E. Garmire, "An empirical rule for band offset between III–V alloy compounds," *J. Appl. Phys.*, vol. 74, pp. 2112–2113, 1993.

102 B. L. Stein, E. T. Yu, E. T. Croke, A. T. Hunter, T. Laursen, A. E. Bair, J. W. Mayer and C. C. Ahn, "Band offsets in $Si/Si_{1-x-y}Ge_xC_y$ heterojunctions measured by admittance spectroscopy," *Appl. Phys. Lett.*, vol. 70, pp. 3413–3415, 1997.

103 R. Hartmann, U. Gennser, H. Sigg, D. Grutzmacher and E. Ensslin, "Bandgap and band alignment of strain reduced $Si/Si_{1-x-y}Ge_xC_y$ multiple quantum well structures obtained by photoluminescence measurements," *Appl. Phys. Lett.*, vol. 73, pp. 1257–1259, 1998.

104 K. Rim, S. Takagi, J. L. Hoyt and J. F. Gibbons, "Capacitance–Voltage Characteristics of p-Si/SiGeC MOS Capacitors," in *Mat. Res. Soc. Symp. Proc.*, vol. 379, pp. 327–332, 1995.

105 S. Maikap, S. K. Ray, S. John, S. K. Banerjee and C. K. Maiti, "Electrical characterization of ultra-thin gate oxides on $Si/Si_{1-x-y}Ge_xC_y/Si$ quantum well heterostructures," *Semicond. Sci. Technol.*, vol. 15, pp. 761–764, 2000.

106 C. L. Chang, A. St. Amour and J. C. Sturm, "Effect of Carbon on the Valance Band Offset of $Si_{1-x-y}Ge_xC_y/Si$ Heterojunctions," in *IEEE IEDM Tech. Dig.*, pp. 257–260, 1996.

107 J. Kolodzey, F. Chen, B. A. Orner, D. Guerin and S. I. Shah, "Energy band offsets of SiGeC heterojunctions," *Thin Solid Films*, vol. 302, pp. 201–203, 1997.

108 J. J. Welser, *The application of strained-silicon/relaxed-silicon germanium heterostructures to metal–oxide–semiconductor field-effect transistors.* PhD Thesis, Stanford University, 1994.

109 J. M. McGregor, T. Manku, J. P. Noel, D. J. Roulston, A. Nathan and D. C. Houghton, "Measured in-plane hole drift and Hall mobility in heavily-doped strained p-type $Si_{1-x}Ge_x$," *J. Electron. Mater.*, vol. 22, pp. 319–321, 1993.

110 A. B. Sproul, M. A. Green and A. W. Stephens, "Accurate determination of minority carrier- and lattice scattering-mobility in silicon from photoconductance decay," *J. Appl. Phys.*, vol. 72, pp. 4161–4171, 1992.

111 S. E. Swirhun, J. A. del Alamo and R. M. Swanson, "Measurement of hole mobility in heavily doped n-type silicon," *IEEE Electron Dev. Lett.*, vol. EDL-7, pp. 168–171, 1986.

112 W. R. Thurber, R. L. Mattis and Y. M. Liu, "Resistivity-dopant density relationship for boron-doped silicon," *J. Electrochem. Soc. : Solid State Sc. and Tech.*, vol. 127, pp. 2291–2294, 1980.

113 J. Dziewior and D. Silber, "Minority-carrier diffusion coefficients in highly doped silicon," *Appl. Phys. Lett.*, vol. 35, pp. 170–172, 1979.

Chapter 4

Gate Dielectrics on Strained Layers

Silicon dioxide and silicon nitride are used in a variety of ways in integrated circuit fabrication. Silicon dioxide films are used as gate dielectrics, for device isolation, as the capacitor material in DRAMs, for tunneling oxides in EPROMs and for final passivation, apart from their common use as a masking material. This chapter will focus on the manufacturing challenges for the formation of gate dielectrics such as oxide and nitride on strained layers.

It is well known that the semiconductor industry is on an aggressive scaling program and the device dimensions are expected to shrink as low as 0.18 μm in 2001, 0.12 μm in 2004 and 0.10 μm in 2007. Along with this, power supply voltage has to drop to 1.5 V in 2004 as indicated in table 4.1. The gate dielectric is of key importance in this scaling effort, since it forms the "heart" of n- and p-channel MOSFETs in CMOS technology, and largely determines the performance of the transistors. Gate dielectrics are getting thinner as device dimensions shrink: now about 60 Å, gate thickness is expected to shrink to about 40 Å in the next five years.

Given that it is always desirable to use the thinnest gate possible to maximize the transistor's drain current, and that lower power supply voltages allow thinner gates, the thickness of the gate is mainly governed by yield and reliability issues. The ultrathin dielectric films that are being used for gate, tunnel and storage applications need to have a high dielectric strength, low defect and interface state density, good radiation hardness, diffusion resistance, resistant to hot carrier degradation and minimum charge trapping.

In this chapter, processing issues of gate dielectric on SiGe and other strained layers in Si integrated circuits are addressed. The dielectrics for a surface channel SiGe MOSFET have been incorporated in a number of ways. One may oxidize the alloy or grow a Si cap on the alloy and oxidize it or

162

Table 4.1. Principal device and electrical characteristics for high performance processors. Source: National Technology Road Map 1997, Semiconductor Industries Association, San Jose, Calif., USA

Year	1998	2001	2004	2007	2010
Min. feature (μm)	0.25	0.18	0.12	0.10	0.07
Logic V_{dd} (V)	2.5/1.2–1.8	1.2–1.8	1.2–1.5	< 1.2	< 1.2
Substrate	Si	Si	Si	SOI	SOI
Max I_{off} (nA/mm) at 100 °C)	10–20	20	30	30	40
Nominal L_{eff} (μm)	0.20–0.25	0.14–0.18	0.10–0.13	< 0.10	
V_T variation ($\pm$mV)	50	40	30	25	20
t_{ox} effective (nm)	4–6	4–5	4–5	< 4	< 4
S/D depth (μm)	0.1–0.15	0.07–0.13	0.05–0.1	< 0.07	< 0.05
On-chip clock (MHz)					
Capability (DSP)	600	800	1100	1500	1900
Capability (μP)	450	600	800	1000	1100
Off-chip freq. (MHz)					
Chip-to-chip speed	200	300	350	450	550
Chip-to-board speed	133	166	200	200	200

deposit the dielectric on the alloy. Deposited oxides do not provide adequate quality for device fabrication. The problems examined in this chapter concern formation of ultrathin dielectrics on strained SiGe, strained Si and SiGeC layers. The present status of silicon dioxide formation on SiGe and related films using various techniques such as thermal, rapid thermal, and microwave/ECR plasma-assisted growth is reviewed. Results of studies on the low temperature (150–200 °C) growth of ultrathin oxides using microwave/ECR plasma in O_2 and N_2O/NO ambient, compositional analysis, electrical characterization and interfacial properties of the oxides are presented.

4.1   Characterization Techniques

Rutherford backscattering analysis, x-ray photoelectron spectroscopy, spectroscopy ellipsometry, and high resolution x-ray diffraction techniques are used to find the crystalline quality, chemical compositions and the thickness of films. Atomic force microscopy is used to determine the surface roughness.

RBS provides depth distribution without requirements for destruction of the sample by layer removal as in etching of films by sputtering used in secondary ion mass spectroscopy or Auger electron spectroscopy (AES). The relative concentrations of Ge and Si can be estimated from the relationship [1]

$$\frac{x}{1-x} = \frac{H_{Ge}}{H_{Si}} \frac{d\sigma_{Si}}{d\sigma_{Ge}} \frac{\epsilon_{Si}^{Si_{1-x}Ge_x}}{\epsilon_{Ge}^{Si_{1-x}Ge_x}} \tag{4.1}$$

where x and $1-x$ are the fractional atomic concentrations of Ge and Si, respectively. H, σ and ϵ are the edge heights, scattering cross-section (Z^2 dependent) and resultant stopping power, respectively, of Ge and Si in SiGe films.

Ion channeling is often used in conjunction with RBS. It allows the investigation of the crystalline perfection of the sample. When the incident direction of an ion beam is aligned along the crystallographic axis or plane of a single-crystal sample the majority of the ions penetrate into the crystal. Channeled particles cannot get close enough to the atomic nucleus to undergo large angle Rutherford scattering and hence the backscattering is drastically reduced.

X-ray photoelectron spectroscopy is a powerful tool for understanding chemical compositions in a multicomponent system. Though other analytical techniques, namely Auger electron spectroscopy, Rutherford backscattering spectroscopy, secondary ion mass spectroscopy, are more useful for yielding absolute atomic compositions with higher precision, XPS offers great potential for the determination of chemical bonding of the film constituents. To calculate the stoichiometry of the specimen studied, areas under the important photoelectron peaks need to be considered. Peak areas are measured on the assumption that they are superimposed on a linear background [2]. For quantitative estimation, published values of the sensitivity factors should be used [3]. This gives an estimated measurement accuracy within 10 at.%.

Ellipsometry is an optical characterization method typically used to measure the layer thickness. In ellipsometry, light of known polarization is reflected from the surface of interest and then analyzed for the change in polarization state. In ellipsometric measurements two variables ψ and Δ are usually measured and are given by

$$\tan \psi e^{i\Delta} = \frac{r_p}{r_s} \tag{4.2}$$

where r_p and r_s are the pth and sth complex amplitude reflection coefficients of a sample, respectively. These components depend on the angle of incidence, complex dielectric constants ($\epsilon = \epsilon_1 - i\epsilon_2$) of the sample and other geometrical parameters, e.g., thickness. However, single wavelength ellipsometric measurements at a single angle of incidence cannot determine more than two unknown parameters. The spectroscopic ellipsometer has been developed to overcome this limitation. It is a nondestructive optical technique extremely sensitive to thickness, alloy composition and surface and interfacial conditions of the sample structure. The composition of SiGe films can be estimated from the relationship using best fit parabola method [4]

$$1 - x = (6.538E_1 - 0.0397)^{1/2} - 3.707 \qquad (4.3)$$

where E_1 is the critical point of the energy band in eV.

4.2   Oxidation of $Si_{1-x}Ge_x$ Films

The interest in SiGe material results from its basic process compatibility with the well established Si processing technology which dominates the present microelectronics industry. However, there is a need for some adaptations of the Si technology to SiGe materials and to Si/SiGe heterostructures [5]. A critical step for successful integration of SiGe materials in conventional Si process technology is to grow a high quality oxide layer which could be used for gate, mask and device isolation.

Integrating SiGe into a Si process raises a number of new processing issues, primarily associated with the lattice mismatch between Si and Ge ($\approx 4.17\%$). As discussed in Chapter 2, due to lattice mismatch, misfit dislocations tend to form at SiGe/Si interface during growth and processing. For a given $Si_{1-x}Ge_x$ composition, equilibrium theory predicts the thickness at which it is energetically favorable for misfit dislocations to exist. In the SiGe/Si system, it is generally believed that there are kinetic barriers to the formation of misfit dislocations. A number of investigations have shown that at low temperatures ($\leq 650\ ^oC$), films with low misfit dislocation density can be grown to thicknesses many times the theoretical equilibrium critical thickness. These films are metastable, and misfit dislocation formation and strain relaxation can occur upon annealing. Conventional high temperature thermal oxidation of the $Si_{1-x}Ge_x$ layer adversely affects the properties of the film.

The interfacial and transition regions of Si/SiO_2 comprise a nonstoichiometric monolayer (due to incomplete oxidation) followed by a layer of strained SiO_2 [6]. Therefore, the strained interface is prone to damage by high field stress and radiation which causes device reliability problems. This is more so in the case of strained layer oxidation. Chemical modification of the SiO_2/Si

interface can reduce this strain. Considerable work has been done to improve the dielectric integrity of conventional thermal oxides in the form of chemical modifications of the SiO_2 bulk and SiO_2/Si interface by incorporating small amounts of nitrogen or fluorine within the framework of MOS processing.

It was reported in the early 1980s that furnace annealing of SiO_2 in NH_3 resulted in an improvement in barrier properties against impurity penetration during subsequent processing steps [7]. However, this technique is not popular for industrial applications because it incorporates a significant amount of hydrogen throughout the SiO_2 layer, which in turn is responsible for reduced reliability [8, 9]. For this reason, rapid thermal or classical furnace annealing of SiO_2 layers in N_2O ambient has been proposed as a plausible hydrogen-free process [10, 11]. The use of rapid thermal oxidation of Si wafers in N_2O gas ambient in the temperature range 900–1100 °C has also been reported [12, 13]. The growth pressure is 760 Torr leading to oxynitride films with thicknesses ranging from 20 to 450 Å for different growth times. A relatively high thermal budget is one of the limitations for this N_2O oxynitridation process. Oxidation of strained Si and SiGe films at low temperature using N_2O plasma is worth investigating.

4.2.1   Thermal Oxidation

Considerable work has been done to study the thermal oxidation of SiGe alloy in order to get a stable oxide/substrate interface. Balk [14] was the first to study the oxidation of SiGe < 111 > wafers. The rate of dry oxidation with a low Ge concentration (< 1%) was the same as that of bulk Si in the temperature range 700–1100 °C. Ge was found to be present in the oxide but the concentration of Ge was lower than that in the substrate. An oxidation study was made on SiGe < 111 > single crystals, epitaxial layers, and polycrystals by Margalit *et al.* [15]. Wet oxidation studies at 1063 °C with a Ge concentration of 4%–37% showed that the rate of wet oxidation was unaffected by the presence of Ge. The segregation of Ge into the substrate was also pointed out.

Steam oxidation of Ge-implanted n-Si < 100 > with different Ge concentrations was carried out by Fathy *et al.* [16] at 900 °C and 1000 °C for various oxidation times. The Ge pile-up, due to the rejection of Ge from the oxidation front, led to the formation of a Ge-rich layer at the oxide–substrate interface. Oxidation rates were enhanced over the intrinsic values due to the presence of a Ge-rich layer which modified the interfacial reaction rate. The authors suggested that the increased rate was related to lower Si–Ge binding energy compared to Si binding energy.

Prokes *et al.* [17, 18] proposed the formation of epitaxial strained SiGe/Si heterostructures by wet oxidation of amorphous SiGe at 900 °C. Reflected high energy electron diffraction and Rutherford backscattering were used to examine the quality of the SiGe layer. Results indicated the formation of an epitaxial SiGe layer after oxidation whereas vacuum or nitrogen annealing produced a polycrystalline layer.

LeGoues *et al.* [19, 20] observed different oxidation rates of epitaxial SiGe film in dry and wet oxidation processes. Results showed an enhancement of wet oxidation rate in the presence of Ge at the SiO_2/Si interface but the rate of dry oxidation remained unaffected. In both processes, Ge did not diffuse into the substrate but was found to pile up at the interface during oxidation.

Nayak *et al.* [21] performed wet oxidation studies on strained SiGe layers with different Ge concentrations in the temperature range 870–960 °C. The authors suggested that Ge acted like a catalytic agent. During wet oxidation, GeO was formed which was less stable than SiO_2. Subsequently, it was reduced by Si to form Ge and SiO_2. As the Ge content in the SiGe layer was increased, more Ge could take part in the catalytic reaction which in turn increased the wet oxidation rate. Also the thickness of the Ge-rich layer increased with oxidation time.

Several common features are found in both dry and wet oxidation of $Si_{1-x}Ge_x$ on Si(100). First, the Ge phase segregates or "piles up" at the oxide/$Si_{1-x}Ge_x$ interface [19, 21–26]. Second, Ge acts as a catalyst for the formation of SiO_2 because the presence of Ge enhances the wet oxidation rate without the formation of a Ge oxide [22]. Third, in wet oxidation, the presence of Ge increases the oxidation rate in the linear regime by a factor of 2–3 but it has been reported to have no effect on the diffusion limited parabolic regime [19, 21–23]. During dry oxidation, Ge does not affect the oxidation rates at all [19, 25]. The mechanism of thermal oxidation of $Si_{1-x}Ge_x$ alloys has not been clarified and Ge segregation at the oxide/$Si_{1-x}Ge_x$ interface produces a high density of fixed charges and interface traps which inhibit device applications.

4.2.2   Rapid Thermal Oxidation

Rapid thermal processing is attractive due to its ability to provide reduced thermal budgets and better thermal control with added benefits of single wafer processing. The following process steps can be incorporated into a CMOS process flow using RTP:

- gate electrode: RTCVD/LPCVD amorphous and polycrystalline Si,
- gate dielectric: dry RTO,

- n- and p- wells: RTO for initial oxide and RTP oxynitridation enhanced diffusion for well formations,
- source–drain junctions: S/D rapid thermal annealing,
- silicided contacts: RTP TiN/TiSi$_2$ react,
- poly-buffered LOCOS or PBL isolation: RTCVD/LPCVD silicon nitride on LPCVD polysilicon on dry RTP oxide masks,
- oxide spacers: RTCVD/LPCVD silicon dioxide,
- RTA forming-gas anneal, and
- double-level metal system: includes LPCVD tungsten.

In thin $Si_{1-x}Ge_x$ strained layers, a short high temperature anneal is preferred to a long low temperature anneal in order to avoid propagation of misfit dislocations which can degrade the electronic properties of the heterostructures [27]. RTO is preferred to furnace oxidation because it provides better growth control of a thin oxide, and also produces better quality oxide in a thin oxide regime [28]. The dependence of oxidation rates on alloy composition and oxidizing temperature in RTO of strained SiGe layers for dry and wet ambients has been studied by Nayak *et al.* [29].

Figure 4.1 shows a comparison of the dry oxidation rate of Si with that of an undoped $Si_{0.92}Ge_{0.08}$ sample (780 Å) at two temperatures, 905 °C and 1010 °C. There is no noticeable difference in the dry oxidation rates. After a short initial time (10 s), the oxide thickness increases linearly with time, which suggests that the oxidation kinetics are limited by surface reaction controlled mechanisms. The appearance of a fast oxidation rate during the initial period (< 10 s) is due to the growth of a thin oxide layer during the ramp-up process. The dry oxidation rates of $Si_{1-x}Ge_x$ layers with different compositions (780 Å of $Si_{0.92}Ge_{0.08}$, 220 Å of $Si_{0.85}Ge_{0.15}$, and 180 Å of $Si_{0.80}Ge_{0.20}$, all undoped) are also found to be the same as that of Si.

In the case of wet oxidation, however, the oxidation rate of $Si_{1-x}Ge_x$ is found to be higher than that of Si. Figure 4.2 shows a comparison of the oxidation rates of Si and undoped $Si_{0.80}Ge_{0.20}$ (200 Å) at 870 °C. The oxidation rate of $Si_{0.8}Ge_{0.2}$ is 2–4 times higher than that of Si, and the rate decreases with the oxidation time. The wet oxidation behavior at a higher temperature, 960 °C, and for different Ge compositions (780 Å of $Si_{0.92}Ge_{0.08}$, and 200 Å of $Si_{0.8}Ge_{0.2}$, both undoped) is shown in figure 4.3. The wet oxidation results for Si and $Si_{0.8}Ge_{0.2}$ at 960 °C exhibit behavior similar to that found at a lower temperature of 870 °C. Unlike dry RTO results, the wet RTO rate of $Si_{1-x}Ge_x$ depends strongly on the Ge concentration.

Liou *et al.* [26] extended the study for varying Ge concentrations. Oxidation in dry oxygen at 880 °C was studied using Auger spectroscopy.

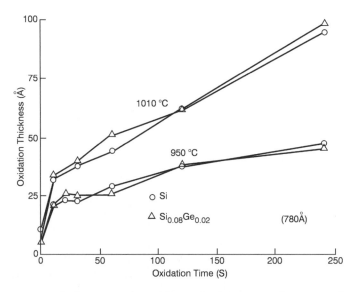

*Figure 4.1. Rapid thermal dry oxidation thickness as a function of oxidation time for Si and $Si_{0.92}Ge_{0.08}$ at two temperatures, 905 °C and 1010 °C. After D. K. Nayak et al., IEEE Trans. Electron Dev., **39**, 56–63 (1992), copyright ©1992 IEEE.*

Si was selectively oxidized and Ge pile-up was observed at the oxide/substrate interface for SiGe films with Ge concentrations below 50%. On the other hand, a mixed SiGe oxide layer was formed at the oxide surface region in addition to the enhanced Ge concentration at the alloy–oxide interface for Ge concentrations above 50%. Close to the surface of this layer Ge was fully oxidized, but at larger depth the state of oxidation was an intermediate one where the Ge was in elemental form at the oxide/substrate interface region.

Xiang *et al.* [30] have studied the wet oxidation of epitaxial SiGe films with Ge concentration up to 85% using Rutherford backscattering. Figure 4.4 shows the 2.0 MeV RBS spectra of three $Si_{1-x}Ge_x$ alloy thin films before (Δ) and after ($\bullet$) a 2 h dry oxidation at 1000 °C for Ge fractions prior to oxidation equal to (i) $x = 0.045$ (C49), (ii) $x = 0.24$ (C50), and (iii) $x = 0.48$ (C54), respectively. RBS detects the incorporation of oxygen into Si on the top of the $Si_{1-x}Ge_x$ layer, and the segregation of Ge underneath the oxide layer. When the Ge fraction x is "low" (i.e., $x < 0.30$), Ge is clearly piled up at the $SiO_2/Si_{1-x}Ge_x$ interface after oxidation (figures 4.4a and 4.4b). However, if

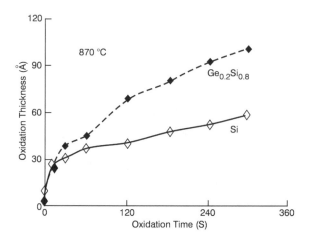

*Figure 4.2. Comparison of oxidation rates of Si and $Si_{0.80}Ge_{0.20}$ in rapid thermal wet oxidation at 870 °C. After D. K. Nayak et al., IEEE Trans. Electron Dev., **39**, 56–63 (1992), copyright ©1992 IEEE.*

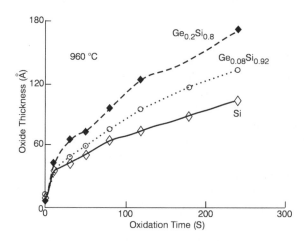

*Figure 4.3. Comparison of oxidation rates of bulk Si, $Si_{0.8}Ge_{0.2}$ and $Si_{0.92}Ge_{0.08}$ in rapid thermal wet oxidation at 960 °C. After D. K. Nayak et al., IEEE Trans. Electron Dev., **39**, 56–63 (1992), copyright ©1992 IEEE.*

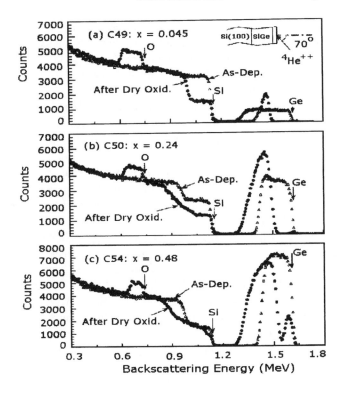

Figure 4.4. 2.0 MeV RBS spectra for binary $Si_{1-x}Ge_x$ alloy thin films before (Δ) and after (•) the 2 h dry oxidation at 1000 °C. The sample normal is tilted 70° with respect to the incoming beam of $^4He^{2+}$ ions. The Ge fraction of the as-deposited film (marked as Δ) is: (a) x = 0.045, (b) x = 0.24, and (c) x = 0.48, respectively. After J. Xiang et al., J. Appl. Phys., 80, 1857–1866 (1996).

the Ge fraction x is close to 0.50, a large surface signal for Ge is detected in the top layer of the oxide and no Ge segregation is observed by RBS at the $SiO_2/Si_{1-x}Ge_x$ interface (figure 4.4c). This indicates that a Ge mole fraction $x < 0.20$ does not affect the dry oxidation rate, which agrees with the results reported on dry oxidation of $Si_{1-x}Ge_x$ alloys by other workers [19, 25]. In addition, a surface signal of segregated Ge is clearly detected for all oxidized $Si_{1-x}Ge_x$ samples. It is seen by comparing figures 4.4a–4.4c that, with the increase in x, the Ge signal at the surface of the oxidized sample becomes higher and wider, while the area of the oxygen signal decreases.

The amount of oxygen incorporated during dry oxidation on $Si_{1-x}Ge_x$ thin films with $x = 0.24$ is 85% of the amount measured on 1000 Å SiO_2 films grown on Si(100) simultaneously in the same furnace. It decreases to 40% for a Ge fraction $x = 0.44$ and further to 55% for a Ge fraction $x = 0.48$. These observations demonstrate that the dry oxidation rate, interpreted by the amount of oxygen incorporated during dry oxidation, starts to decrease when the Ge fraction x is larger than 0.20, and is reduced by about a factor of 2 when the Ge fraction x exceeds 0.30. The total amount of Ge, obtained by measuring the aerial density of the signal of segregated Ge at the surface and the signal of Ge beneath the oxide, is found to remain constant before and after dry oxidation within the precision of RBS analysis ($\pm 5\%$). In other words, no significant loss of Ge due to the formation and thermal desorption of GeO_2 is found to occur.

A Rutherford backscattering study by Eugene *et al.* [23] on wet oxidation at 900 °C of samples containing an epitaxial SiGe film with Ge concentration up to 75% showed similar results. Manukutta *et al.* [31] reported anomalous behavior for the dry oxidation of SiGe film at 800 °C. The growth rate was found to be 10%–20% higher than that of Si while a decrease by 60% was reported for relaxed films. The reason for this discrepancy with the earlier reports is not clear, however.

From the analysis of the depth profiles by SIMS for Ge redistribution after dry oxidation (see figure 4.5), one observes that a layer of SiO_2 does form during the 2 h dry oxidation and Ge separates from the oxide phase and segregates at the oxide/$Si_{1-x}Ge_x$ interface. Figure 4.5 shows that the Ge remaining in the $Si_{1-x}Ge_x$ film is uniformly distributed. A small amount of Ge segregates at the oxide surface which is not detected by SIMS (figures 4.5a and 4.5b) because SIMS lacks sensitivity in the first 100 Å of the surface due to the transitional regime in the initial stages of sputtering. However, when x is increased to 0.5, SIMS can detect the segregation (figure 4.5c).

Figure 4.6 shows the high frequency $C - V$ characteristics of thermally oxidized (700 °C, 10 min) samples with three different Ge concentrations. No hysteresis is found in the $C - V$ plots. However, a little dispersion at high frequency is observed in the accumulation region. The values of minimum capacitance were found to be different for different Ge concentrations (curves a, b and c). This is due to the varying contribution of the heterojunction depletion layer capacitance which is in series with the MOS accumulation and inversion layer capacitances. The variation of oxide thickness values during processing is also a contributing factor in the variation of C_{min} [32].

Using thermal oxidation for SiGe films, the lowest interface state density achieved so far is much higher than that of single-crystalline Si. Negative fixed

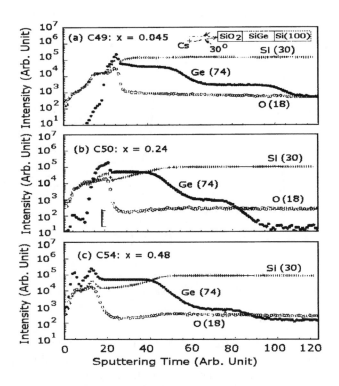

*Figure 4.5. SIMS profiles of the three $Si_{1-x}Ge_x$ alloys depicted in figure 4.4 after 2 h dry oxidation at 1000 °C: (a) x = 0.45, I_{Cs^+} = 53 nA; (b) x = 0.24, I_{Cs^+} = 52 nA; (c) x = 0.48, I_{Cs^+} = 52 nA. The 10 keV Cs^+ primary ions are incident at an angle of 30° to the sample normal. After J. Xiang et al., J. Appl. Phys., **80**, 1857–1866 (1996).*

oxide charges have also been found in SiGe samples which are absent in silicon dioxide. In addition, the thermal oxidation at higher temperature generates dislocations and stacking faults and gives rise to impurity redistribution in the layer. Also, high temperature processing is particularly detrimental for the SiGe strained layers as it causes strain relaxation.

It is very difficult to fabricate MOSFETs directly on SiGe. This problem has so far been avoided by using a Si cap layer so that oxidation can be confined to the Si cap layer in order to get a high quality interface [33]. The interface state density has been found to decrease with increasing Si cap layer thickness and is independent of the Ge content in the film.

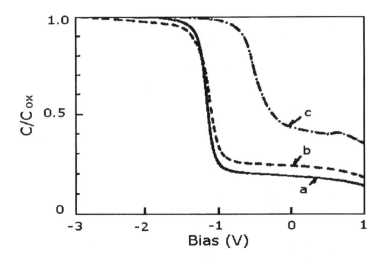

Figure 4.6. 1 MHz $C - V$ characteristics of thermally oxidized SiGe samples of varying Ge concentrations: (a) $Si_{0.82}Ge_{0.18}$, (b) $Si_{0.91}Ge_{0.09}$, and (c) $Si_{0.74}Ge_{0.26}$.

It is evident that the success of electronic devices based on the SiGe/Si system depends mainly on low temperature processing as well as optimization of the thermal budget with respect to the electrical and structural properties of these metastable strained layers. In fact, the nonavailability of a low temperature process for the growth of gate quality oxide has been a concern for the fabrication of MOSFETs directly on SiGe. Low temperature oxidation of SiGe is thus essential not only for submicron devices with ultrashallow junctions, but also to grow a high quality dielectric layer without Ge segregation and to maintain the pseudomorphic nature of the alloy.

4.2.3 Alternative Techniques

Several alternative techniques such as ultraviolet (UV) oxidation, ion and plasma-assisted growth and deposition of dielectric films have been proposed. Agarwal *et al.* [34] showed the formation of a two-phase oxide consisting of SiO_2 and GeO_2 using UV light in air at room temperature. The increased oxidation rate (by a factor of ~ 2) with increasing Ge content is attributed to weak Si–Ge bond energy in comparison with Si–Si bond energy. A further study on strained $Si_{0.8}Ge_{0.2}$ layers using UV-assisted dry oxidation at 550 ^{o}C showed similar growth rate behavior [35].

Two research groups have experimented with nonequilibrium oxidation of $Si_{1-x}Ge_x$ alloys such as high pressure oxidation (HPO) [36] and ion beam oxidation (IBO) to chemically incorporate Ge into a ternary oxide phase and form an atomically sharp oxide/$Si_{1-x}Ge_x$ interface [37]. But the main problems facing $Si_{1-x}Ge_x$ based device technology are the constraints that the metastable nature of $Si_{1-x}Ge_x$ alloy thin films places on Ge content, layer thickness, subsequent thermal cycling, and effective process integration.

Low temperature oxidation using ion beams [37] and ECR plasma [38, 39] without significant Ge pile-up has also been reported. Low energy ion beam oxidation with energies ranging from 100 eV to 1 keV have been reported to result in fully oxidized SiGe alloy at room temperature. Due to preferential sputtering and decomposition, these films tend to contain less Ge than in the SiGe alloy. But the Ge content approaches the bulk content at the lowest ion energies. However, the range of ion energies used in the oxidation may often induce damage or defects on the surface. Oxides with a reasonably good quality have been reported by using rf plasma anodization [40, 41] of MBE grown SiGe alloy at low temperature. The oxidation rate was found to be about twice that of bulk Si. Negative fixed oxide charges were not observed which confirms that there is no Ge pile-up at the interface.

Li and Yang [39] reported the formation of fully oxidized Si and Ge without significant Ge pile-up using ECR oxygen plasma. The Ge content in the oxide was lower than the original Ge content in the SiGe alloy. The oxide was found to be stoichiometric and did not lose its GeO_2 component even after vacuum annealing at 450 °C. The measured fixed oxide charge density was 1×10^{11} cm^{-2} and the interface state density was found to be 7×10^{11} cm^{-2}/eV after 30 min annealing in forming gas at 450 °C, which are acceptable for device fabrication. Li *et al.* [42] also demonstrated the potential use of low temperature ECR plasma oxidation by fabricating MOSFET devices directly on SiGe surfaces. Some of the reported results obtained from different oxidation techniques are summarized in table 4.2.

4.3   Plasma Oxidation of SiGe Films

Plasma discharge has been employed as a useful tool for low temperature film formation in IC technology. Although plasma processing has been a subject of extensive research work since the 1960s, commercial applications in the field of microelectronics started during the 1970s. Plasma may be excited by dc [43, 44], rf [45-47], or microwave discharge [48-51] in different gases at pressures ranging from 10^{-3} Torr to a few Torr. RF plasma is the most widely used technique for film formation and etching in IC processing. However,

Table 4.2. Comparison of electrical and interfacial properties of oxides grown using various techniques on SiGe layers.

Oxidation technique	Ge pile-up	Q_f/q (cm^{-2})	D_{it} (cm^{-2}/eV)
Thermal [22]	Yes	1×10^{12}	1×10^{12}
Rapid thermal [29]	Yes	7.4×10^{11}	2×10^{12}
UV ozone induced [34]	No	–	–
Ion beam [37]	No	–	–
rf plasma [40]	No	3×10^{11}–1×10^{12}	1.6×10^{12}
ECR plasma [39]	No	1×10^{11}	7×10^{11}

work carried out with microwave excited discharges show the creation of high density plasma. This has been used for the growth of ultrathin dielectrics on Si/SiGe films [52].

4.3.1   Microwave Plasma Oxidation

ECR and non-ECR mode microwave plasma have been used for the growth and deposition of various gate dielectrics for Si technology [53]. Because of very low thermal budget (150–200 ^{o}C, 2–3 min) and low self-bias in microwave plasma, the technique has been extended for the growth and deposition of ultrathin gate dielectrics on strained SiGe [54–57], strained Si [58–60] and strain-compensated SiGeC layers [61–63]. In the following, the microwave plasma oxidation of strained SiGe films is considered.

Using GSMBE, nominally undoped strained Si$_{1-x}$Ge$_x$/Si epitaxial films were grown on $< 100 >$ p-Si (5–10 Ω cm) substrates at 700 ^{o}C. An undoped Si buffer layer, typically 500 Å, was used to provide a good quality surface for subsequent growth of the SiGe epilayer. The SiGe epilayers grown had Ge concentrations of 9% (500 Å), 18% (500 Å) and 26% (300 Å). The oxide films were grown using a microwave (700 W, 2.45 GHz) cavity discharge system [64] as shown in figure 4.7. All the samples were subjected to a standard cleaning schedule followed by dipping in dilute HF just before loading into the chamber. In a typical deposition process, the system was first evacuated to 10 mTorr and then O_2 gas (99.99%) and N_2O gas (purity 99.999%) was introduced into the chamber. No external heating of the substrate was done during the experiment. The plasma discharge itself produced a temperature in the range 150–200 ^{o}C, depending upon the growth conditions.

The growth pressure was 1 Torr with 12 sccm gas flow in the chamber.

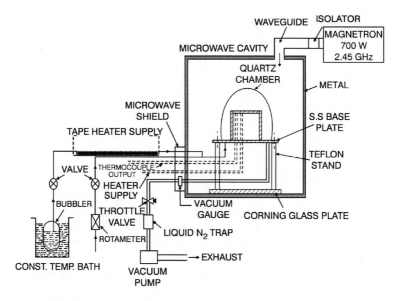

Figure 4.7. Microwave plasma cavity discharge system.

Post-oxidation annealing of grown oxides was done at different temperatures ranging from 400 °C to 650 °C in an O_2 ambient for 30 min. A low temperature *in situ* hydrogen plasma treatment of the SiGe surface preceded oxidation in some cases to improve the interfacial properties [55]. Thermal oxidation was also carried out at 700 °C for 10 min with 1000 sccm oxygen flow in the furnace for comparison. Aluminum, typically 1000–2000 Å thick, was evaporated through a metal mask to form electrodes with diameters of 0.5 and 1.0 mm. A post-metallization annealing using forming gas $(90\%N_2+10\%H_2)$ at 400 °C for 15 min was also carried out to reduce interface trap charges.

4.3.2   Growth Rate

The variation of oxide thickness as a function of growth time for $Si_{0.74}Ge_{0.26}$ films has been studied by Mukhopadhyay *et al.* [65]. It has been shown that there is a fast growth regime followed by a slow growth regime for longer oxidation times. At the initial stage of growth, the oxidation involved a reaction zone, indicating a direct reaction of oxygen species with the SiGe. Later oxidation is limited by the diffusion of oxygen species through the growing oxide to the SiGe. The oxidation without a sample bias becomes self-limiting. This implies that the positive ions may not be the main reaction species and the developed self-bias provides a barrier to negative ions for further oxidation. Similar results have been previously reported from studies of the ECR plasma oxidation of Si [66] and SiGe [38]. An observed enhancement of the oxidation rate of SiGe compared to that of Si may be explained as the result of the breaking of the weak Ge–Si bond due to bombardment by electrons and ions during plasma oxidation.

4.3.3   Stoichiometry

Detailed XPS analysis may be performed to understand the nature of the oxides grown on epitaxial $Si_{1-x}Ge_x$ layers [57]. The line shape and binding energy of the x-ray photoelectron spectra provide information about the composition and the presence of sub-oxide phases in the film. A chemically cleaned SiGe film with well defined Si 2p and Ge 2p peaks positioned at 99.0 eV and 1217.1 eV, respectively, is generally used as a reference. Figure 4.8 shows the corresponding XPS peaks of Si and Ge for different times of exposure to a microwave O_2 plasma.

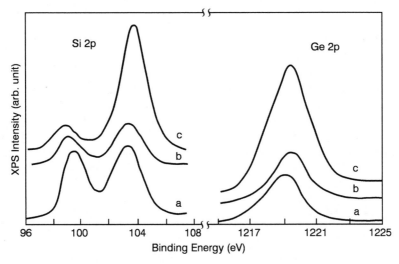

Figure 4.8.   High resolution XPS spectra of a $Si_{0.82}Ge_{0.18}$ surface: plasma oxidized for (a) 2 min, (b) 4 min, and (c) 10 min.

The chemically shifted Si 2p binding energy values of 103.4 eV (curves a and b) and 103.3 eV (curve c) in figure 4.8 coincide with those for stoichiometric SiO_2. The observed Ge 2p peaks are characterized by core-level chemical shifts of 3.9 eV (curve a), 4.3 eV (curve b) and 4.0 eV (curve c) towards higher binding energies. The observations indicate complete oxidation of Ge into a Ge^{4+} state and Si into a Si^{4+} state even with an exposure of 2 min to microwave plasma. An additional peak around 99 eV is attributed to the Si^{0+} state from the underlying film. This may occur since the escape depth of Si 2p photoelectrons (77 Å) is comparable to the thickness of the oxide. The spectra (figure 4.8) reveals that Si and Ge are simultaneously oxidized in microwave O_2 plasma forming a mixed oxide of SiO_2 and GeO_2 [55].

Resolved XPS spectra of Si 2s for thermally and plasma oxidized SiGe alloy at different steps of etching show two distinct peaks of Si 2s. The peaks at 154.0 eV correspond to Si^{4+} states with peak shifts of 3.7 eV from bulk Si towards higher binding energies. The nature of the curve remains the same throughout the oxide layer. The results indicate complete oxidation of Si at different depths for both the oxide samples. The signal contribution of Si^{0+} states from the underlying substrate is also observed for thin oxide samples.

Figure 4.9 shows XPS spectra for Ge 2p electrons for the case of plasma oxidized $Si_{0.82}Ge_{0.18}$ samples for the same etching condition as that for thermal oxide. Curve (a) with a binding energy shift of 3.9 eV from bulk Ge exhibits the complete oxidation of Ge at the surface of the oxide layer. However, Ge is found to be present in the intermediate oxidation state below the surface (curves b and c). But Ge segregation is found to be absent throughout the oxide layer. At a greater depth (curve d), Ge is again found to be in the Ge^{4+} state along with the GeO phase.

At a particular depth, the XPS signal contribution due to Ge^{4+} relative to Ge^{2+} is found to be more in the case of the plasma oxidized sample than the thermally oxidized one [32]. This suggests that there is no significant pile-up of Ge during plasma oxidation compared to thermal oxidation. Similar observations have also been reported for wet oxidation at 700 °C [24, 67] and for low temperature plasma oxidation of SiGe films [38]. The strain relaxation is also negligible at a low process temperature as has been reported by Nayak *et al.* [29] and van de Walle *et al.* [68].

Figure 4.10 shows the backscattering spectra of an as-grown plasma oxide (4 min). The Si to Ge atomic ratio in the oxide layer is estimated using RBS to examine the rejection of Ge, if any, from the oxide. The spectra indicate that the interface of the SiGe alloy layer is sharp with uniform Si and Ge content throughout the layer. The compositions of the alloys are found to be $Si_{0.83}Ge_{0.17}$ and $Si_{0.84}Ge_{0.16}$ for the as-grown and annealed samples, respectively. No Ge rejection is observed from the plasma grown oxides. The results are similar to XPS studies described in an earlier section.

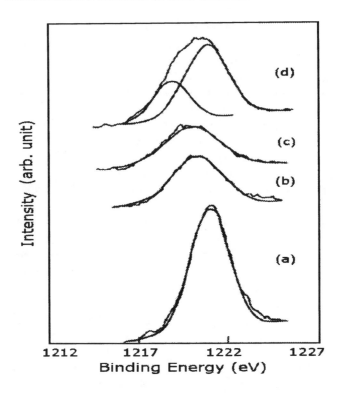

Figure 4.9. Deconvoluted XPS spectra of Ge 2p electrons with different etching time for microwave plasma oxidized (10 min) $Si_{0.82}Ge_{0.18}$ surface: (a) without etching, (b) after first etching, (c) after second etching, and (d) after third etching.

The observations from RBS analysis reveal that SiGe can be oxidized at low temperature using the plasma oxidation technique without a significant Ge pile-up. The oxidation states, observed from the XPS spectrum, confirm the formation of fully oxidized states of Si and Ge.

4.3.4   Electrical Characterization

Fixed oxide charge densities (Q_f/q) for different SiGe oxides are calculated from the flat-band voltage shift (V_{FB}) using high frequency $C - V$ characteristics. The conductance–voltage $(G - V)$ curves are used to calculate the interface state density (D_{it}) at mid-gap using Hill's method [69]. The high

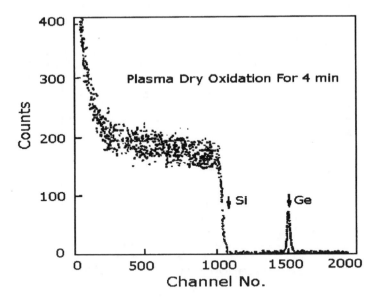

Figure 4.10. 3.049 MeV $^4He^{2+}$ backscattering spectra of a pseudomorphic $Si_{0.82}Ge_{0.18}$ sample after 4 min dry plasma oxidation.

frequency $C - V$ characteristics of MOS capacitors with as-grown oxides on SiGe having Ge concentrations of 18% and 26% show flat-band voltages of –0.96 V and –0.88 V, respectively. The occurrence of a negative flat-band voltage indicates the presence of positive oxide charges in the films [54]. This is attributed to the presence of hole trapping centers in the oxide rather than the water related electron traps observed by Li and Yang [39]. The hole trapping has also been confirmed by ramp and constant voltage stressing of the oxide film. Though as-grown oxides exhibit stable MOS characteristics, a post-oxidation annealing step is found to be useful to reduce the leakage current and interface trap density [54].

The variation of V_{FB} and mid-gap D_{it} values for $Si_{0.74}Ge_{0.26}$ samples with different post-oxidation annealing temperatures in an oxygen ambient for 30 min is shown in figure 4.11. The results indicate that annealing at a low temperature is adequate to achieve a low D_{it} value. The D_{it} value of 2.9 × 10^{11} cm^{-2}/eV obtained at the post-oxidation annealing temperature of 400 °C is lower than the values reported for ECR plasma oxidized SiGe films. Also evident from figure 4.11 is that the interface property degraded slightly on increasing the annealing temperature. It is observed that V_{FB} decreases with

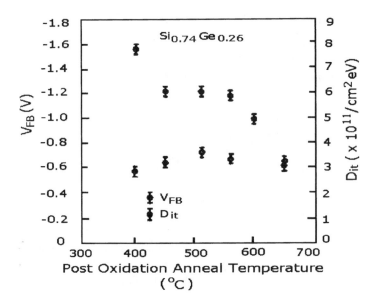

Figure 4.11. Variation of flat-band voltage (V_FB) and mid-gap interface trap density with different post-oxidation anneal temperature for Si_{0.74}Ge_{0.26} MOS capacitors.

increase in post-oxidation annealing temperature. Since GeO_2 is volatile with a high vapor pressure and strained SiGe on Si can relax at higher process temperatures causing misfit dislocations and interdiffusion, post-oxidation annealing at higher temperatures ($> 650\ ^oC$) should be avoided [55].

4.3.5   Trapping Characteristics

Information regarding the trapping characteristics of the oxide can be obtained from dynamic ramp current–voltage $(I - V)$ characteristics measured either in accumulation or depletion mode on previously unstressed capacitors with a fixed ramp rate. It is observed that the displacement current increases with applied bias following the Fowler–Nordheim relation. A trapping ledge is observed before breakdown. The electric field near the injecting interface is found to be almost constant during the extent of the trapping ledge [57].

Assuming the centroid of trapped charge distribution in the oxide layer is located at the middle of the oxide, the capture cross-section (σ), the total number of traps per unit area (N_t), and the trapping probabilities ($N_t \times \sigma$)

are calculated following the work of DiMaria *et al.* [70] by

$$N_t = 2\epsilon_{ox}\frac{\Delta V_{gl}}{qt_{ox}} \tag{4.4}$$

$$\sigma = q\frac{dV_g/dt}{\Delta V_{gl}J} \tag{4.5}$$

where ϵ_{ox} is the permittivity of the oxide, ΔV_{gl} is the voltage width of the ledge, q is the electronic charge, t_{ox} is the oxide thickness, dV_g/dt is the ramp rate, and J is the current density.

Table 4.3. Calculated values of capture cross-section (σ), number of traps/area (N_t) and trapping probability ($N_t \times \sigma$) for different SiGe oxides.

Sample	N_t (cm^{-2})	σ (cm^2)	$N_t \times \sigma$
Thermal oxide			
Si$_{0.82}$Ge$_{0.18}$	2.70×10^{12}	10.9×10^{-16}	2.9×10^{-3}
Si$_{0.74}$Ge$_{0.26}$	5.70×10^{12}	3.00×10^{-16}	1.7×10^{-3}
Plasma oxide			
Si$_{0.91}$Ge$_{0.09}$	6.80×10^{12}	0.68×10^{-16}	4.6×10^{-4}
Si$_{0.82}$Ge$_{0.18}$	15.2×10^{12}	0.12×10^{-16}	1.8×10^{-4}
Si$_{0.74}$Ge$_{0.26}$	6.20×10^{12}	0.84×10^{-16}	5.2×10^{-4}

Table 4.3 lists the calculated values for N_t and σ for plasma as well as thermally grown SiGe oxides. The values are comparable in magnitude with the results obtained from ramp $I - V$ measurements for Si MOS devices using a dual dielectric layer [70]. As seen from table 4.3, the total number of traps per unit area is slightly higher for the microwave plasma oxidized sample compared to the thermally oxidized sample. This may be attributed to the defects created in the oxide interface due to very high power microwave plasma used in the study. However, the trapping probabilities are found to be lower for plasma oxide samples compared to their thermal counterpart [55].

To find out the nature of the trapping in plasma grown SiGe oxides, ramp and dc voltage stressing in the F–N injection region was carried out. It was seen that the stressing causes a negative flat-band voltage shift indicating a build-up of positive charge. The net positive charge build-up by injection is known to be directly related to the change in the flat-band voltage. The oxygen deficiency [71] at the Si/SiGe oxide interface may be responsible for the creation of hole traps.

Oxides grown on $Si_{0.82}Ge_{0.18}$ exhibited good static $I - V$ characteristics with a moderately low value $(10^{-8}$ A cm$^{-2})$ of low field leakage current density. The resistivity value of the post-oxidation annealed and H_2 plasma cleaned oxide films is found to be in the range of 5–9 $\times$ 10^{14} Ω cm. The breakdown field strength is calculated to be 6–9 MV/cm. However, the resistivity (1–2 $\times$ 10^{13} Ω cm) and the breakdown strength (3–5 MV/cm) for thermal oxide $Si_{0.82}Ge_{0.18}$ samples are found to be lower than those of plasma oxide samples.

4.3.6   Oxidation Using N_2O

The N_2O plasma oxide films were grown using a microwave (700 W, 2.45 GHz) cavity discharge system described earlier (see figure 4.7). The growth time (typically 5 min) was kept constant for a particular set of observations. The growth pressure was 1 Torr with 12 sccm N_2O flow in the chamber. Oxide thicknesses, calculated from the oxide capacitance in accumulation, were in the range of 75–100 Å.

Fixed oxide charge density and interface trap density for the samples are extracted from the MOS $C - V$ and conductance–voltage characteristics. The minimum values of Q_f/q and D_{it} obtained for $Si_{0.74}Ge_{0.26}$ samples are 3.8 $\times$ 10^{11} cm^{-2} and 2.7 $\times$ 10^{12} cm^{-2}/eV, respectively. In order to obtain a good quality interface, a two-step oxidation can be carried out using oxygen plasma after N_2O oxidation. In doing so the interface quality was found to be improved. For $Si_{0.74}Ge_{0.26}$ samples, the minimum values of Q_f/q and D_{it} were found to be 1.5 $\times$ 10^{10} cm^{-2} and 4.2 $\times$ 10^{11} cm^{-2}/eV, respectively [56].

4.4   Oxidation of Strained Si Films

Oxidation of ultrathin strained Si layers is a key processing issue as the gate oxide thickness and strained Si channel thickness (remaining strained Si after oxidation) are determined by this gate oxidation step. High temperature processing ($>$ 700 °C) is not desirable for strained Si/SiGe structures to prevent strain relaxation and dislocation propagation from the buffer layers. Many of the reported device structures used a MODFET configuration with a Schottky gate to avoid high temperature gate oxide growth [72–74]. Several workers have used conventional thermal (both dry and wet) oxidation at 750–850 °C for the fabrication of strained Si MOSFETs [75–78].

4.4.1 Thermal Oxidation

Device quality oxides usually result from thermal oxidation of Si. Dry oxygen gas or steam is used at high temperature (typically 800 to 1000 ^{o}C) in the furnace, and Si reacts in the oxidizing ambient to form SiO_2 on the surface of the wafer. During oxidation, Si is consumed by the growing oxide, resulting in a loss of approximately $0.44t_{ox}$ of Si for an oxide thickness t_{ox}.

To investigate possible relaxation of strained Si during thermal oxidation, Welser *et al.* [75] processed three samples (grown using LRPCVD) with graded $Si_{1-x}Ge_x$ buffer layers ($x = 0.30$), a constant Ge buffer 1 μm thick, and varying thicknesses (160, 445, and 1020 Å) of strained Si grown on p-type substrates. The strained Si layers were grown at 700 ^{o}C and the buffer layers were over 95% relaxed. The samples were doped 7×10^{16} cm^{-3} p-type by flowing 0.5 sccm of diborane during the growth. The samples were oxidized (both dry and wet) at temperatures ranging from 750 to 900 ^{o}C in a standard oxidation furnace to yield approximately 130 Å of oxide.

Figure 4.12 shows a Raman scan before and after oxidation at 850 ^{o}C. It is noted that the Si–Si peak from the strained Si layer is diminished due to Si consumption during oxidation, but unaltered peak positions of the Si–Si from the strained Si and buffer $Si_{0.7}Ge_{0.3}$ layer indicate that strain has not been reduced in the strained Si and relaxation remains unchanged in the buffer $Si_{0.7}Ge_{0.3}$ layer.

By measuring the wavenumber shift of the Si–Si from the Raman scan for the strained Si, the strain before and after oxidation was found and is plotted as a function of temperature in figure 4.13. It is seen that for all oxidation temperatures, the thinnest strained Si layer (160 Å) appears to remain fully strained. The 445 Å strained Si layer is fully strained as-grown, but shows some evidence of partial strain relaxation during oxidation. Finally, the thickest strained Si layer (1020 Å) is partially relaxed even as-grown, and continues to relax more as a function of oxidation temperature.

Since the oxide thickness is relatively thin (≈ 13 nm), oxide growth rate is expected to be linear and precisely that was reported by Welser *et al.* [75]. Figure 4.14 shows a plot of linear rate constants (B/A) for all three strained Si samples and a control Si sample as a function of temperature. The values extracted for all samples at all temperatures are very similar, well within the expected accuracy of the measurement. In addition, the activation energy for B/A was found to be of the order of 1.7, which is comparable to the value of 1.96 for bulk Si. Although there was no measured difference between the average growth rate of oxide on strained Si and bulk Si, the authors reported that the oxidation rate decreases slightly as strain in the Si layer increases.

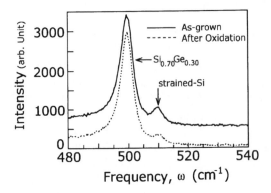

Figure 4.12.   High resolution Raman spectra of a strained Si layer before and after wet oxidation at 850 °C. After J. J. Welser, The application of strained-silicon/relaxed-silicon germanium heterostructures to metal–oxide–semiconductor field-effect transistors, *PhD Thesis, Stanford University, 1994.*

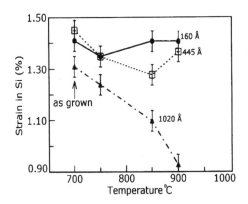

Figure 4.13. Strain in Si layers after thermal oxidation. Si thicknesses refer to as-grown thickness, prior to oxide growth. Wet oxidation times were t = 40, 8 and 2.5 min for 750, 850, and 900 °C, respectively. After J. J. Welser, The application of strained-silicon/relaxed-silicon germanium heterostructures to metal–oxide–semiconductor field-effect transistors, *PhD Thesis, Stanford University, 1994.*

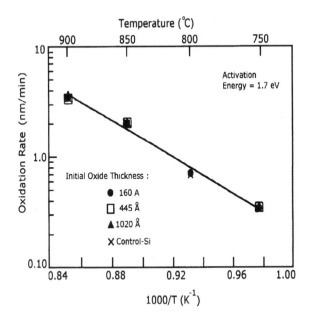

Figure 4.14. *Oxide growth rate of strained Si in wet oxidation. The linear rate constant (B/A) vs. inverse oxidation temperature is plotted. Experiments show very little dependence of oxide growth rate on the initial thicknesses of strained Si. After J. J. Welser,* The application of strained-silicon/relaxed-silicon germanium heterostructures to metal–oxide–semiconductor field-effect transistors, *PhD Thesis, Stanford University, 1994.*

Thermal (dry) oxidation of strained Si film (300 Å thick on a fully relaxed step-graded SiGe buffer layer grown using GSMBE on a p-type Si substrate at 700 °C) and quantum hole confinement in accumulation at the strained-Si/SiGe interface have been reported by Bera *et al.* [58] as shown in figure 4.15. The simulated hole confinement in accumulation in a SiO_2/strained-Si/SiGe heterostructure using a one-dimensional numerical Poisson solver is also shown. All simulations were performed for strained-Si/SiGe channel widths exceeding 200 Å, and quantum effects were neglected. At the SiO_2/strained-Si interface, a trap density at the middle of the bandgap of 2×10^{12} cm^{-2}/eV was assumed in simulation. A good agreement between the simulation and experimental results is observed.

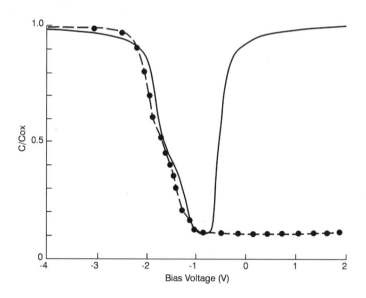

Figure 4.15. $C-V$ characteristics of thermally (700 °C, 140 min, dry) grown oxides on strained Si samples: filled circle: experimental data; solid line: computed low frequency; and (- - - - -): computed high frequency $C-V$ plot.

4.4.2 Plasma Oxidation

As discussed above, there is evidence of oxidation-induced partial strain relaxation in relatively thicker strained Si films, which continues to increase as a function of thermal oxidation temperature. Low temperature oxidation of strained Si on relaxed SiGe films is thus important to achieve a high quality gate oxide while maintaining the pseudomorphic nature of the film. It has also been reported in the literature that the mobility of strained Si degrades due to the presence of high fixed oxide charge ($\sim 4 \times 10^{11}$ cm^{-2}) [75–77]. Bera *et al.* [58, 59] have reported the growth of ultrathin oxides (< 100 Å) on strained Si layers at a very low temperature (< 200 °C) using microwave N_2O plasma and microwave-ECR O_2 plasma. The values of fixed oxide charge density are found to be lower in plasma grown oxides than in thermal oxides [75, 77]. The interface state densities at mid-gap of the oxide grown using thermal (dry) and microwave O_2 plasma are higher compared to that of microwave N_2O plasma grown oxide. The breakdown field strength of the N_2O plasma grown oxide is comparable to that of the mainstream thermal oxide.

4.5   Oxidation of $Si_{1-x-y}Ge_xC_y$ Films

Wet oxidation of $Si_{1-x-y}Ge_xC_y$ films grown on Si(100) by chemical vapor deposition has been reported [79]. The effect of carbon incorporation upon thermal oxidation of ion and MBE grown $Si_{1-x}Ge_x$ alloys and its role on strain compensation in $Si_{1-x}Ge_x$ alloys with Ge fraction in the range $0 \leq x \leq 0.50$ and carbon fraction y in the range $0 \leq y \leq 0.085$ has been studied by Xiang *et al.* [30].

The thickness and the composition of all samples before and after oxidation were measured by RBS combined with ion channeling at 2.0 MeV and carbon nuclear resonance analysis (NRA) at 4.3 MeV using $^4He^{2+}$ ions. The authors reported Ge segregation towards the oxide surface and at the interface between the oxide and the thin film. The amount of oxygen incorporated during dry oxidation was measured as a function of Ge fraction x. The dependence of dry oxidation rates of $Si_{1-x}Ge_x$ alloys upon strain and microstructure, as well as Ge and C fractions, was also analyzed.

2.0 MeV RBS analysis showed that a layer of SiO_2 is formed on the top surface of both $Si_{1-x}Ge_x$ and $Si_{1-x-y}Ge_xC_y$ films, while Ge segregates at the $SiO_2/Si_{1-x}Ge_x$ and $SiO_2/Si_{1-x-y}Ge_xC_y$ interfaces. Carbon within the film is found to modify strain and local bonding, thereby affecting the rate of dry oxidation. Ion channeling analysis and strain measurements indicate that the incorporation of C rather than the amount of C itself affects the dry oxidation mechanism because of its strong influence on film strain and crystalline quality. These results are discussed in conjunction with observations by secondary ion mass spectrometry, high resolution transmission electron microscopy, Fourier transform infrared spectrometry, and tapping mode atomic force microscopy.

The effect of carbon incorporation on the dry oxidation of three heteroepitaxial $Si_{1-x-y}Ge_xC_y/Si(100)$ thin films ((i) $y = 0$, $T_{growth} = 500$ ^oC (C55); (ii) $y = 0.006$, $T_{growth} = 560$ ^oC (C32); and (iii) $y = 0.005$, $T_{growth} = 450$ ^oC (C34)) is shown in figure 4.16. The figure shows the 2.0 MeV RBS spectra of the three samples before (Δ) and after ($\bullet$) a 2 h dry oxidation at 1000 ^oC. The difference in the area under the oxygen signal in figures 4.16a–4.16c, demonstrates that even a small amount of C can significantly affect the oxygen incorporated during dry oxidation.

Ge segregates both towards the $SiO_2/Si_{1-x-y}Ge_xC_y$ interface and towards the SiO_2 surface. The front edge of the segregated Ge signal underneath SiO_2 only widens from 19 to 26 keV in figure 4.16a, to 23 keV in figure 4.16b, and remains 19 keV wide in figure 4.16c. These values match the width of the front edge of the Ge signal at the surface prior to the oxidation within less than 7 keV. In other words, the broadening can be contributed entirely to

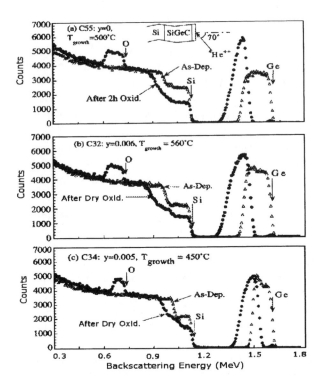

*Figure 4.16. 2.0 MeV RBS spectra for three $Si_{1-x-y}Ge_xC_y$ samples: before (Δ) and after ($\bullet$) 2 h dry oxidation at 1000 °C. The sample normal is tilted 70° with respect to the incoming beam of $^4He^{2+}$ ions. (a) $x = 0.22$, $y = 0$, $T_{growth} = 500$ °C; (b) $x = 0.27$, $y = 0.006$, $T_{growth} = 560$ °C; and (c) $x = 0.28$, $y = 0.005$, $T_{growth} = 450$ °C. After J. Xiang et al., J. Appl. Phys., **80**, 1857–1866 (1996).*

straggling which means a sharp compositional interface between SiO_2 and the $Si_{1-x-y}Ge_xC_y$ film as was observed in binary $Si_{1-x}Ge_x$ films.

Heteroepitaxial $Si_{1-x-y}Ge_xC_y$ thin films were depth profiled by SIMS after dry oxidation. Figure 4.17 shows the SIMS depth profiles corresponding to the three oxidized $Si_{1-x-y}Ge_xC_y$ films (C55, C32, and C34) whose RBS spectra are shown in figure 4.16. The SIMS depth profile of Ge in figure 4.17 matches closely the Ge profiles obtained for oxidized binary $Si_{1-x}Ge_x$ films. This confirms RBS observations on Ge phase segregation.

Microwave plasma oxidation (below 200 °C) of partially strain-

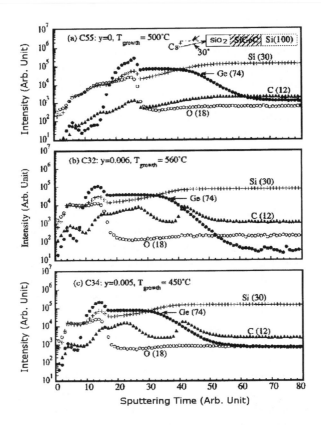

*Figure 4.17. SIMS profiles of the three $Si_{1-x-y}Ge_xC_y$ alloys depicted in figure 4.16 after 2 h dry oxidation at 1000 °C: (a) x = 0.22, y = 0.0, I_{Cs^+} = 51 nA; (b) x = 0.27, y = 0.006, I_{Cs^+} = 54 nA; (c) x = 0.28, y = 0.005, I_{Cs^+} = 53 nA. The 10 keV Cs^+ primary ions are incident at an angle of 30° to the sample normal. After J. Xiang et al., J. Appl. Phys., **80**, 1857–1866 (1996).*

compensated $Si_{1-x-y}Ge_xC_y$ with and without a Si cap layer has been reported [61, 62]. The electrical properties of grown oxides have been characterized using a metal oxide semiconductor structure. Fixed oxide charge density and mid-gap interface trap density are found to be 2.9×10^{11} cm^{-2} and 8.8×10^{11} cm^{-2}/eV, respectively, for directly oxidized $Si_{0.79}Ge_{0.2}C_{0.01}$ film. The oxide on samples with low C < 0.5% concentration exhibits hole trapping, whereas electron trapping is observed for oxides on alloys containing 1% C.

4.6   Nitridation of Strained Layers

The dielectric used for fabrication of the gate is a key factor in determining the long-term reliability of a device. In the quest for higher reliability, research has moved to sandwiches of oxides and nitrides like ONO (oxide/nitride/oxide) and more recently to oxynitrides (SiO_xN_y). As gate material for MOS VLSI, the use of oxynitride has shown clear advantages over SiO_2 in electrical properties and hot-carrier reliability. Optimized process conditions and the use of nitrogen sources allow for the formation of ultrathin gates where it is possible to localize the nitrogen at both the poly-Si/SiO_2 and the SiO_2/Si interfaces.

Several techniques have been used for the formation of oxynitrides. Depending on the N source used, these can be grouped into three categories: NH_3, N_2O and NO. The use of NH_3 is advantageous in introducing large amounts of N at the Si/SiO_2 interface. Work in oxynitrides first focused on growing the oxide in an ammonia (NH_3) ambient, but the excess hydrogen was found to create electron traps. Now, instead of ammonia, nitrous oxide (N_2O) is the nitrogen source of choice. Good results (see figure 4.18 for mobility enhancement when oxynitride is used) have been obtained by growing thermal oxides in a N_2O ambient, but the growth rate is low. The kinetics of this process have been shown to be highly dependent on temperature, pressure and furnace geometry.

Another advantage of N_2O usage is the low thermal budget required to obtain an oxide of desired thickness in reasonable processing times. This lowering of the thermal budget is desirable during processing. NO grown oxides require even lower thermal budgets than those used in N_2O processing, to introduce the same quantity of N into the gate. In addition, NO grown oxides also exhibit a higher amount of N incorporation than is the case with N_2O, with similar processing conditions. This higher amount of incorporation is probably the consequence of completely different growth kinetics, as demonstrated by the much slower growth rate of the oxynitride layer in a NO ambient, compared with N_2O. It is postulated that the higher amount of N introduced at the Si/SiO_2 layer has a retarding effect on the growth of the oxide itself, as it is the limiting factor. Excellent electrical properties are demonstrated by the NO grown gates, confirming the importance of N introduction into the SiO_2 gate material and at the interface [80].

The important application of N incorporation at the interface is to effectively create a barrier to boron penetration to improve short-channel performance of p-MOSFETs. The advantage obtained by N accumulation in the interfacial region is somewhat mitigated by a decrease in gate reliability

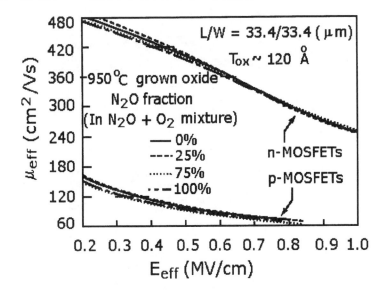

Figure 4.18. Effective mobility vs. effective electric field for MOSFETs with SiO₂ and N₂O oxides. After G. D. Leonarduzzi and D.-L. Kwong, Semicond. Intl., 225 (July 1998).

because of the accumulation of boron within the gate itself. This effect is schematically illustrated in figure 4.19, which shows the effect of the oxynitride interface layer in stopping boron penetration and producing an accumulation of boron atoms within the gate. Such accumulation results in degradation of the gate charge trapping properties, as confirmed by charge trapping studies [81]. However, while such degradation is certainly an undesired effect, NO grown gate dielectric film is always superior to a SiO₂ gate in overall performance and reliability behavior. The performance behavior of oxynitrides can be explained by considering the four cases exemplified in figure 4.20:

(i) Figure 4.20a represents the control, where boron can diffuse easily into the gate during activation, resulting in degradation of oxide reliability.

(ii) Figure 4.20b shows the blocking effect of an N barrier at the SiO_2/Si interface, which does reduce (V_{FB}) shift, but accumulation of boron atoms within the gate causes charge trapping, provoking Q_{bd} degradation.

(iii) Figure 4.20c shows an improved scenario, where the N barrier positioned at the poly-Si/Si interface decreases boron penetration into the gate oxide.

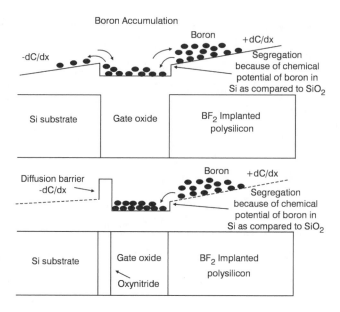

Figure 4.19. The effect of the oxynitride interface layer (bottom) is compared to a gate without such a layer (top). The oxynitride layer stops boron penetration and produces an accumulation of boron atoms within the gate. After G. D. Leonarduzzi and D.-L. Kwong, Semicond. Intl., 225 (July 1998).

(iv) Finally, figure 4.20d portrays the behavior in the presence of the "double wall" N barrier. This renders it impossible for boron atoms, which penetrated the first barrier (at the poly-Si/Si interface), to penetrate into the gate, yielding the smallest (V_{FB}) shift. However, boron atoms within the gate are now effectively "trapped" leading to decreased Q_{bd} performance compared with the case presented in figure 4.20c. In this case, Q_{bd} performance is better than the cases of figures 4.20a and 4.20b, as much less boron can penetrate the gate because of the presence of two blocking "walls".

Fluorine also appears to be a beneficial addition to the gate dielectric, in that it enhances oxidation rate, relaxes the oxide stress and suppresses the hot electron induced generation of interface traps. Research is under way to understand the effects of fluorine on dielectric breakdown strength, junction leakage currents, radiation hardness and hot electron immunity [82].

In Si, various types of oxynitride processes such as reoxidized-nitrided oxide (ROXNOX) and N_2O treatment have been employed to retard the boron penetration and to improve the oxide integrity [11, 85–87]. For strained Si

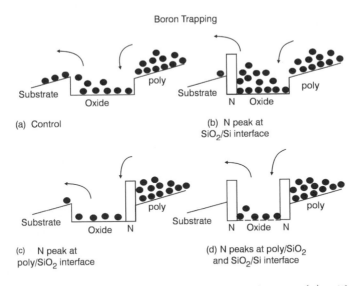

Figure 4.20. The impact on boron penetration of gates (a) without any interface layers, (b) with nitrogen at the SiO₂/Si interface, (c) with nitrogen at the poly-Si/SiO₂ interface and (d) with nitrogen at both interfaces. After G. D. Leonarduzzi and D.-L. Kwong, Semicond. Intl., 225 (July 1998).

channel MOSFET applications, it is also important to study the growth of gate quality oxynitride films on strained Si. However, the growth temperature should be much lower than the conventional nitridation or ROXNOX process (above 900 °C) to prevent strain relaxation in strained Si. Bera *et al.* [60] have studied the growth and properties of a low temperature (< 200 °C) oxynitride film on strained Si using N_2O and NH_3 plasma. Two different types of oxynitride films were grown using a microwave (700 W, 2.45 GHz) plasma cavity discharge system [64].

Table 4.4 shows the processing condition for various samples. High resolution XPS studies were performed on as-grown and Ar^+ ion etched sample surfaces. No nitrogen signal was observed in the XPS spectra from the as-grown film surfaces. Figure 4.21 shows Si 2p and N 1s photoelectron spectra for etched N_2 and N_2O+NH_3 treated samples. The observed Si 2p peaks are attributed to signals from the Si substrate (~ 98.8 eV) and SiO_xN_y layer (~ 101–103 eV). N 1s spectra of the N_2O grown films exhibit two peaks corresponding to the Si $\equiv$ N bond at 396.2 eV and the $Si_2 =$ N–O bond

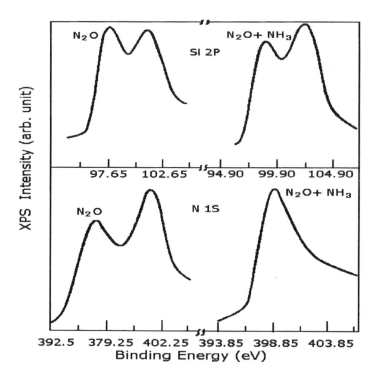

Figure 4.21. High resolution XPS spectra of Si 2p and N 1s signals after 1 min Ar⁺ ion etching for N₂O and N₂O+NH₃ nitrided oxide on strained Si.

at 400.8 eV. However, the corresponding spectra for NH_3-nitrided N_2O oxides show a single peak at 397.9 eV with a relatively greater broadening having a full width at half maximum of 2.7 eV. The position (397.9 eV) and line shape of the N 1s spectrum in this case suggests that the peak is due to the convoluted spectra due to $Si \equiv N$, $Si_2 = N-O$ and $Si-N = H_2$ bonds with the most significant contribution from $Si \equiv N$ bonds in the sample [88].

It may be noted that the nitrogen atoms accumulate at or near the oxide/substrate interface forming a silicon oxynitride layer. The fixed insulator charge density (Q_f/q) for two samples (NH_3 plasma treated and N_2O–NH_3 plasma treated) is shown in table 4.4. The minimum value of Q_f/q for the N_2O–NH_3 plasma treated sample is comparable to that of mainstream thermal oxide (2.5×10^{10} cm^{-2}) [89, 90] and shows that the growth of a good quality oxynitride film, necessary for conventional Si processing technology, is possible on strained Si at a low temperature.

Table 4.4. Electrical properties of various dielectric films grown on strained Si on a relaxed SiGe layer.

Processing condition	Temp. (^{o}C)	Q_f/q (cm^{-2})	D_{it} (cm^{-2}/eV)	Breakdown field (MV/cm)
Thermal oxidation				
Wet oxidation [83]	850	2.0×10^{11}	$5.5\text{--}10 \times 10^{10}$	–
Dry oxidation [77]	–	4.1×10^{11}	1.3×10^{12}	–
Microwave plasma oxidation				
N$_2$O plasma [58]	150	6.0×10^{10}	1.2×10^{11}	5.0–7
ECR-O$_2$ plasma [59]	150	2.6×10^{10}	2.2×10^{12}	2.7–3
Mainstream dry oxidation (Si) [84]	1000	2.5×10^{10}	$1.6\text{--}6 \times 10^{10}$	6–8
Nitridation				
N$_2$O+NH$_3$ plasma (2 min + 2 min) [60]	150	6.1×10^{10}	6.5×10^{11}	10–12
NH$_3$ plasma (2 min) [60]	150	6.0×10^{11}	4.7×10^{11}	5–8

4.7 Summary

Recent progress made in dielectric formation on strained layers has been discussed. The thermal oxidation of heteroepitaxial $Si_{1-x}Ge_x$/Si, strained Si on relaxed SiGe buffer and $Si_{1-x-y}Ge_xC_y$ alloy films has been presented in detail. There is no clear correlation between the total amount of C and dry oxidation rates. The incorporation of C into $Si_{1-x-y}Ge_xC_y$ alloys either in substitutional or in interstitial sites modifies strain and epitaxial quality and it can be correlated with the change in dry oxidation rates: at low carbon fraction, epitaxial quality is better when carbon is substitutional and the dry oxidation rate is lower; at high carbon fraction, most carbon is interstitial, the films become highly defective, and the dry oxidation rate is high.

Rapid thermal dry oxidation rate of SiGe layers has been found to be the same (up to 20% Ge) as that of Si, but in the case of rapid thermal wet oxidation, the rate is 2–4 times higher than that of Si. The wet oxidation rate of SiGe alloy depends strongly on Ge concentration.

SiGe films can be oxidized at a temperature as low as 200 ^{o}C using microwave oxygen plasma and also at 700 ^{o}C by the thermal technique.

Table 4.5. Oxidation conditions of published oxides on SiGe. Values are calculated when not stated. After L. S. Riley and S. Hall, J. Appl. Phys., 85, 6828–6837 (1999).

Reference	Oxidation type	Temp. (°C)	Growth rate (nm/min)	Ge segregation	GRE[c]
[19, 22, 91]	Dry	800	0.06	Yes	No
[16, 92][a]	Dry	1000	0.4	Yes	–
[30]	Dry	900	0.3	Yes	No
[30]	Dry	1000	–	Yes	No
[29]	RTP dry	905	∼ 0.66 (8% Ge)	Yes	No
		1010	1.5 (8% Ge)	Yes	No
[21]	RTP wet	960	2.6 (8% Ge)	Yes	∼ 1.5
		870	2.2 (20%Ge)	Yes	∼ 3.0
		960	4.1 (20%Ge)	Yes	∼ 2.5
[19, 22, 91]	Wet	800	0.6	Yes	2–3
[93]	Wet	1000	8.0 (50%Ge)	Intermediate	∼ 3
[94]	Wet	700/1000	1.2–3.0	Intermediate	Yes
[23]	Wet	900	25%–75%	Intermediate	Yes
[24]	Wet	700	3.0 at 3.5 h[b]	No	Yes[b]
[95]	Plasma anodization	400	∼ 2.7	No	Yes
[40]	Plasma anodization	80	3.0	No	2–3
[32, 54]	Plasma	150–200	1.5	No	–
[38, 39]	ECR plasma	0–500	0.1 (at 300 K) > 0.1 (at 500 K)	No	–
[96]	Plasma anodization	65	2.7	No	2.5–3
		80	1.6	No	–
[35]	UV assisted dry	550	0.133	Yes	Yes
[97]	Atomic oxygen assisted dry	500	0.135 (5%) 0.195 (10%) 0.270 (20%)	Intermediate	1.1 1.5 2.1

[a] Implanted Ge.

[b] Parabolic regime only.

[c] GRE, growth rate enhancement.

Table 4.6. Fixed charge, interface state densities and breakdown voltages for various oxides and nitrides formed below 850 °C. After U. Konig and J. Hersener, Solid State Phenomena, **47-48**, 17–32 (1996).

Process	Temp (°C)	Annealing (min, °C)	Fixed charge (cm^{-2}) $(\times 10^{11})$	D_{it} (cm^{-2}/eV) $(\times 10^{11})$	V_{bd} (MV/cm)
Thermal					
dry RTO	850		5	20	8
wet RTO	850		10	20	7.5
HIPOX	700		5	1	13
		30, 800	4	0.4	13
Plasma					
ECR	400		> 1	> 10	8
		20, 850	0.4	2	11
PECVD	350		50	30	8
ECR CVD	< 100	5, 450, N_2/H_2	2.5	3	9.5
SPOX	140	5, 450, N_2/H_2	-0.7–2.5	0.8	13
Anodization					
ANOX	45	30, 250, N_2	1.5	0.3	13
ANOX	20	20, 400, N_2	0.25	1	12
Nitridation					
ECR CVD	< 100	5, 450, N_2/H_2	-6	1	9.5
PECVD	350		35	30	10

The reported chemical state for SiGe oxides grown by both techniques have been discussed. Thermally oxidized samples show the formation of an intermediate oxidation state for Ge at the surface, while plasma oxidation results in a complete oxidation of Ge to GeO_2. XPS spectra also confirm the absence of Ge segregation at the oxide–substrate interface. The electrical and interfacial properties of the oxide after post-oxidation and post-metal annealing treatments exhibit low values of fixed oxide and interface trap densities. The plasma grown oxides have relatively better properties than thermally grown oxides. A controlled in situ H_2 plasma treatment of the samples prior to oxide growth is useful to improve the electrical properties. In tables 4.5 and 4.6, reported oxidation conditions of published oxides on SiGe and electrical properties such as fixed charge, interface state densities and breakdown voltages for various oxides and nitrides formed below 850 °C have been summarized.

The results of plasma-enhanced oxidation of strained epitaxial SiGe films in N_2O ambient have been presented. The XPS study has demonstrated the incorporation of nitrogen below the SiO_2 surface. The electrical properties of the oxides show moderately low values of fixed oxide charge density and high breakdown strength. Different process sequences and various gaseous N sources allow one to "engineer" N profiles within an oxynitride gate, leading to improved performance and reliability.

4.8 References

1 W. K. Chu, J. W. Mayer and M. A. Nicolet, *Backscattering Spectrometry.* Academic Press, New York, Chapters 2, 4 and 5, 1987.

2 T. B. Ghosh and M. Sreemany, "On the selection of an integration limit for quantitative XPS analysis," *Appl. Surf. Sci.*, vol. 64, pp. 59–70, 1993.

3 D. Briggs and M. P. Seah, *Practical Surface Analysis by Auger and X-ray Photoelectron Spectroscopy.* John Wiley & Sons, New York, 1983.

4 J. Humlicek, M. Garriga, M. I. Alonso and M. Cardona, "Optical properties of $Si_{1-x}Ge_x$ alloy," *J. Appl. Phys.*, vol. 65, pp. 2827–2832, 1989.

5 U. Konig and J. Hersener, "Needs of Low Thermal Budget Processing in SiGe Technology," *Solid State Phenomena*, vol. 47–48, pp. 17–32, 1996.

6 F. J. Grunthaner, B. F. Lewis, N. Zamini and J. Maserjian, "XPS studies of structure induced radiation effects at the Si/SiO_2 interface," *IEEE Trans. Nucl. Sci.*, vol. NS-27, pp. 1640–1646, 1980.

7 T. Ito, T. Nakamura and H. Ishikawa, "Advantages of thermal nitride and nitroxide gate films in VLSI process," *IEEE Trans. Electron Dev.*, vol. ED-29, pp. 498–502, 1982.

8 T. Hori, H. Iwasaki and K. Tsuji, "Electrical and physical characteristics of ultrathin reoxidized nitrided oxides prepared by rapid thermal processing," *IEEE Trans. Electron Dev.*, vol. 36, pp. 340–350, 1989.

9 G. J. Dunn and J. T. Krick, "Channel hot-carrier stressing of reoxidized nitrided oxide p-MOSFET's," *IEEE Trans. Electron Dev.*, vol. 38, pp. 901–906, 1991.

10 V. K. Mathews, R. L. Maddox, P. C. Fazan, J. Rosato, H. Hwang and J. Lee, "Degradation of junction leakage in devices subjected to gate oxidation in nitrous oxide," *IEEE Electron Dev. Lett.*, vol. 13, pp. 648–650, 1992.

11 Z. H. Liu, H. J. Wann, P. Ko, C. Hu and Y. C. Cheng, "Effects of N_2O anneal and reoxidation on thermal oxide characteristics," *IEEE Electron Dev. Lett.*, pp. 402–404, 1992.

12 H. T. Tang, W. N. Lennard, M. Zinke-Allmang, I. V. Mitchell, L. C. Feldman, M. L. Green and D. Brasen, "Nitrogen content of oxynitride films on Si (100)," *Appl. Phys. Lett.*, vol. 64, pp. 3473–3475, 1994.

13 N. Bellafiore, F. Pio and C. Riva, "Thin SiO_2 films nitrided in N_2O," *Microelectron. J.*, vol. 25, pp. 495–500, 1994.

14 P. Balk, "Surface Properties of Oxidized Germanium-doped Silicon," *J. Electrochem. Soc.*, vol. 118, pp. 494–495, 1971.

15 S. Margalit, A. Bar-Lev, A. B. Kuper, H. Aharoni and A. Neugroschel, "Oxidation of Silicon–Germanium Alloys," *J. Cryst. Growth*, vol. 17, p. 288, 1972.

16 D. Fathy, O. W. Holland and C. W. White, "Formation of epitaxial layers of Ge on Si substrates by Ge implantation and oxidation," *Appl. Phys. Lett.*, vol. 51, pp. 1337–1339, 1987.

17 S. M. Prokes, W. F. Tseng and A. Christou, "Formation of epitaxial $Si_{1-x}Ge_x$ films produced by wet oxidation of amorphous Si–Ge layers deposited on Si(100)," *Appl. Phys. Lett.*, vol. 53, pp. 2483–2485, 1988.

18 S. M. Prokes and A. K. Rai, "The mechanism of epitaxial Si–Ge/Si heterostructure formation by wet oxidation of amorphous Si–Ge thin films," *J. Appl. Phys.*, vol. 67, pp. 807–813, 1990.

19 F. K. LeGoues, R. Rosenberg and B. S. Meyerson, "Kinetics and mechanism of oxidation of SiGe: dry versus wet oxidation," *Appl. Phys. Lett.*, vol. 54, pp. 644–646, 1989.

20 F. K. LeGoues, R. Rosenberg and B. S. Meyerson, "Dopant redistribution during oxidation of SiGe," *Appl. Phys. Lett.*, vol. 54, pp. 751–753, 1989.

21 D. K. Nayak, K. Kamjoo, J. S. Park, J. C. S. Woo and K. L. Wang, "Wet oxidation of GeSi strained layers by rapid thermal processing," *Appl. Phys. Lett.*, vol. 57, pp. 369–371, 1990.

22 F. K. LeGoues, R. Rosenberg, T. Nguyen, F. Himpsel and B. S. Meyerson, "Oxidation studies of SiGe," *J. Appl. Phys.*, vol. 65, pp. 1724–1728, 1989.

23 J. Eugene, F. K. LeGoues, V. P. Kesan, S. S. Iyer and F. M. d'Heurle, "Diffusion versus oxidation rates in silicon–germanium alloys," *Appl. Phys. Lett.*, vol. 59, pp. 78–80, 1991.

24 W. S. Liu, E. W. Lee, M.-A. Nicolet, V. Arbet-Engels, K. L. Wang, N. M. Abuhadba and C. R. Alta, "Wet oxidation of GeSi at 700 ^{o}C," *J. Appl. Phys.*, vol. 71, pp. 4015–4018, 1992.

25 D. K. Nayak, K. Kamjoo, J. C. S. Woo, J. S. Park and K. L. Wang, "Rapid Thermal Oxidation of GeSi Strained Layers," *Appl. Phys. Lett.*, vol. 56, pp. 66–68, 1990.

26 K. K. Liou, P. Mei, U. Gennser and E. S. Yang, "Effects of Ge concentration on SiGe oxidation behavior," *Appl. Phys. Lett.*, vol. 59, pp. 1200–1202, 1991.

27 J. C. Bean, "Silicon-Based Semiconductor Heterostructures: Column IV Bandgap Engineering," *Proc. IEEE*, vol. 80, pp. 571–587, 1992.

28 J. Nulman, J. P. Crusius and A. Gat, "Rapid Thermal Processing of Thin Gate Dielectrics Oxidation of Silicon," *IEEE Electron Dev. Lett.*, vol. 6, pp. 205–207, 1985.

29 D. K. Nayak, K. Kamjoo, J. S. Park, J. C. S. Woo and K. L. Wang, "Rapid Isothermal Processing of Strained GeSi Layers," *IEEE Trans. Electron Dev.*, vol. 39, pp. 56–62, 1992.

30 J. Xiang, N. Herbots, H. Jacobsson, O. Ye, S. Heame and S. Whaley, "Comparative study on dry oxidation of heteroepitaxial $Si_{1-x}Ge_x$ and $Si_{1-x-y}Ge_xC_y$ on Si(100)," *J. Appl. Phys.*, vol. 80, pp. 1857–1866, 1996.

31 L. V. Manukutla, M. D. Faltys, M. R. Moore, R. Santhanam, M. H. Liaw, K. Evans and T. A. Anderson, "Manufacturability studies of SiGe MOSFETs: Oxidation study," in *Proc. NSF design and manufacturing systems conference*, pp. 287–290, 1991.

32 M. Mukhopadhyay, S. K. Ray, T. B. Ghosh, M. Sreemani and C. K. Maiti, "Interface properties of thin oxides grown on strained SiGe layers at low temperatures," *Semicond. Sci. Technol.*, vol. 11, pp. 360–365, 1996.

33 S. S. Iyer, P. M. Solomon, V. P. Kesan, A. A. Bright, J. L. Freeouf, T. N. Nguyen and A. C. Warren, "A Gate-quality Dielectric System for SiGe Metal–Oxide–Semiconductor Devices," *IEEE Electron Dev. Lett.*, vol. EDL-12, pp. 246–248, 1991.

34 A. Agarwal, J. K. Patterson, J. E. Greene and A. Rockett, "Ultraviolet ozone induced oxidation of epitaxial $Si_{1-x}Ge_x(111)$," *Appl. Phys. Lett.*, vol. 63, pp. 518–520, 1993.

35 V. Craciun, I. W. Boyd, A. H. Reader, W. J. Kersten, F. J. G. Hakkens, P. H. Oosting and D. E. W. Vandenhoudt, "Microstructure of oxidized layers formed by the low temperature UV assisted dry oxidation of strained $Si_{0.8}Ge_{0.2}$ on Si," *J. Appl. Phys.*, vol. 75, pp. 1972–1975, 1994.

36 D. C. Paine, C. Caragianis and A. F. Schwartzman, "Oxidation of $Si_{1-x}Ge_x$ Alloys at Atmospheric and Elevated Pressure," *J. Appl. Phys.*, vol. 70, pp. 5076–5084, 1991.

37 O. Vancauwenberghe, O. C. Hellman, N. Herbots and W. J. Tan, "New SiGe dielectrics grown at room temperature by low-energy ion beam oxidation and nitridation," *Appl. Phys. Lett.*, vol. 59, pp. 2031–2033, 1991.

38 P. W. Li, H. K. Liou, E. S. Yang, S. S. Iyer, T. P. Smith, III and Z. Lu, "Formation of stoichiometric SiGe oxide by electron cyclotron resonance plasma," *Appl. Phys. Lett.*, vol. 60, pp. 3265–3267, 1992.

39 P. W. Li and E. S. Yang, "SiGe gate oxide prepared at low-temperatures in an electron cyclotron resonance plasma," *Appl. Phys. Lett.*, vol. 63, pp. 2938–2940, 1993.

40 I. S. Goh, J. F. Zhang, S. Hall, W. Eccleston and K. Werner, "Electrical Properties of Plasma Grown Oxide on MBE-Grown SiGe," *Semicond. Sci. Technol.*, vol. 10, pp. 818–828, 1995.

41 I. S. Goh, S. Hall, W. Eccleston, J. F. Zhang and K. Werner, "Interface quality of SiGe oxide prepared by rf plasma anodization," *Electronics Lett.*, vol. 30, pp. 1988–1989, 1994.

42 P. W. Li, E. S. Yang, Y. F. yang, J. O. Chu and B. S. Meyerson, "SiGe pMOSFET's with Gate Oxide Fabricated by Microwave Electron Cyclotron Resonance Plasma Processing," *IEEE Electron Dev. Lett.*, vol. 15, pp. 402–405, 1994.

43 M. A. Copeland and R. Pappu, "Comparative Study of Plasma Anodization of Silicon in a Column of a dc Glow Discharge," *Appl. Phys. Lett.*, vol. 19, pp. 199–201, 1971.

44 K. K. Ng and J. R. Ligenza, "The Mechanism of Plasma Oxidation on Floating Silicon Substrates," *J. Electrochem. Soc.*, vol. 131, pp. 1968–1970, 1984.

45 D. L. Pulfrey and J. J. Reche, "Preparation and properties of plasma anodized silicon dioxide films," *Solid-State Electron.*, vol. 17, pp. 627–632, 1974.

46 V. Q. Ho and T. Sugano, "An Improvement of the Interface Properties of Plasma Anodized SiO_2/Si System for the Fabrication of MOSFET's," *IEEE Trans. Electron Dev.*, vol. ED-28, pp. 1060–1065, 1981.

47 A. K. Ray and A. Reisman, "Plasma Oxide FET devices," *J. Electrochem. Soc.*, vol. 128, pp. 2424–2428, 1981.

48 J. R. Ligenza, "Silicon oxidation in an oxygen plasma excited by microwaves," *J. Appl. Phys.*, vol. 36, pp. 2703–2707, 1965.

49 J. Kraitchman, "Silicon Oxide Films Grown in a Microwave Discharge," *J. Appl. Phys.*, vol. 38, pp. 4323–4330, 1967.

50 J. Asmussen and J. Root, "Characteristics of a Microwave Plasma Disk Ion Source," *Appl. Phys. Lett.*, vol. 44, pp. 396–400, 1984.

51 S. K. Ray, C. K. Maiti and N. B. Chakrabarti, "Low Temperature Oxidation of Silicon in Microwave Oxygen Plasma," *J. Mat. Sci.*, vol. 25, pp. 2344–2348, 1990.

52 M. Moisan, C. Barbeau, R. Claude, C. M. Ferreira, J. Margot, J. Paraszczak, A. B. Sa, G. Sauvi and M. R. Wertheimer, "Radio frequency or microwave plasma reactors? Factors determining the optimum frequency of operation," *J. Vac. Sci. Technol. B*, vol. 9, pp. 8–25, 1991.

53 S. K. Ray, *Studies on Microwave Plasma Processing for Microelectronic Applications*. PhD Thesis, Indian Inst. of Technology, Kharagpur, 1990.

54 M. Mukhopadhyay, S. K. Ray, C. K. Maiti, D. K. Nayak and Y. Shiraki, "Electrical properties of oxides grown on strained SiGe layer at low temperatures in a microwave oxygen plasma," *Appl. Phys. Lett.*, vol. 65, pp. 895–897 (see also Vol. 66(12), p.1566, March 20, 1995 issue), 1994.

55 M. Mukhopadhyay, S. K. Ray, C. K. Maiti, D. K. Nayak and Y. Shiraki, "Properties of SiGe oxides grown in a microwave oxygen plasma," *J. Appl. Phys.*, vol. 75, pp. 6135–6140, 1995.

56 M. Mukhopadhyay, S. K. Ray, D. K. Nayak and C. K. Maiti, "Ultrathin Oxides Using N_2O on Strained $Si_{1-x}Ge_x$," *Appl. Phys. Lett.*, vol. 68, pp. 1262–1264, 1996.

57 M. Mukhopadhyay, L. K. Bera, S. K. Ray, D. K. Nayak and C. K. Maiti, "Ultrathin Oxides on Strained Epitaxial $Si_{1-x}Ge_x$ Films at Low Temperature," *IETE J. Research*, vol. 43, pp. 165–177, 1997.

58 L. K. Bera, S. K. Ray, D. K. Nayak, N. Usami, Y. Shiraki and C. K. Maiti, "Electrical properties of oxides grown on strained Si using microwave N_2O plasma," *Appl. Phys. Lett.*, vol. 70, pp. 66–68, 1997.

59 L. K. Bera, M. Mukhopadhyay, S. K. Ray, D. K. Nayak, N. Usami, Y. Shiraki and C. K. Maiti, "Oxidation of strained Si in a microwave electron cyclotron resonance plasma," *Appl. Phys. Lett.*, vol. 70, pp. 217–219, 1997.

60 L. K. Bera, S. K. Ray, M. Mukhopadhyay, D. K. Nayak, N. Usami, Y. Shiraki and C. K. Maiti, "Electrical Properties of N_2O/NH_3 Plasma Grown Oxynitride on strained Si," *IEEE Electron Dev. Lett.*, vol. 19, pp. 273–275, 1998.

61 S. K. Ray, L. K. Bera, C. K. Maiti, S. John and S. K. Banerjee, "MOS capacitor characteristics of plasma oxide on partially strained SiGeC films," *Thin Solid Films*, vol. 332, pp. 375–378, 1998.

62 S. K. Ray, L. K. Bera, C. K. Maiti, S. John and S. K. Banerjee, "Electrical Characteristics of Plasma Oxidized $Si_{1-x-y}Ge_xC_y$ MOS Capacitors," *Appl. Phys. Lett.*, vol. 72, pp. 1250–1253, 1998.

63 S. Maikap, L. K. Bera, S. K. Ray, S. John, S. K. Banerjee and C. K. Maiti, "Electrical characterization of $Si/Si_{1-x}Ge_x/Si$ quantum well heterostructures using MOS capacitor," *Solid-State Electron.*, vol. 44, pp. 1029–1034, 2000.

64 S. K. Ray, C. K. Maiti, S. K. Lahiri and N. B. Chakrabarti, "Properties of silicon dioxide films deposited at low temperatures by microwave plasma enhanced decomposition of tetraethylorthosilicate," *J. Vac. Sci. Technol. B*, vol. 10, pp. 1139–1150, 1992.

65 M. Mukhopadhyay, S. K. Ray and C. K. Maiti, "Oxidation Studies on Strained SiGe Layer Using Microwave Oxygen Plasma," in *Proc. of the International Workshop on Physics of Semiconductor Devices, NPL, New Delhi*, pp. 353–357, 1995.

66 Y. Z. Hu, J. Joseph and E. A. Irene, "*In-situ* spectroscopic ellipsometry study of the ECR plasma oxidation of silicon and interfacial damage," *Appl. Phys. Lett.*, vol. 59, pp. 1353–1355, 1991.

67 S. G. Park, W. S. Liu and M.-A. Nicolet, "Kinetics and mechanism of wet oxidation of Ge_xSi_{1-x} alloys," *J. Appl. Phys.*, vol. 75, pp. 1764–1770, 1994.

68 G. F. A. van de Walle, L. J. van Ijzendoorn, A. A. van Gorkum, R. A. van den Heuvel and A. M. L. Theunissen, "Thermal stability of strained $Si/Si_{1-x}Ge_x/Si$ structures," *Semicond. Sci. Technol.*, vol. 5, pp. 345–347, 1990.

69 W. A. Hill and C. C. Coleman, "A single frequency approximation for interface state density determination," *Solid-State Electron.*, vol. 23, pp. 987–993, 1980.

70 D. J. DiMaria, R. Ghez and D. W. Dong, "Charge trapping studies in SiO_2 using high current injection from Si-rich SiO_2 films," *J. Appl. Phys.*, vol. 51, pp. 4830–4841, 1980.

71 C. H. Seager and W. K. Schubert, "Hole trapping in oxides grown by rapid thermal processing," *J. Appl. Phys.*, vol. 63, pp. 2869–2871, 1988.

72 S. F. Nelson, K. Ismail, T. N. Jackson, J. J. Nocera and J. O. Chu, "Systematics of electron mobility in Si/SiGe heterostructures," *Appl. Phys. Lett.*, vol. 63, pp. 794–796, 1993.

73 K. Ismail, B. S. Meyerson, S. Rishton, J. Chu, S. Nelson and J. Nocera, "High-Transconductance n-Type Si/SiGe Modulation-Doped Field-Effect Transistors," *IEEE Electron Dev. Lett.*, vol. 13, pp. 229–231, 1992.

74 U. Konig, A. J. Boers, F. Schaffler and E. Kasper, "Enhancement mode n-channel Si/SiGe MODFET with high intrinsic transconductance," *Electronics Lett.*, vol. 28, pp. 160–162, 1992.

75 J. Welser, J. L. Hoyt and J. F. Gibbons, "Growth and Processing of Relaxed-Si$_{1-x}$Ge$_x$/Strained-Si Structures for Metal–Oxide–Semiconductor Applications," *Jap. J. Appl. Phys.*, vol. 33, pp. 2419–2422, 1994.

76 J. Welser, J. L. Hoyt and J. F. Gibbons, "Growth and Processing of Relaxed-Si$_{1-x}$Ge$_x$/Strained-Si Structures for MOS Applications," in *Proc. SSDM*, pp. 377–379, 1993.

77 K. Rim, J. Welser, J. L. Hoyt and J. F. Gibbons, "Enhanced Hole Mobilities in Surface-channel Strained Si p-MOSFETs," in *IEEE IEDM Tech. Dig.*, pp. 517–520, 1995.

78 D. K. Nayak, K. Goto, A. Yutani, J. Murota and Y. Shiraki, "High-Mobility Strained Si PMOSFETs," *IEEE Trans. Electron Dev.*, vol. 43, pp. 1709–1715, 1996.

79 Z. Atzmon, A. E. Bair, T. L. Alford, D. Chandrasekhar, D. J. Smith and J. W. Mayer, "Wet oxidation of amorphous and crystalline Si$_{1-x-y}$Ge$_x$C$_y$ alloys grown on (100) substrates," *Appl. Phys. Lett.*, vol. 66, pp. 2244–2246, 1995.

80 L. K. Han, M. Bhat, D. Wristers, H. H. Wang and D. L. Kwong, "Recent Developments in Ultra Thin Oxynitride Gate Dielectrics," *Microelectronic Engineering*, vol. 28, pp. 89–96, 1995.

81 D. Wristers, L. K. Han, T. Chen, H. H. Wang, D. L. Kwong, M. Allen and J. Fulford, "Degradation of oxynitride gate dielectric reliability due to boron diffusion," *Appl. Phys. Lett.*, vol. 68, pp. 2094–2096, 1996.

82 J. G. Huang and R. J. Jaccodine, "Fast Growth on Thin Gate Dielectrics by Thermal Oxidation of Si in N$_2$O Gas Ambient with Low Concentrations of NF$_3$ Addition," *J. Electrochem. Soc.*, vol. 140, pp. L15–L16, 1993.

83 J. J. Welser, *The application of strained-silicon/relaxed-silicon germanium heterostructures to metal–oxide–semiconductor field-effect transistors.* PhD Thesis, Stanford University, 1994.

84 B. Maiti, *Novel Thin Dielectric Devices and Processing Using Low-Pressure Chemical Vapor Deposited Films.* PhD Thesis, The University of Texas at Austin, 1993.

85 B. Maiti, M. Y. Hao, I. Lee and J. C. Lee, "Improved ultrathin oxynitride formed by thermal nitridation and low pressure chemical vapor deposition process," *Appl. Phys. Lett.*, vol. 61, pp. 1790–1792, 1992.

86 M. Mukhopadhyay, S. K. Ray and C. K. Maiti, "Microwave Plasma Grown Oxynitride Using Nitrous Oxide (N_2O)," *Electronics Lett.*, vol. 22, pp. 1953–1954, 1995.

87 H. Hwang, W. Ting, D. L. Kong and J. Lee, "Electrical and reliability characteristics of ultrathin oxynitride gate dielectric prepared by rapid thermal processing in N_2O," in *IEEE IEDM Tech. Dig.*, pp. 421–424, 1990.

88 M. Bhat, G. W. Yoon, J. Kim and D. L. Kwong, "Effects of NH_3 nitridation on oxides grown in pure N_2O ambient," *Appl. Phys. Lett.*, vol. 64, pp. 2116–2118, 1994.

89 V. R. Rao, I. Eisele, R. M. Patrikar, D. K. Sharma, J. Vasi and T. Grabolla, "High-Field Stressing of LPCVD Gate Oxides," *IEEE Electron Dev. Lett.*, vol. 18, pp. 84–86, 1997.

90 B. Maiti and J. C. Lee, "Electrical and Reliability Characteristics of Silicon-Rich Oxide for Non-Volatile Memory Applications," *Mat. Res. Soc. Symp. Proc.*, vol. 265, pp. 255–260, 1992.

91 F. K. LeGoues, R. Rosenberg, T. Nguyen and B. S. Meyerson, "The mechanism of oxidation of SiGe," in *Mat. Res. Symp. Proc.*, vol. 105, pp. 313–318, 1988.

92 O. W. Holland, C. W. White and D. Fathy, "Novel oxidation process in Ge^+-implanted Si and its effect on oxidation kinetics," *Appl. Phys. Lett.*, vol. 51, pp. 520–522, 1987.

93 J. P. Zhang, P. L. F. Hemment, S. M. Newstead, A. R. Powell, T. E. Whall and E. H. C. Parker, "A Comparison of the Behavior of $Si_{0.5}Ge_{0.5}$ Alloys During Dry and Wet Oxidation," *Thin Solid Films*, vol. 222, pp. 141–144, 1992.

94 H. Tsutsu, W. J. Edwards, D. G. Ast and T. I. Kamins, "Oxidation of polycrystalline-SiGe alloys," *Appl. Phys. Lett.*, vol. 64, pp. 297–299, 1994.

95 S. Hall, J. F. Zhang, S. Taylor, W. Eccleston, P. Beahen, G. T. Tatlock, C. J. Gibbings, C. Smith and C. Tuppen, "Assessment of plasma grown oxide on Si:Ge substrate," *Appl. Surf. Sci.*, vol. 39, pp. 57–64, 1989.

96 L. S. Riley and S. Hall, "X-ray photoelectron spectra of low temperature plasma anodized $Si_{0.84}Ge_{0.16}$ alloy on Si(100): Implications for SiGe oxidation kinetics and oxide electrical properties," *J. Appl. Phys.*, vol. 85, pp. 6826–6837, 1999.

97 C. Tetelin, X. Wallart, J. P. Nys, L. Vescan and D. J. Gravesteijn, "Kinetics and mechanism of low temperature atomic oxygen-assisted oxidation of SiGe layers," *J. Appl. Phys.*, vol. 83, pp. 2842–2846, 1998.

Chapter 5

SiGe Heterojunction Bipolar Transistors

A SiGe heterojunction bipolar transistor uses a silicon–germanium alloy layer as the active base material. The introduction of Ge causes a reduction of the bandgap and thus an increase in the intrinsic carrier concentration, and at the same time creates a valence band offset which impedes the flow of holes into the emitter. SiGe-HBTs fabricated using MBE in 1987 exhibited a remarkable enhancement of current gain. There have been many technological developments and innovations since then. Some recent developments, such as IBM epi-base technology, SiGe-BiCMOS technology and SiGe-HBTs on SOI, are briefly described in this chapter. DC performance of SiGE-HBTs and their behavior at low temperature are presented. Incorporation of carbon into the SiGe alloy enables attainment of many improvements in performance. This is discussed. Special techniques and processes need to be used for fabricating HBTs for high voltage and high power applications. IC processes for incorporation of SiGe-HBTs in a standard Si IC receive special attention.

RF applications are known to be the special niche for SiGe-HBTs because of their excellent high frequency performance. For application in rf circuits, several analog characteristics become important. The cutoff frequency and maximum frequency of oscillation are important figures to specify the high frequency capability. It is known that the f_T value must be 10 times higher than the operating frequency. The noise figure is another factor for rf circuit design especially for low noise amplifier applications. Flicker noise or $1/f$ noise is important as it results in phase noise in voltage controlled oscillators and nonlinear rf circuits. Flicker noise in HBTs is two orders of magnitude smaller than with CMOS technology. Applications of SiGe-HBTs in MMIC and rf circuits have been discussed. The choice of technology, bipolar or CMOS or

210

their combination, remains a live issue. This receives the consideration due. To indicate the level of maturity of SiGe-HBT technology, some available products are listed.

5.1 Principle of Operation

The idea of a bipolar transistor with a wide gap emitter is almost as old as the bipolar junction transistor. The performance advantages of this type of device, commonly known as the heterojunction bipolar transistor, have been known for many years. The concept of inserting heterojunctions into bipolar transistors was suggested by Kroemer in 1957 [1] but it was not until 1987 that a SiGe-HBT was demonstrated [2]. The introduction of Ge into the base of an npn Si-BJT alters the performance of a transistor in several fundamental ways. The most obvious change is a reduction in the bandgap of the SiGe alloy in the p-doped base, and an increase in the intrinsic carrier concentration compared to Si in the n-doped emitter and collector regions. The bandgap discontinuity creates the heterojunctions needed for the enhanced performance of a SiGe-HBT.

Some of the advantages of the HBT can be illustrated by analyzing the current components in an npn HBT. Figure 5.1 shows the current components in an npn HBT in the forward active mode of operation. Following Kroemer [1], the major dc current components in the transistor are: the electron current I_n injected from the emitter into the base; the hole current I_p injected from the base into the emitter; electron–hole recombination I_s in the forward biased emitter–base space charge layer; and the current I_r due to injected electrons recombining with holes in the quasi-neutral base. Current components I_p, I_r, and I_s should be minimized in order to maximize the device performance. The terminal currents can be expressed in terms of these four components:

$$I_e = I_n + I_p + I_s \qquad (5.1)$$

$$I_b = I_p + I_r + I_s \qquad (5.2)$$

$$I_c = I_n - I_r \qquad (5.3)$$

An important figure of merit for bipolar transistors is the dc current gain:

$$\beta = \frac{I_c}{I_b} = \frac{I_n - I_r}{I_p + I_r + I_s} \leq \frac{I_n}{I_p} \approx \beta_{max} \qquad (5.4)$$

By definition, β_{max} is the value of the current gain for which the recombination currents I_r and I_s are negligible. The electron and hole injection currents in a forward biased p–n junction can be expressed as

$$I_n = \frac{qAD_{nB}}{L_{nB}} n_{p0} \left[\exp\left(\frac{qV_{be}}{kT} \right) - 1 \right] \tag{5.5}$$

$$I_p = \frac{qAD_{pE}}{L_{pE}} p_{n0} \left[\exp\left(\frac{qV_{be}}{kT} \right) - 1 \right] \tag{5.6}$$

where V_{be} is the applied bias, A is the area of the junction, D_{nB} and D_{pE} are the minority carrier diffusion constants, L_{nB} and L_{pE} are minority carrier diffusion lengths, and n_{p0} and p_{n0} are the equilibrium minority carrier concentrations in the neutral base and emitter.

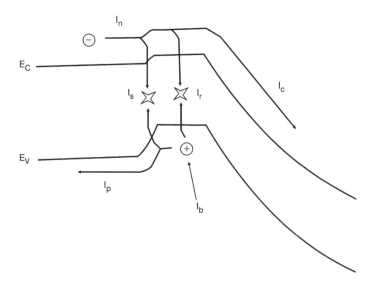

*Figure 5.1. Basic current components in a heterojunction bipolar transistor. After H. Kroemer, Proc. IEEE, **70**, 13–25 (1982), copyright ©1982 IEEE.*

In SiGe-HBTs, a narrow bandgap SiGe alloy base is used. Due to its smaller bandgap, the intrinsic carrier concentration in the SiGe base increases. The difference between the HBT and BJT is that the concentration of the injected electrons is much higher (several orders of magnitude) into the base due to the lower conduction band barrier. This means that the collector current in a HBT will be much higher than in a similarly doped BJT, by a factor of $\exp(\Delta E_g/kT)$, while the base current is not affected. However, the real advantage of the HBT is not to achieve a very high current gain, but to trade it against a high base doping, necessary to reduce the base resistance. Base resistance is an important parameter in determining f_{max}, the maximum frequency of oscillation. Low base resistivity allows a reduction in base width and consequently in the transit time of the minority carriers across the base. The better high frequency performance of HBTs over BJTs is a result of heavier base doping since the base resistance and parasitic junction capacitances are reduced.

In a SiGe-HBT, the bandgap difference $\Delta E_g = E_{g,SiGe}(x) - E_{g,Si}$ can be made much larger than kT. For a Ge fraction $x = 0.2$ in the base, the bandgap difference is more than 170 meV. Therefore, the current gain of the HBT can be made large irrespective of the doping ratio in the emitter and the base. A consequence is that the base in an HBT can be more heavily doped than in a BJT, and a HBT can achieve a higher current gain than a BJT with a lighter base doping.

In a heterostructure bipolar transistor with $Si_{1-x}Ge_x$ base, the bandgap of the emitter is larger and therefore the injection efficiency can be made very high, even if the base is doped more heavily than the emitter [1, 3]. In an optimized transistor, a value of 50 for β is usually sufficient. Therefore the high injection efficiency can be traded for heavy doping in the base and reduced base resistance.

High values of the oscillation frequency (for microwave transistors) and low values of gate delay τ_d (for digital switching applications) can be obtained in HBTs. Heavy doping also avoids punch-through in the thin base. Tunneling currents can be avoided in the heterostructures by reducing the doping concentration in the emitter. In double heterostructures, the collector and emitter can be interchanged, providing an additional advantage in some digital circuits. The design of a Si homojunction bipolar transistor or a SiGe heterojunction bipolar transistor is optimized by varying geometrical parameters of the device such as the region widths and doping profiles in the emitter, base and collector regions. Additionally, in a SiGe-HBT the bandgap of the base layer is tailored by grading the Ge concentration.

There are four commonly used figures of merit of a bipolar transistor: current gain, cutoff frequency, maximum oscillation frequency, and gate delay τ_d in digital circuits. For high frequency microwave operation, the design must be optimized so that f_{max} is as high as possible. For digital switching transistors, the gate propagation delay τ_d should be reduced to a minimum. In addition to high frequency or high speed, the transistor must have low noise and high junction breakdown voltage. The product βV_A of current gain β and Early voltage V_A is an important index of performance for analog applications.

5.2   Fabrication of SiGe-HBTs

Work on the growth of SiGe layers for SiGe-HBTs started in the early 1980s and initial HBTs were demonstrated using MBE [2]. The base layer which has an induced strain due to the presence of Ge of the same order of magnitude as stresses due to conventional device isolation or metallization, is unconditionally stable. Ge is buried so that to all processing tools the material looks and behaves like Si.

In order to meet the stringent defect density requirements for large volume manufacturing in a Si process line, a conservative graded Ge profile is used in the base. UHVCVD techniques used for base layer deposition have been refined to provide reproducible and uniform layers, as well as to place the dopant deeper in the structure to reduce the surface electric field and improve reliability, and to extend the process from 125 mm to 200 mm wafers [4–6]. Excellent base profile control is achieved compared to advanced implanted-base bipolar transistors, where a majority of the implanted dose is buried under the emitter dopant during the emitter drive-in.

A move towards commercial manufacturing was the transition from MBE to CVD layer growth, which has the uniformity, reproducibility and basic tool commonality to support volume manufacturing in a Si process line [7]. Epitaxial layer growth is well established at high temperatures for CMOS and bipolar substrates, and extension to lower temperatures as well as the addition of Ge has been optimized for over a decade.

Over the last few years, several research groups have attempted to investigate the potential of high level integration of SiGe in conventional Si processing technology. Some of the processes developed are derivatives of existing III–V processes and will probably result in the same level of yield and integration as III–V ICs. Conversion of these processes to BiCMOS requires major process and device parameter changes, due to the large number of medium and high temperature thermal cycles in a standard Si process. In the BiCMOS process, the fabrication and device advantages of using a polysilicon emitter with thermal oxide at the emitter–base interface are retained, along with the associated two decades of experience in double polysilicon bipolar transistors. BiCMOS also reduces component count and improves overall system performance by combining optimized functional blocks using either bipolar or CMOS technology [8].

In order to achieve the highest flexibility for circuit fabrication and to decrease the cost of production, it is necessary to realize simple implementation of SiGe-HBTs in standard Si processing technology. Aiming towards this, IBM and Temic research groups have attempted various epitaxy techniques for the

SiGe films. The evolution to the manufacturability of the SiGe technology with reference to the SiGe-BiCMOS process has been described in Chapter 1. Table 5.1 shows some of the important process integration issues.

Table 5.1. Types of SiGe-HBTs and process integration issues.

Process integration features	Blanket epitaxy	Differential epitaxy	Selective epitaxy
Precleaning	simple	difficult	difficult
Type of epitaxy	MBE	MBE	UHVCVD
	UHVCVD	UHVCVD	LPCVD
	APCVD	LPCVD	APCVD
Epitaxy process	simple	complex	complex
BiCMOS process integration	HBT/CMOS	HBT/CMOS	HBT/CMOS
Compatibility to CMOS	low	acceptable	acceptable
Research group	DBAG/Temic	DBAG/Temic, IBM	NEC, Siemens

Presently, two different concepts in SiGe-HBT technology exist, namely (i) the high emitter doping concept with graded base doping and graded Ge profiles, which was introduced by the IBM researchers [9], and (ii) the doping inversion concept with high base and low emitter doping mainly used by the Daimler-Benz research group [10]. Both concepts are shown in figures 5.2 and 5.3. The advantage of the second one is the distinct reduction of the base resistance in spite of small base widths down to 100 Å. For example, an MBE grown 150 Å base layer doped with 8×10^{19} cm^{-3} has a sheet resistance of 1 kΩ/sq, whereas the value for standard bipolar and UHVCVD grown films is approximately 10 times higher. The base resistance is very important for attaining high f_{max} values as will be shown later.

One of the advantages of using MBE is that the entire HBT layer structure, i.e., the low doped Si collector, the heavily doped SiGe base embedded in undoped SiGe layers (doping setback), the low doped Si emitter and the heavily doped Si emitter contact layer, can be grown in a single process step. This simplifies the fabrication sequence and is therefore preferable to the UHVCVD concept, where the growth is performed in several pieces of equipment.

The approach of the IBM researchers for their SiGe-HBTs was to modify

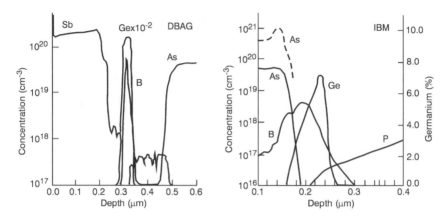

Figure 5.2. Comparison of SiGe-HBT doping profiles of MBE (Daimler-Benz AG) and UHVCVD (IBM) grown samples. SIMS profile for DBAG (a) and IBM (b). After A. Schuppen et al., J. Mater. Sci.: Mater. Electron., 6, 298–305 (1995).

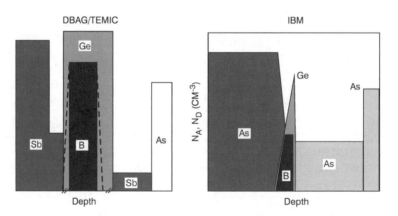

Figure 5.3. Comparison of SiGe-HBT doping profiles of MBE (Daimler-Benz AG) and UHVCVD (IBM) grown samples. Schematic profile exhibiting doping setback layers on the collector and/or the emitter side of the base preventing outdiffusion. Profile for DBAG (a) and IBM (b). After A. Schuppen et al., J. Mater. Sci.: Mater. Electron., 6, 298–305 (1995).

an existing poly-emitter bipolar transistor process. A limited number of fabrication steps were changed to allow quick development and the use of existing circuit designs. Non-self-aligned SiGe-HBTs grown by UHVCVD have reached an f_T of 75 GHz [9] and self-aligned HBTs have exhibited f_T values of 113 GHz [11].

In this chapter we shall consider the different processes: (i) a non-passivated self-aligned double mesa process [12], (ii) a passivated double mesa process, and (iii) a passivated self-aligned single mesa process, which leads to the so-called differential SiGe-HBT [13]. All three technologies start with low doped p-type 4 inch wafers, in which buried collector layers are implanted (100 keV As, 2×10^{16} cm^{-2}) and subsequently diffused and activated over 4 h at 1150 °C, resulting in a 5 Ω/sq sheet resistance.

5.2.1   Non-passivated Double Mesa Technology

After the formation of the buried layer, the complete MBE layer sequence is grown. The first lithography step defines the emitter contact, which is fabricated by a lift-off process with Pt/Au (200/3000 Å). Then the emitter is etched in a modified KOH solution down to the SiGe base as shown in figure 5.4a. It stops at the SiGe layer. A slight overetch provides a self-alignment of the following 200 Å Pt and 1500 Å Au metal contacts with respect to the emitter (figure 5.4b). In order to produce the collector contact, the base and the low doped collector are dry etched in an SF_6/O_2 plasma. Ti–Pt–Au (200/200/2000 Å) is patterned using a lift-off process (figure 5.4c). The essential step of this technology is the groove etching, which separates the active transistor from the contact pads (figure 5.4d). Hence, it is possible to measure dc and rf characteristics directly on the wafer without any de-embedding. In addition, the advantage of this HBT technology is not only the zero thermal budget, but also the simplicity and quick turnaround time. A disadvantage is the unpassivated mesa surface. However, these structures show very high f_{max} values due to very small base resistance and small base–collector capacitance.

5.2.2   Passivated Double Mesa Technology

In order to obtain passivated SiGe-HBTs a double mesa process has been implemented by using a self-aligned spacer technology. After buried layer formation the MBE layers are grown. The thick 2300 Å emitter contact layer is reduced to 1000 Å. Then a 200 Å low thermal budget oxide is grown and a 3000 Å nitride layer is deposited and patterned by a plasma etch step defining

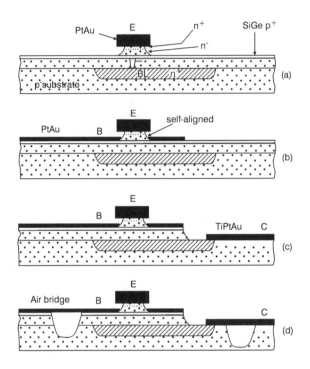

Figure 5.4. Fabrication process of double mesa-type SiGe-HBTs. (a) Emitter mesa etching, (b) self-aligned base contact, (c) collector mesa and collector contact, (d) transistor separation by isotropic undercutting. After A. Schuppen et al., J. Mat. Sci.: Mater. Electron., 6, 298–305 (1995).

the emitter. Subsequently the high doped emitter contact layer and 200 Å of the emitter are removed by plasma etching with the same mask. A first oxide spacer prevents short circuit between the emitter and the subsequently implanted (BF_2) base region. After spacer removal the formation of the collector mesa follows. A thermal oxidation at 750 °C and a second spacer process passivates the double mesa structure. After removing the nitride, titanium is evaporated and $TiSi_2$ is formed self-aligned at 750 °C for 30 s by RTA on the emitter, base and collector contact. After removal of the unreacted Ti an RTA step at 820 °C improves the contact and sheet resistance of the $TiSi_2$ layer. PECVD oxide layers serve as passivation for the two TiAl metallization.

5.2.3 Differential SiGe-HBT Technology

For integrated circuits the differential SiGe-HBT offers a technology with reduced parasitic base–collector capacitance. The process starts with a buried layer on a 20 Ω cm substrate. The collector is grown by Si low pressure CVD. Subsequently a recessed LOCOS technology defines the active HBT area. The collector contact regions are implanted with phosphorus. The base and emitter layers are deposited by MBE leading to monocrystalline growth in the defined emitter and collector contact windows and to polycrystalline Si and SiGe on the field oxide. A similar process as described above with emitter–base mesa passivation by thermal oxidation and spacer technique has been developed. This technology includes three important advantages: (i) reduction of the base–collector capacitance, (ii) lowering of the base lead resistance by using the polycrystalline Si/SiGe as base interconnect, and (iii) isolation between the transistors provided by the field oxide.

Use of differential or selective epitaxy for the fabrication of SiGe-HBTs using different deposition techniques has been reported by Behammer *et al.* [14]. Figure 5.5 shows the desired Ge mole fraction and doping in the base for a device optimized for high f_T with very steep donor–acceptor profiles and Ge trapezoidal distribution. Depending on the type of epitaxy, such as MBE, RT-LPCVD and APCVD, and process features (for instance *in situ* cleaning, dopants, Si and Ge sources used), quite different interface and layer qualities, and doping and contamination levels (mainly O, C and metals) were obtained and are shown in figures 5.6, 5.7, and 5.8, respectively.

5.2.4 IBM Epi-base Technology

Generally, one of the difficulties associated with the fabrication of self-aligned heterojunction bipolar transistors using SiGe is the need to avoid oxidation of the SiGe layer in the intrinsic device region. IBM's epi-base technology [15] stated below allows fabrication of self-aligned SiGe epitaxial base (epi-base) bipolar transistors without detrimental oxidation of SiGe in the active device region. Figures 5.9a–5.9h show the fabrication process flow for the emitter definition and extrinsic base self-alignment process. The steps required in the process are outlined below.

1. Epi-base deposition (figure 5.9a) consists of: (i) an arbitrary SiGe base profile in n- or p-type; (ii) a thin ($\sim$ 100 Å), heavily doped ($p^{++} > 2 \times 10^{20}$ cm^{-3}), conductive etch stop layer; and (iii) an intrinsic Si buffer region for consumption during isolation ($\sim$ 300–400 Å).

2. Oxide/nitride deposition and patterning for the emitter opening stack.

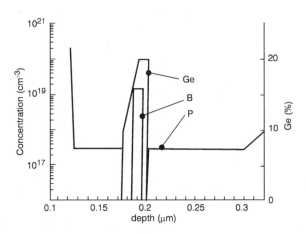

Figure 5.5. Desired profiles for Ge mole fraction and doping in the base for a device optimized for high f_T with very steep donor–acceptor profiles and Ge trapezoidal distribution. After D. Behammer et al., Solid-State Electron., 39, 471–480 (1996).

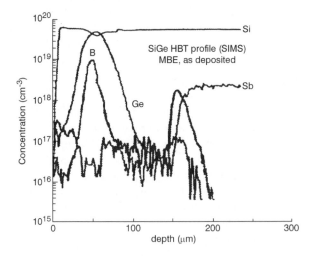

Figure 5.6. Measured base profiles using MBE for Ge mole fraction and doping for a device optimized for high f_T. After D. Behammer et al., Solid-State Electron., 39, 471–480 (1996).

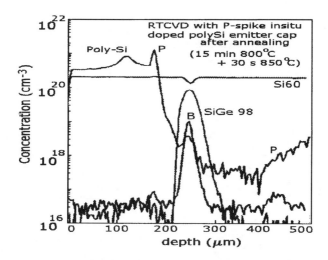

*Figure 5.7. Measured base profiles using RTCVD for Ge mole fraction and doping for a device optimized for high f_T. After D. Behammer et al., Solid-State Electron., **39**, 471–480 (1996).*

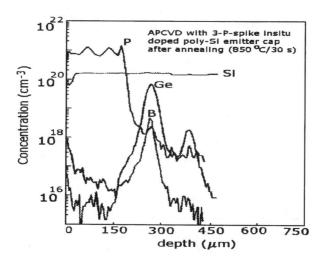

*Figure 5.8. Measured base profiles using APCVD for Ge mole fraction and doping for a device optimized for high f_T. After D. Behammer et al., Solid-State Electron., **39**, 471–480 (1996).*

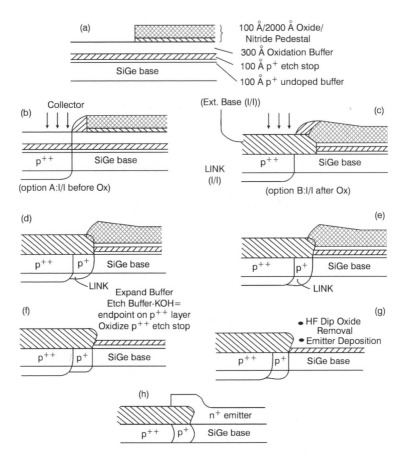

Figure 5.9.   Fabrication process flow of IBM's epi-base technology for SiGe-HBTs. After IBM Technology Disc. Bull., 179–182 (1992).

There are two options at this stage:

- Implant before oxidation (figure 5.9b) – sidewall formation and removal for extrinsic base ion implantation; 700 °C low temperature oxidation, e.g., HIPOX, for isolation; link implant defined by the nitride stack (figure 5.9d).

- Implant after oxidation (figure 5.9c) – 700 °C low temperature oxidation, e.g., HIPOX, for isolation; sidewall formation and removal for extrinsic base ion implantation; link implant defined by the nitride stack, if desired (figure 5.9e).

3. Removal of the nitride stack.

4. Wet chemical (KOH/IPA) etch of the intrinsic Si buffer in the emitter opening.

5. Oxidation of the p^{++} conductive etch stop (figure 5.9f).

6. HF dip for removal of the converted (oxidation) etch stop layer (figure 5.9g).

7. Emitter deposition (figure 5.9h).

5.2.5   IBM SiGe-BiCMOS Technology

It is instructive to look at the state-of-the-art Si bipolar technology. For reasons of base-width control (difficulty in vertical scaling), it is unlikely that the Si-only process will be able to achieve well-controlled ac and dc parameters (achievable in epitaxial-base SiGe technology) which are important to the robustness of the circuit design [16–18].

A comparison of identical Si-only and SiGe technologies (as shown in table 5.2) shows that there is a considerable advantage in both speed and power for SiGe technology [19]. It is also important to use the same tool-set (as far as possible) and to keep the actual physical process steps the same (see figure 5.10). The measured cutoff frequency vs. breakdown voltage for devices from various technologies is shown in figure 5.11.

Table 5.2. Process technology comparison. After S. Subbanna et al., ISSCC Tech. Dig., 66–67 (1999).

	Imp Si	Epi Si	Epi SiGe
Sub-collector and epitaxial collector	Y	Y	Y
Shallow trench isolation	Y	Y	Y
Base process	Imp.	Si epi	SiGe epi
Emitter process	Y	Y	Y
Emitter As implant	Y	Y	Y
Emitter anneal	RTA	RTA	RTA
Base dopant loss in emitter	High	Low	Low
Current gain	Medium	Medium	High
Mask levels	20	20	20

SiGe-BiCMOS technology has now matured from the laboratory level to a standard manufacturing process technology. Several manufacturers have attempted to investigate the potential of high level integration in SiGe-

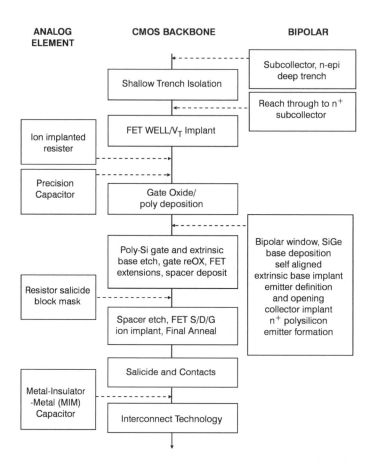

ANALOG ELEMENT   CMOS BACKBONE   BIPOLAR

*Figure 5.10. CMOS process flow with required additions to produce SiGe-based BiCMOS and additional analog devices. After D. Ahlgren and B. Jagannathan, Solid State Technol., **43**, 53(January 2000).*

BiCMOS using conventional Si processing with minimum changes in the process [6, 20]. Among them IBM has been the most successful and has fabricated a 1.8 million transistor CMOS ASIC test site (die size: 8.06 mm × 8.06 mm) along with various SiGe heterojunction bipolar transistor circuits suitable for addressing both high performance and low power system requirements. It has been shown that the incorporation of high speed bipolar devices does not degrade the operation and the reliability of VLSI CMOS

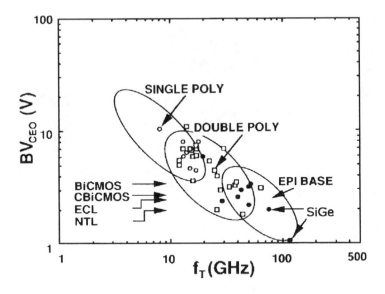

*Figure 5.11. Measured cutoff frequency vs. breakdown voltage for devices from various technologies. After T. Nakamura and H. Nishizawa, IEEE Trans. Electron Dev., **42**, 390–398 (1995), copyright ©1995 IEEE.*

circuits. A single-chip radio for wireless communications has been fabricated using SiGe-BiCMOS technology [21].

IBM's SiGe-BiCMOS technology is a double polysilicon self-aligned process [22]. A cross-section schematic of the IBM SiGe-HBT in a BiCMOS process is shown in figure 5.12. It offers CMOS ($L_{\text{eff}} = 0.36 \ \mu$m), standard ($f_T = 50$ GHz, $BV_{\text{ceo}} = 3.4$ V) and high-breakdown voltage ($f_T = 28$ GHz, $BV_{\text{ceo}} = 5.7$ V) npn-HBTs, a lateral pnp-BJT and a Schottky diode, as well as a complement of passive components such as resistors, capacitors and inductors. Dc and ac parameter values and characteristics of SiGe-HBTs measured from the production SiGe-HBT BiCMOS technology are shown in table 5.3 and figure 5.13, respectively. The dc electrical characteristics of the CMOS devices are shown in table 5.4. These characteristics are equivalent to those of the CMOS-only technology to which the SiGe-HBTs have been added. The properties of passive components such as resistors, capacitors and inductors realized in the SiGe-BiCMOS technology are shown in table 5.5. The technology used is described briefly below.

The starting wafer is a relatively high resistivity substrate (boron-doped

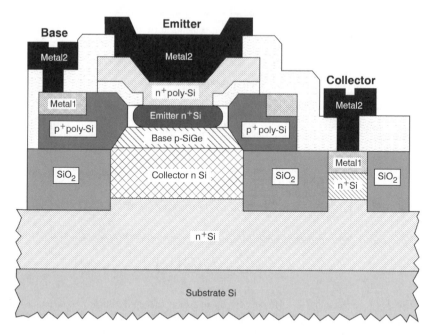

*Figure 5.12. A cross-section schematic of the IBM SiGe-HBT in a BiCMOS process. After D. L. Harame et al., IEEE Trans. Electron Dev., **42**, 469–482 (1995), copyright ©1995 IEEE.*

15 Ω cm) and is used to improve the performance of rf components fabricated on it. The process begins with an n^+-subcollector implanted under the HBTs. This implant is also used as part of the N-wells. An n^+ epitaxial layer is then grown followed by deep trench and shallow trench isolation. The subcollector buried layer is contacted with a phosphorus ion implant. The wells are then formed with arsenic and boron implants. A 78 Å gate oxide is capped with a thin undoped polysilicon layer to protect the gate oxide. The polysilicon is then patterned and etched to open the bipolar window. A single-crystal SiGe epitaxial base is grown in the bipolar active region and the remaining area, including the CMOS gates, is covered with polysilicon. This has been termed the "base equals gate" scheme [16].

A series of films are deposited and etched to form a pedestal for the self-aligned extrinsic base implant. This pedestal is then removed to form the emitter opening. A polysilicon layer is deposited, implanted and etched to form

Table 5.3. DC characteristics of npn SiGe-HBTs ($A_E = 0.5~\mu m \times 2.5~\mu m$) fabricated in a BiCMOS process. After S. Subbanna et al., ISSCC Tech. Dig., 66–67 (1999).

Parameter	Value
β ($V_{BE} = 0.72$ V)	100
BV_{cbo} (nom)	14.4 V
BV_{cbo} (min)	12.5 V
V_{Early}	66 V
βV_{Early}	> 6000 nom
R_{bi}	9.0 kΩ/sq
R_E	10.2 Ω
R_B (low current)	$< 120~\Omega$
C_{jeo}	10.8 fF
C_{jco}	3.0 fF
C_{jbo}	5.8 fF

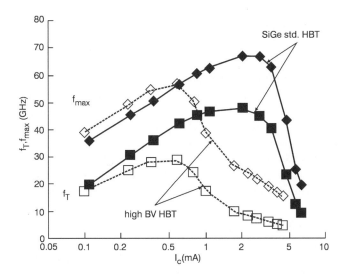

Figure 5.13. AC characteristics of npn SiGe-HBTs fabricated in a BiCMOS process. After S. Subbanna et al., ISSCC Tech. Dig., 66–67 (1999), copyright ©1999 IEEE.

Table 5.4.   DC characteristics of field effect devices fabricated in a SiGe-BiCMOS process. After S. Subbanna et al., ISSCC Tech. Dig., 66–67 (1999).

3.3 V LDD MOSFETs
n-FET $L_{\text{eff}} = 0.39$ μm, $I_{\text{D,sat}} = 485$ mA/mm
p-FET $L_{\text{eff}} = 0.36$ μm, $I_{\text{D,sat}} = 213$ mA/mm

Table 5.5. Passive components fabricated in a BiCMOS process. After S. Subbanna et al., ISSCC Tech. Dig., 66–67 (1999).

Diodes	Schottky Barrier
	Varacter
	p-i-n
	ESD
Capacitors	MIM 0.7 fF/μm^2
	poly-n$^+$ Silicon 1.5 fF/μm^2
Resistors	poly-Si Low and Medium
	Single-crystal Si Low, Medium, High
Inductor	Q 12 at 2 GHz (4 nH)

the emitter. The extrinsic base and CMOS gates are then formed using a single polysilicon etch. The HBT is now completely formed and only needs a limited thermal cycle for dopant redistribution and activation. This thermal cycle will come from the remaining CMOS processing which begins with a sidewall oxidation. The LDDs are implanted followed by a nitride/oxide spacer and source/drain/gate ion implants. A final RTA is done to activate the dopants followed by a titanium salicide process for low resistance contacts. The devices are interconnected using a fully planarized 5-level metallization process.

5.2.6   SiGe-HBTs on SOI

In Si bipolar technology, the two well-known disadvantages are a high power dissipation and low density.   High power dissipation is a result of the high parasitic junction capacitance associated with the use of Si as the substrate. Silicon-on-insulator has been used for high performance MOSFET microelectronics, primarily for applications requiring radiation hardness. The advantages of utilizing a composite substrate comprising a monocrystalline semiconductor layer, such as Si, epitaxially deposited on a supporting insulating substrate are well recognized.   Several advantages include the substantial reduction of parasitic capacitance between charged active regions and the substrate, and the effective elimination of leakage currents flowing between adjacent active devices.

Modern communication system components present greater difficulties in high level integration because they require digital computing capability (logic and memory) along with analog and rf circuitry. The need to reduce power consumption in battery powered wireless communication systems is a need which has not previously been met.   Bipolar transistors fabricated on SOI substrates offer the possibility of lower parasitic capacitances, but have a greater susceptibility to self heating. The objective of fabrication of high speed SiGe-HBTs on insulators is to develop ultra-low power circuits, which are suitable for mixed analog/digital circuit performance up to and beyond 5.5 GHz. A cross-sectional view of the device is shown in figure 5.14.   The Southampton University research group have fabricated SiGe-HBTs on SOI using two alternative approaches:

(i) shallow trench isolation and non-selective growth of the SiGe base and Si emitter, and
(ii) selective growth of the Si collector, combined with non-selective growth of the SiGe base and Si emitter.

A silicon-on-insulator bipolar structure has also been reported [23] with a very low resistance silicided extrinsic base contact which is self-aligned to the collector window, and a very low resistance buried collector contact (metal or refractory silicide). The structure is especially effective for reducing the parasitic resistance and capacitance of Si or SiGe epitaxial-base bipolar devices. The structure can also be extended to SOI BiCMOS.

The unique features of the structure are a low resistance extrinsic base contact which is self-aligned to the collector pedestal by a spacer and a highly conductive buried collector contact and/or local interconnect. Refractory metal or refractory metal silicide is formed directly under the device to serve as a very low resistance collector contact and buried local interconnect. This metal layer also helps in heat removal and chip cooling. It is encapsulated by the surrounding Si and insulator layers during subsequent device fabrication steps. It is now feasible to achieve high performance circuits combining analog circuits with high speed ECL circuits (gate delay < 10 ps) and high density and low power integrated injection logic (I^2L) circuits (gate delay approaching 30 ps and power–delay product less than 30 fJ).

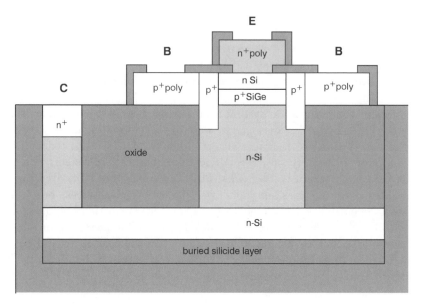

Figure 5.14. Schematic structure of a SiGe-HBT fabricated on a SOI substrate. Source: Southampton University, UK website.

5.3   Performance of SiGe-HBTs

Silicon–germanium based heterojunction bipolar transistors (SiGe-HBTs) are npn bipolar transistors with a thin ($\sim$ 500 Å) pseudomorphically grown $Si_{1-x}Ge_x$ alloy layer as the base. Ge can be incorporated up to a mole fraction of 0.5. Compared to standard Si-BJTs, the base may be as thin as 50 to 100 Å. In addition, the doping in the base may be high, above 10^{20} cm^{-3}, which reduces the base sheet resistance. The highest maximum oscillation frequency from an HBT has been achieved with a base sheet resistance below 1 kΩ/sq.

The dc performance of a SiGe-HBT is basically governed by the high current gain due to suppressed hole reinjection because of emitter–base valence band offset [1] This is shown in figure 5.15 where the Gummel plots at room temperature and 77 K have been plotted. Nearly constant current gain over 9 decades are observed with a high gain of 2000 at 77 K (80 at room temperature). As shown in the Gummel plot, the SiGe-HBT can operate at current densities in excess of 1.5 mA/μm^2 and with near perfect ideality. Unique to the SiGe-HBT is the fact that β also remains virtually flat over a broad temperature range. Because of its large peak f_T, the SiGe-HBT

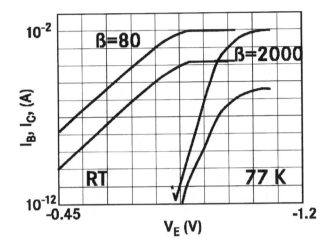

Figure 5.15. Gummel plots at room temperature and 77 K for a passivated SiGe-HBT fabricated in a Si production line. After H. Presting and U. Konig, Proc. Current Developments of Microelectronics, 139–150 (1999).

retains significant high frequency performance even at low currents, allowing the designer the choice of tradeoff between speed and low power.

The potential of high-β HBTs is that they can further raise the dynamic input resistance for low noise amplifiers in the low GHz range. The high frequency properties of various SiGe-HBTs are summarized in terms of the transit and maximum oscillation frequency plots vs. base thickness for various manufacturers (Daimler-Benz, Hitachi, NEC, Siemens, Philips, and TEMIC) as shown in figures 5.16 and 5.17 [24].

f_T values as high as 116 GHz and f_{max} of 160 GHz have been achieved by the Daimler-Benz research group with base thicknesses of 70 and 370 Å. The maximum transit frequencies at lower base width however are associated with lower f_{max} values due to higher base resistance. The f_T/f_{max} ratio depends strongly on the collector design. A thin, heavily doped collector favors higher f_T values. Presently HBTs with equal f_T and f_{max} of about 70/80 GHz can be fabricated. The frequencies increase with collector current up to a maximum after which there is a rapid fall-off, attributed to the Kirk effect, which can be also seen for passivated devices [25]. By increasing the collector–emitter voltage V_{ce}, f_{max} can be further increased by utilizing the reduction of the base–collector capacitance.

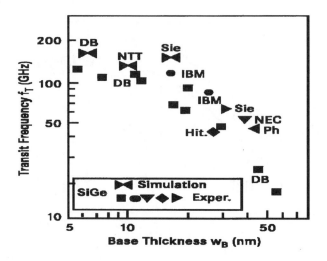

Figure 5.16. Reported cutoff frequencies of SiGe-HBTs by various research laboratories. Simulation results are also shown. After H. Presting and U. Konig, Proc. Current Developments of Microelectronics, 139–150 (1999).

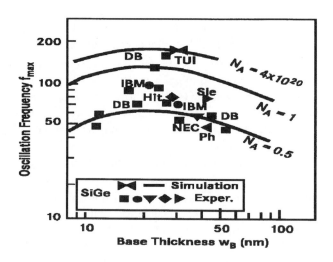

Figure 5.17. Reported maximum oscillation frequencies of SiGe-HBTs by various research laboratories. Simulation results are also shown. After H. Presting and U. Konig, Proc. Current Developments of Microelectronics, 139–150 (1999).

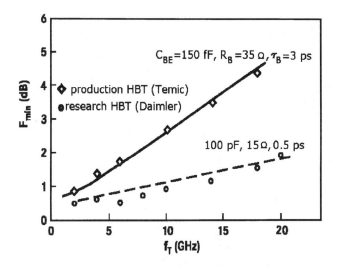

Figure 5.18.   High frequency noise behavior of a mesa on-chip HBT and packaged HBT. Simulated influence of emitter–base capacitance, base resistance and base transit time. After H. Presting and U. Konig, Proc. Current Developments of Microelectronics, 139–150 (1999).

Low frequency noise is another important device parameter of SiGe-HBTs. Figure 5.18 compares the noise figure F_{min} for frequencies above 2 GHz for on-wafer HBTs and packaged devices. The noise figure from a production line HBT is found to be below 1 dB at 10 GHz. This can be seen in figure 5.18 which compares the noise figures of a research mesa-HBT with a planar passivated production HBT. The corner frequency f_c, which marks the transition point from $1/f$ noise to shot noise, is below 300 Hz, even for packaged HBTs [26].

The maximum f_T and f_{max} values were achieved at current densities of 0.65 mA/μm^2 and 0.3 mA/μm^2 for the selectively implanted collector (SIC) and the non-SIC devices, respectively, as shown in figure 5.19. For large signal applications, e.g., power amplifiers (PAs), the linearity of the gain over a voltage swing is of great interest. The voltage dependence of the maximum frequency of oscillation and of the transit frequency for the non-SIC HBT revealed a constant value for V_{ce} voltages above 1.5 V. Hence for power amplifiers high PAE values and good linearity can be expected.

For low noise amplifiers, a low high-frequency noise of SiGe-HBTs in the 1–5 GHz range is a real advantage over homojunction transistors. A typical

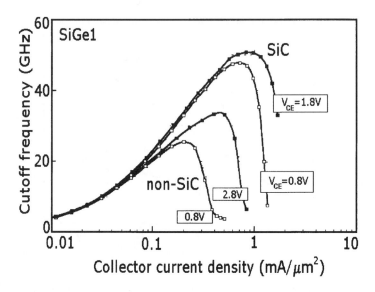

Figure 5.19. f_T values of single emitter devices with and without SIC. After A. Schuppen et al., Proc. Current Developments of Microelectronics, 151–162 (1999).

minimum noise figure of SiGe-HBTs at 2 GHz is about 0.8 dB, owing to the low base sheet resistance and the high current gain of 180. The associated gain is in the region of 17 dB.

5.3.1   SiGe-HBTs at Low Temperature

Currently the vast majority of integrated circuits are fabricated using either CMOS or bipolar (e.g., ECL) technologies. One way of improving CMOS circuit performance, which has been well known for many years, is operation at low temperatures [27]. Advantages include increased device speed and reliability, as well as the virtual elimination of latch-up. Low temperature operation has also been used successfully in the design of high speed digital supercomputers. At low temperatures, the device transconductance increases, metal resistance decreases, interconnect delays are reduced and signal swings become smaller [28, 29]. Built-in drift field in the base is more effective at low temperatures. It compensates for the degradation in the base diffusivity and results in an improvement in the cutoff frequency of the HBT. There is great potential for enhanced performance and improved reliability of computer systems at liquid nitrogen temperature.

Compared to MOS transistors, bipolar transistors have higher transconductance and therefore better current drive capability. For these reasons, BiC-MOS has received considerable attention as an attractive room-temperature VLSI technology. Use of BiCMOS circuits at liquid nitrogen temperature would provide all the advantages of CMOS and low-resistance interconnects, along with bipolar drivers for large capacitive loads. While bipolar technology is available for such applications at room temperature, it was not practical until recently for cryogenic operation because of unacceptably low current gain at low temperatures.

The dominant physical reason for this current gain decrease is bandgap narrowing (ΔE_{ge}) in the heavily doped emitter edge. Since the functional dependence of current gain β is given by

$$\beta \propto \exp\left(-\frac{\Delta E_{\text{ge}}}{kT}\right) \tag{5.7}$$

β is much smaller at lower temperatures. In addition, carrier freezeout in the base can cause the intrinsic base resistance to increase. It has also been suggested that freezeout will degrade the base storage time as well, due to trapping of carriers by impurity atoms. The built-in electric field in the base caused by the impurity gradient will prevent, to a certain degree, the freezeout of carriers due to high field effects. However, since the magnitude of this electric field is to a first order proportional to temperature and is $\approx 10^3$ V/cm at 77 K, freezeout is expected to remain important at reduced temperatures. It is well known that the amount of carrier freezeout depends on the level of compensation. Injected minority carriers in the base can also show freezeout behavior, i.e., be trapped in impurity sites. This effect, which is directly proportional to the compensating dopant concentration, is negligible at room temperature but can be significant at 77 K. Finally, surface state and bulk traps can play a much more important role at 77 K, increasing the nonideal component of the base current–voltage relation substantially.

The outstanding performance advantages of SiGe-HBTs for low temperature operation have been demonstrated in the state-of-the-art Si bipolar process [30]. Several research groups have presented detailed experimental investigations on the low temperature properties of SiGe-HBTs [31–33]. However, the design and optimization issues associated with the low temperature operation of SiGe-HBTs remain unclear.

Because of its bandgap engineered base, the SiGe-HBT is particularly suitable for operation at cryogenic temperatures [29, 31–36]. It has been demonstrated that present SiGe technology is capable of providing transistors with higher current gain at liquid nitrogen temperature than at room temperature and unloaded ECL circuits which are as fast at 77 K as they are at room temperature.

A high performance liquid nitrogen temperature BiCMOS (CRYO-BiCMOS) technology with Si/SiGe heterojunction bipolar transistors has been reported [37]. The newly developed HBT, which has an n^+-poly-Si/n-type Si epitaxial layer emitter structure on a p-type SiGe base layer, shows a high current gain of 50 at liquid nitrogen temperature. Under the conditions of 3.3 V and 83 K, the driving capability of CRYO-BiCMOS gates is twice as large as that of the CRYO-CMOS gate. At 3.3 V and a load capacitance of 1 pF, the gate delay of the CRYO-BiCMOS gate with pull-up HBT is 480 ps. The CRYO-BiCMOS with Si/SiGe-HBTs is very promising for the future progress of BiCMOS LSIs.

Problems encountered for low temperature operation of Si-BJTs can be solved effectively by using heterojunction technology. In an HBT, the bandgap of the emitter is larger, and therefore the current gain increases at low temperatures. Since doping in the base is large, carriers do not freeze at low temperatures. The emitter doping can also be kept sufficiently high to avoid freezing of the carriers. Patton *et al.* [38] have studied the low temperature operation of a SiGe-HBT fabricated in a poly-emitter bipolar process. The devices showed improved low temperature behavior with an extremely high current gain of 1600 at 77 K.

Crabbe *et al.* [29] investigated the low temperature behavior of Si-BJTs and SiGe-HBTs fabricated and optimized for room temperature operation. The authors demonstrated that the degradation in both current gain and base transit time can be eliminated in a SiGe base with a graded Ge profile. The effect of introducing an spacer layer in the emitter–base junction was studied in detail. The spacer layer reduced the low-level parasitic emitter–base tunneling (leakage current) at low temperature but gave rise to carrier freezeout and increase of base resistance at liquid nitrogen temperature. The current gains were 20 to 40 for the Si-BJT and 100 to 140 for the HBT for the temperature range 77–300 K. The graded Ge profile in the base improved both the low temperature current gain and base transit time resulting in a peak cutoff frequency of 94 GHz at 85 K compared to 75 GHz at 298 K.

A much improved low temperature SiGe-HBT was reported by Cressler *et al.* [39]. The devices were specifically designed for low temperature operation and were fabricated using self-aligned epitaxial base technology [40]. Lightly doped spacers were used at both the junctions to reduce the electric field. The base width was approximately 59 nm and the peak concentration of the graded

Ge profile was 9%. ECL circuits built with this transistor performed well at 84 K. For a high power design (about 10 mW), the gate delay at 84 K was 28.1 ps, roughly the same as at 310 K. This value is a factor 2 better than the best value obtained with low temperature Si-BJTs. However, the emitter area of the transistor used in this circuit was 0.4 μm $\times$ 4.3 μm. The circuit was operated under full logic swing of 500 mV at an average switch current density of 0.88 mA/μm^2 at both temperatures. Low power ECL circuits showed switching speeds as fast as 51 ps at 2.2 mW (112 fJ power–delay product) at 84 K. The measured gate delays are in reasonable agreement with the theoretical predictions [41]. These results represent unprecedented performance levels for silicon-based bipolar technology at 77 K.

Cressler *et al.* [31, 33, 35] have reported performances of SiGe base transistors with further improvements in the design and technology optimized for low temperature operation. Reduced temperature processing with a novel *in situ* doped polysilicon contact, a lightly doped epitaxial cap layer and a graded SiGe base was employed. A low thermal budget allowed a sharp transition from low doped emitter to heavy doped base, making the base immune to carrier freezeout at liquid nitrogen temperature. At 84 K, transistors showed a current gain of 500, a cutoff frequency of 61 GHz and a gate delay of 21.9 ps, 3.5 ps faster than at room temperature. Typical parameters and performance of the transistors (on wafer) at 310 K and 84 K for the epitaxial emitter cap SiGe-HBT design and an i-p-i SiGe-HBT design with emitter areas $A_{\rm E} \sim 0.72$ μm $\times$ 4.42 μm and $A_{\rm E} \sim 0.6$ μm $\times$ 4.3 μm, respectively, are given in table 5.6.

The effect of introducing lightly doped spacer layers at both the emitter–base and base–collector junctions was studied in detail [33]. The spacer layer reduced the low-level parasitic base leakage but gave rise to carrier freezeout and increase of base resistance at liquid nitrogen temperature. However, it was shown that thin abrupt base profiles attainable with epitaxial processing are particularly useful for low temperature operation since they yield profiles less sensitive to base freezeout than ion-implanted profiles. The authors have shown that properly designed homojunction transistors may have sufficient current gain and switching speed at liquid nitrogen temperature for many digital applications. In several applications, however, the flexibility offered by using SiGe for the base layer yields great benefits.

Gruhle *et al.* [34] have reported a high performance SiGe-HBT (fabricated using MBE) having a base doping of 2×10^{19} cm^{-3} largely exceeding the emitter impurity level and a base sheet resistance of about 1 kΩ/sq. The device exhibited an Early voltage of 500 V and a maximum room temperature current gain of 550, rising to 13,000 at 77 K. Devices built on buried-layer

Table 5.6. Typical SiGe-HBT transistor parameters at 310 K and 84 K at the wafer level. After J. D. Cressler et al., IEEE Electron Dev. Lett., 15, 472–474 (1994).

SiGe Profile	Emitter Cap Design		i-p-i Design	
Temperature	310 K	84 K	310 K	84 K
β_{max}	102	498	105	82
β at 1.0 mA	94	99	96	34
Peak g_m (mS)	62	113	74	83
R_{bi} (kΩ/sq)	7.7	11.0	8.2	15.9
R_e (Ω)	14.3	11.0	82	15.9
I_{be} (nA)	8.44×10^4	1.91×10^3	2.89	1.11
BV_{ceo} (V)	3.1	2.1	3.2	3.2
BV_{cbo} (V)	10.8	9.6	10.8	9.5
C_{be} (fF/μm^2)	5.47	5.13	6.30	5.90
C_{cbx} (fF/μm^2)	0.46	0.40	1.04	0.93
Peak f_T (GHz)	43	61	53	59
Peak f_{max} (GHz)	40	50	37	48
Min. ECL delay (ps)	25.4	21.9	26.0	30.4

substrates exhibited an f_{max} of 40 GHz and an f_T of 42 GHz.

Sturm *et al.* [42] also fabricated high quality SiGe-HBTs using rapid thermal chemical vapor deposition. Both graded-base and uniform Ge profiles in the base were considered. In a transistor with 20% uniform Ge concentration in the base, current gains of about 2000 at room temperature and 11,000 at 133 K were observed. The performance of SiGe-HBTs at liquid He temperature has been reported [32]. The current gain of a self-aligned, UHVCVD grown SiGe-HBT showed an increase in current gain from 110 at 300 K to 1045 at 5.85 K, although parasitic base current leakage limits the useful operating current to about 1.0 μA at 5.84 K. A very high base doping (peak at 8×10^{18} cm^{-3}) was used to suppress the base freezeout at 4.48 K and resulted in an R_{bi} of 18.3 kΩ/sq.

5.3.2   Parasitic Energy Barriers

Two different approaches used in designing and fabricating SiGe-HBTs have been discussed in section 5.2. When a graded Ge profile is used in the base, the grading provides an aiding drift-field for the minority carriers and considerably reduces the base transit time [43, 44]. However, in these transistors the Ge concentration at the emitter end of the base is zero, and to maintain high injection efficiency, base doping has to be kept low, which makes the base resistance high. In the other approach [34, 45], the Ge concentration in the base is constant and high. In this design, the doping concentration in the base can be made very high, making the base resistance low. Noise characteristics of these transistors are expected to be superior.

At high current densities or high forward bias, the transport of carriers is strongly influenced by the potential barrier that develops due to the grading potential of the heterojunction. A retrograde Ge profile near the collector junction also creates a barrier to the flow of minority carriers [46]. Valence band offsets in SiGe are much larger than the conduction band offsets. However, the conduction band offset in an npn transistor gives rise to a conduction band spike that can reduce the enhancement of the current gain and the effect of the spike will be much larger in SiGe pnp transistors.

Harame *et al.* [46] studied experimentally and theoretically the barrier to the flow of minority carriers in a SiGe base pnp transistor with a retrograde Ge profile in the base. The knee current was found to increase with applied base–collector bias more than can be explained by the Kirk effect. Similarly the f_T vs. collector current density also showed a strong dependence on the base–collector bias. In the retrograde portion of the Ge profile, the Ge concentration changes from 18% to 0% over a distance of 250 Å. This grading causes an electric field at the base–collector junction that opposes the flow of holes. If the grading is linear, the field is constant. Initially this barrier is in the depletion region but close to the neutral base. As the emitter–base voltage increases and at high injection when base widening sets in, the net electric field decreases and the barrier due to the retrograde Ge profile becomes important.

The parasitic barriers in npn transistors have been studied extensively. Prinz *et al.* [47, 48] pointed out that it is important that the boron profile be confined to within the SiGe strained layer in an npn transistor. If the boron profile tails extend into the emitter or the collector Si layers, barriers to the flow of electrons, designated as parasitic barriers, are formed. Parasitic barriers at both the emitter–base junction and the base–collector junction have been studied in detail [49, 50]. The effect of parasitic barriers at the base–collector junction on the behavior of the npn transistor is discussed below.

The SIMS doping profile of the transistor investigated is shown in the upper part of figure 5.20. The Ge fraction x is 0.20. Note that the box-like Ge profile does not extend up to the base–collector pn junction and ends within the p-type base. The calculated conduction band edge is shown for different values of base–collector reverse bias in the lower part of figure 5.20. It is seen that at zero collector bias, a strong parasitic barrier exists at the base–collector

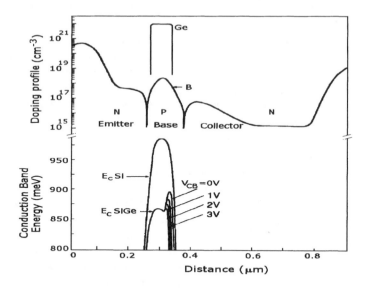

Figure 5.20. SIMS doping profile and conduction band energy in the base of a SiGe npn transistor. After J. W. Slotboom et al., IEEE Electron Dev. Lett., 12, 486–488 (1991), copyright ©1991 IEEE.

junction. The height of the barrier decreases as the collector bias increases and becomes relatively negligible at $V_{bc} = 3$ V. The parasitic barriers suppress the collector current, increase considerably the concentration of minority carriers in the base and also increase the base transit time τ_b.

The effect of parasitic barriers at both the emitter–base junction and the base–collector junction in the npn RTCVD transistors has been studied [47, 48]. The Ge fraction x in the base was 0.20 and the boron doping in the 300 Å thick base layer was either 1×10^{19} cm^{-3} or 1×10^{20} cm^{-3}. Although boron doping was box-like and was confined to the base layer, some mixing and outdiffusion of boron occurred that moved the junctions to the neighboring emitter and collector Si layers. The n-type doping in the emitter was 1×10^{17} cm^{-3} and that in the collector was 5×10^{17} cm^{-3}. In some devices, 80 Å thick undoped SiGe spacers were added to eliminate the effect of outdiffusion and mixing of boron into adjacent Si emitter and collector layers.

For SiGe-HBTs fabricated using MBE with a very narrow (214 Å) base width, doped with a boron concentration of 5×10^{19} cm^{-3} and a Ge concentration of 15% (which gives a valence band offset of 65 meV), a collector current enhancement of 13 was reported by Shafi *et al.* [51]. However, the base current was also found to increase sixfold. The base current was a strong function of the reverse bias at the base–collector junction and decreased rapidly as the bias increased. The authors attributed this large increase to a very low lifetime (3×10^{-13}s) of minority carriers near the collector region in the base. These results can also be explained by assuming parasitic barriers at the base–collector junction. The minority carrier concentration in the base will increase due to the barriers and this will increase the recombination and base current, irrespective of the lifetime of the minority carriers. Since the barriers decrease in height as the reverse bias increases, this would also explain the observed decrease of the base current at higher bias voltages.

5.4   Effect of Carbon Incorporation

The addition of a small amount of carbon to silicon–germanium thin films leads to a new class of semiconducting materials [52]. This new material can alleviate some of the constraints on strained $Si_{1-x}Ge_x$ and may help to open up new fields of device applications for heteroepitaxial Si-based systems. The incorporation of carbon can be used (i) to improve thermal stability of the SiGe layer, (ii) to obtain layers with new properties, and (iii) to control dopant diffusion in microelectronic devices. The phenomenon of suppressed boron diffusion in carbon-rich epitaxial layers can be used to increase the performance of SiGe-HBTs.

In SiGe-HBTs the outdiffusion of the base dopant (B) out of the SiGe region into the emitter or collector strongly degrades device performance. The incorporation of a small amount of carbon within the SiGe base region of a HBT can significantly suppress boron outdiffusion caused by a variety of subsequent processing steps. The reduction of boron outdiffusion by carbon doping as a function of carbon concentration and position within the base has been studied by Gruhle *et al.* [53]. It was found that carbon is only effective when it is placed within the doped base region. C incorporation also provides greater flexibility in process design and wider latitude in process margin. At a carbon level of 2×10^{19} cm^{-3} the allowable anneal time may be increased by a factor of 3.

5.4.1   Performance of SiGeC-HBTs

Osten *et al.* [54] demonstrated almost ideal base current characteristics. Cutoff and maximum oscillation frequencies of more than 70 GHz, and delays per stage down to 15 ps for ring oscillators with integrated SiGeC-HBTs

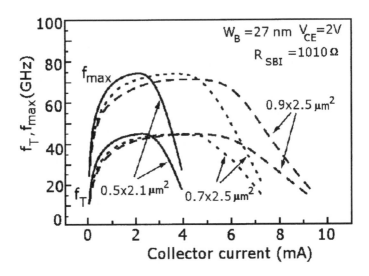

Figure 5.21.   Cutoff frequencies and maximum oscillation frequencies as functions of the collector current for SiGeC HBTs. After H. J. Osten et al., IEEE Trans. Electron Dev., **46**, *1910–1912 (1999), copyright ©1999 IEEE.*

have been achieved.   For a similarly processed SiGe-HBT, a lateral range of transient enhanced diffusion of more than 4000 Å on each side of the emitter has been observed [55]. Figure 5.21 shows the cutoff frequency and the maximum oscillation frequency of SiGeC-HBTs with different emitter widths. Even for emitter widths as low as 0.5 μm, no degradation in high frequency performance was observed. The overall reduction in f_T compared to the results shown earlier is caused by a decrease of base resistance without optimizing the transistor design, especially the boron position within the Ge profile.

Figures 5.22a and 5.22b show Gummel plots and common emitter characteristics of transistors following As implantation and annealing at 647 °C for SiGe- and SiGeC-HBTs. The decrease in collector current and reduced Early voltage in the transistors without carbon show that boron has outdiffused even though the thermal budget of the annealing is low. Devices which were not subjected to the As implant but underwent the same thermal annealing cycle did not show any evidence of boron outdiffusion. This confirms that TED effects are responsible for the boron movement. However, the high Early voltages for the transistors with carbon illustrate that carbon in SiGe has suppressed the TED effects of an As emitter implant.

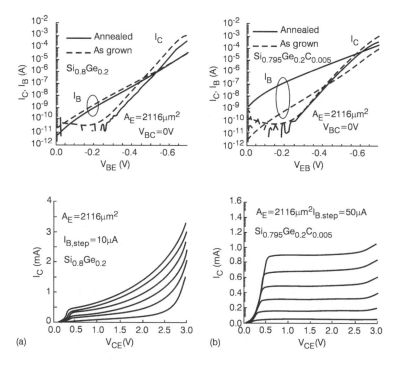

*Figure 5.22.   Room temperature Gummel plots and common emitter characteristics of SiGe- and SiGeC-HBTs following As emitter implantation and anneal at 647 °C. After L. D. Lanzerotti et al., Appl. Phys. Lett., **70**, 3125–3127 (1997).*

5.5   Power SiGe-HBTs

To meet the demands of large signal applications such as power amplifiers, a high breakdown voltage SiGe-HBT having an open-base common–emitter breakdown of 5.5 V and a common–base breakdown of 15 V (at the cost of reduction of peak f_T) has been fabricated [56]. In order to demonstrate the performance of multi-finger power devices, HBTs with different numbers of emitter fingers were investigated. The effective size of each finger is 1.6 μm × 30 μm. On each side of the emitter is a base contact in order to achieve a low base resistance, and a collector finger contact is placed between a group of base and emitter contacts. To achieve homogeneous current distribution for all emitters, a 4 Ω ballast resistance is inserted in series to each emitter. The Gummel plot and the output characteristics of the multi-emitter finger HBTs show good dc performance. The ideality factors of the base and the collector currents are close to 1 and the current gain is nearly constant over a wide current range. Due to the very high base doping the Early voltage is above 50 V. The current gain is in the region of 150 and the breakdown voltage BV_{ceo} is 6 V, which is well suited for systems with 3.6 V battery power supply. The output characteristics of SiGe-HBTs show no offset voltage, which is a big advantage over III–V single heterostructure HBTs, because the offset voltage reduces the efficiency of the HBT in large signal operation.

A SiGe-HBT with 10 emitter stripes with output matching for the fundamental frequency 1.9 GHz and third harmonic achieved a value of 62% PAE. The results of the 10, 20 and 40 emitter-stripe devices show that the power scales with the number of emitter stripes. With these devices 21 dBm, 24 dBm, and 27 dBm have been achieved in class AB operation at a collector–emitter voltage of 3.6 V. The PAE slightly decreases with the number of emitters from 62% at 21 dBm, to 50% at 32 dBm because the gain also decreases with the number of emitters. Gain reduction in multi-emitter devices is a consequence of the long interconnect line of the devices. The electromagnetic coupling between these lines is responsible for gain and cutoff frequency reduction. In order to demonstrate the power capability of these HBTs, an 80-emitter stripe HBT was tested at 0.9 GHz showing 32 dBm output power and a corresponding PAE of 50%. Measurements on different geometries, finger lengths, and contact arrangements indicate that a further lateral layout improvement is still possible. However, the new results showed that it is possible to achieve PAE values as high as is achieved by GaAs HBTs at 900 MHz and 1.9 GHz. Measured PAEs as high as 72% at 900 MHz and 64% at 1.9 GHz are depicted in figures 5.23 and 5.24 [57, 58].

SiGe power-HBTs intended for 1.9 GHz applications using 60 emitter fingers exhibited a collector–emitter breakdown voltage of 4.5 V with f_T and f_{max} values of 16 and 11 GHz, respectively, at a collector current of 400 mA. Class A PAE of 44% at 1 W output power and a PAE of 72% at 900 MHz for class AB operation have been measured. These data are achieved without any thermal shunt precautions in the contacts and without substrate thinning. This reflects the benefits of a Si substrate.

5.5.1   High Voltage SiGe Process

Traditionally, SiGe devices were perceived as an inherently low voltage technology since the process was developed for RFIC and digital applications.

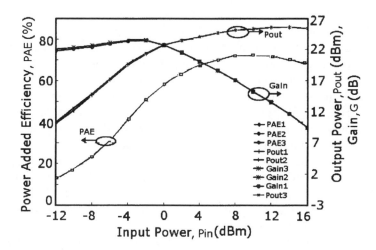

*Figure 5.23. Large signal results of 20 emitter HBT at 900 MHz revealing 72%
PAE. After Temic Semiconductors, Germany.*

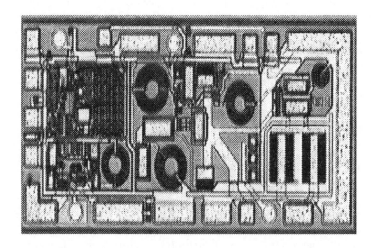

*Figure 5.24. DECT power amplifier with LNA. After Temic Semiconductors,
Germany.*

However, SiGe-HBTs can achieve the same breakdown voltages as Si-BJTs ($V_{cbo} > 80$ V and $V_{ceo} > 26$ V) for rf applications. In addition, performance improvements from SiGe in linearity, efficiency and gain for high voltage wireless power applications have been demonstrated. For Si power transistor applications such as base stations, the typical bipolar process starts with a heavily doped n^+ substrate (collector contact), proceeds with n-collector epitaxy (collector) and concludes with a junction-isolated, traditional implanted base and emitter process. The doping and thickness of the collector (along with the radius of the junction isolation) control the breakdown voltage of the device. A SiGe power process has been developed that replaces the base implant step with SiGe epitaxial growth in a poly-Si emitter. Apart from these minor changes, the process is essentially the same as an implanted bipolar process. A 15 V SiGe power process has been demonstrated using this process flow with $V_{cbo} > 50$ V and $V_{ceo} > 15$ V.

Although researchers from Daimler-Benz and Temic started a feasibility study for SiGe-HBTs as early as 1993, the real transfer into production started through the European SIGMA project only in early 1998 [59]. Temic offers a production technology, called SiGe1, including npn HBTs with and without a selectively implanted collector on the same wafer. The main important issue of the technology is the differential growth of the SiGe layer after a standard recessed LOCOS process. The poly-SiGe is used for the base contact and for two of the three resistor types. The emitter has an inside and an outside spacer and an additional α-Si layer in order to produce the emitter and the collector contact. In addition, spiral inductors, nitride capacitors, three types of poly resistors, a lateral pnp, rf- and dc-ESD protection, and varactor diodes are incorporated in the present technology.

The SiGe1 IC process was developed for RFICs with $V_{cc} = 1.5$–5 V application voltages for an SiGe1 HBT with $BV_{ceo} = 6$ V ($f_T/f_{max} = 30/50$ GHz) and 3 V ($f_T/f_{max} = 50/50$ GHz). Some important device features available from the SiGe1 technology are summarized in table 5.7.

5.6   Si/SiGe-MMICs

In the past, rf systems have been successfully implemented on printed circuit boards combining discrete transistors (MESFETs, HEMTs, HBTs on compound semiconductor substrates), lumped-element (spiral inductors, interdigitated capacitors) or distributed components (microstrips) for impedance matching, dielectric resonator filters, quartz crystal oscillators for frequency synthesis, and other elements. However, emerging consumer and commercial applications require other system aspects for cellular systems, wide

Table 5.7.   Summary of the device parameters available from the SiGe1 technology. After Temic Semiconductors, Germany.

Parameter	non-SIC	SIC
npn transistors		
Transit frequency, f_T (GHz)	30	50
Max. frequency of oscillation f_{max} (GHz)	50	50
Current gain, h_{FE}	180	180
Early voltage, V_A (V)	50	50
Collector emitter breakdown voltage, BV_{ceo} (V)	6.0	3.0
Collector base breakdown voltage, BV_{cbo} (V)	15	12
Noise figure at 2 GHz F_{min} (dB)	1.0	1.0
Lateral pnp		
Collector emitter breakdown voltage, BV_{ceo} (V)	7	
Current gain, h_{FE}	7	
Typical collector current (μA)	40	
Passive Components		
High ohmic poly resistor (poly1) (Ω/sq)	400	
Medium ohmic poly resistor (poly2) (Ω/sq)	110	
Low ohmic poly resistor (poly1-TiSi$_2$) (Ω/sq)	4.5	
Precision MIM capacitor (fF/μm^2)	1.1	
Spiral inductor 4 nH, Q-value at 2 GHz Q	7	
ESD Zener diode, Zener voltage V_Z (V)	6.2	
Zener diode, parasitic capacitance (pF)	5.5	
RF-ESD diode, parasitic capacitance (pF)	0.3	

area networks (WAN), and local area networks (LAN), such as low cost, small size, low power consumption and batch fabrication. These requirements make Si the technology of choice and foster a fully monolithic, single-chip solution, including all rf building blocks (except the antenna) and the intermediate frequency (IF) circuitry [60].

Experience of printed board rf design cannot be directly transferred to Si. The Si substrate is prone to rf losses in contrast to GaAs and printed circuit boards. Different types of active and passive devices have been realized in Si processing technology as shown in figure 5.25. Typical active devices include BJTs, MOS, and diodes and passive devices include resistors, capacitors, inductors and interconnects. The resistivity of Czochralski-grown Si is only about 10 Ω cm and at frequencies beyond 1 GHz the skin depth for 10 Ω cm Si exceeds the typical substrate thickness so that rf losses extend over the entire substrate. Considerable losses are also expected in the interconnects which are typically Al wires combined with W vias, both of limited thickness. Impedance matching by using microstrips at frequencies of 900 MHz to 2.4 GHz is not feasible with Si chip dimensions. These difficulties are the main reasons why the conventional rf design concepts cannot directly be transferred to Si. Some novel rf component structures have to be developed for incorporation in Si technologies to accomplish the desired rf functions, but the techniques of implementing them may be quite different from the established ways.

As a result of a high level of integration, crosstalk and thermal cross-coupling will be increasingly important. These concerns exist not only because of the close device proximity on a single chip, but also because parasitic electromagnetic signal propagation through the substrate becomes an additional issue. RF isolation schemes have to be developed for device integration and it is important to recognize that an oxide or junction isolation structure which works satisfactory at dc signal levels may fail in rf operation.

High resistivity substrates associated with a low substrate doping, however, limit the integration density significantly if junction isolation is used. An interesting innovative type of substrate, which combines full dielectric isolation through a silicon-on-insulator structure with an FZ-Si substrate, has been investigated [61]. Similarly, silicon-on-sapphire (SOS) substrates can be used for even lower rf losses. Another novel approach utilizes bulk micromachining to remove Si material under devices which are particularly prone to electromagnetic substrate losses, such as spiral inductors [62] or transmission lines [63]. Porous Si formation provides another way to reduce the substrate losses [64, 65].

5.6.1 Passive Components

Spiral inductors are performance limiting components for monolithic rf circuits, such as VCOs, low noise amplifiers, and passive filters [66, 67]. The quality factor (Q) of the inductors is limited by resistive losses in the spiral coil and by substrate losses. High Qs can be achieved in state-of-the-art Si fabrication processes [68–70].

Figure 5.26 shows a cross-sectional view along the center line of the spiral inductor structure in an IC process. Besides the conventional planar coil structure, it is possible to build stacked coils in multi-level interconnects. This can lead either to a solenoidal inductor structure, which provides a larger

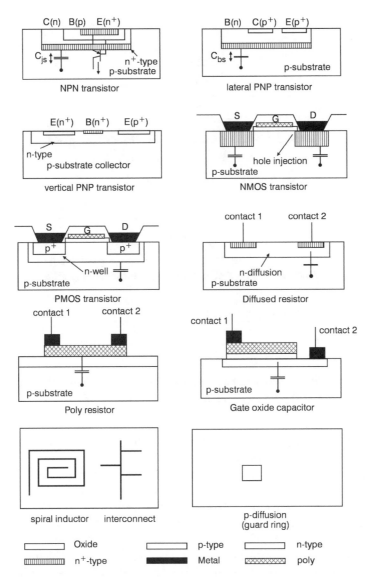

Figure 5.25. Different types of active and passive devices realized in Si processing technology. Typical active devices include BJTs, MOS and diodes and passive devices include resistors, capacitors, inductors and interconnects. After R. Gharpurey, Modeling and Analysis of Substrate Coupling in Integrated Circuits, *PhD Thesis, Univ. of Calif., Berkeley, 1995.*

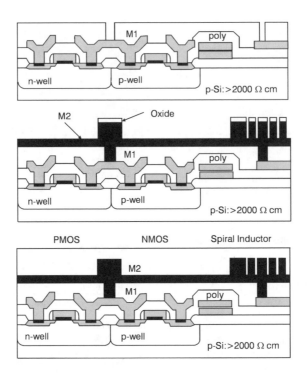

Figure 5.26. Thick metal inductor fabrication process steps. After M. Park et al., IEEE IEDM Tech. Dig., 59–62 (1997), copyright ©1997 IEEE.

inductance for a given area [71], or to transformer structures which are useful in rf circuits for impedance matching [68].

The ohmic losses in the spiral coil can be reduced by replacing aluminum with a less resistive metal, such as gold or copper, and by increasing the thickness of the interconnect layers. Besides generating sufficient spacing of the inductor coil from the substrate, high resistivity Si [69, 72], suspension of the inductor structure [62], sapphire substrates [69], or porous Si [64] can be applied to reduce the substrate losses. The highest Q for inductors fabricated in a manufacturing process to date is 24 [68]. An inductor with the same lateral dimensions but built in a Cu-Damascene process over a sapphire substrate has a Q of 40, which is presently the overall highest Q demonstrated in Si technology [69].

Figure 5.27 shows a summary of various inductor results achieved [68]. It

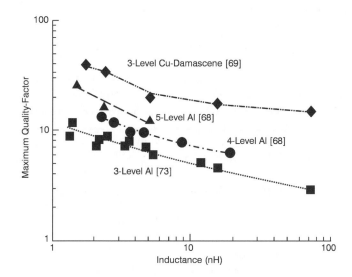

Figure 5.27. Quality factor vs. inductance for inductors fabricated in different IBM Si technologies. After J. N. Burghartz, Proc. ESSDERC, 143–153 (1997).

may be noted that useful Qs have already been achieved with three Al levels, with clearly increasing Qs as metal layers were added. The combination of Cu interconnects and sapphire substrates has resulted in a $4\times$ improvement over an Al structure on bulk Si with the same number of levels.

It is also obvious that large inductances have lower Qs than small inductances. This has been explained as the result of the capacitance between coil and substrate, which does not change in proportion to the number of turns for a given area [68, 73]. Fortunately, high Qs in combination with small inductances are required at high frequencies, whereas smaller Qs are sufficient at low frequencies at which large inductances are needed. Since the maximum Q also appears at higher frequencies for smaller inductances, one can conclude that the spiral inductor on Si is scalable to first order with the operating frequency. Other passive rf components, such as metal–insulator–metal (MIM) capacitors, varactors, and transformers, have been investigated over the past few years as well [68, 73].

In addition to high performance active devices, analog and mixed-signal design depends on having easy access to a wide variety of passive devices. Present SiGe technology fills this need with a comprehensive library of resistors, capacitors, spiral inductors, and diodes such as SBD, PIN and varactor. Table 5.8 shows some of the typical component values achieved in SiGe technology.

Table 5.8. List of devices available in SiGe technology. The main characteristics are provided for each device. These devices are available to the designers to make a full custom design. Source: IBM website.

Device type	Value
Polysilicon resistors	$R = 340 \ \Omega/\text{sq}$
	$R = 220 \ \Omega/\text{sq}$
Implant resistors	
	$1.7 \ \text{k}\Omega/\text{sq}$
	$23 \ \Omega/\text{sq}$
	$8 \ \Omega/\text{sq}$
MIS capacitor	$C = 1.5 \ \text{fF}/\mu\text{m}^2$
MIM capacitor	$C = 0.7 \ \text{fF}/\mu\text{m}^2$
Inductor (2-turn)	$L = 1.5 \ \text{nH}$
	$Q = 10$ at 12 GHz

5.6.2 SiGe RFICs

Present rf circuits operate at frequencies beyond 1 GHz and bipolar appears to be the technology of choice for rf applications. Although the bipolar transistors have dominated the high speed applications so far, with manufacturable devices with 0.1 μm or less feature sizes, CMOS is appearing as a serious candidate for rf applications. If CMOS, the mass production technology today, meets the performance requirements for rf applications, it will be the technology of choice. A 900 MHz transceiver design has been demonstrated and rf CMOS systems operating at 2 GHz have been projected [74]. The implementation of a complete rf transceiver system on a single Si chip is motivated by a rapidly emerging wireless communication market which extends to consumer as well as commercial applications.

While for digital applications the design of devices is primarily focused on large-signal switching speed and a small device size, rf design requires sufficient small-signal (ac) amplification at very high frequency, a low noise figure, a high

PAE and a high breakdown voltage for power amplifiers. Good linearity and other quality aspects of frequency domain device operation are also required.

5.6.3   Bipolar vs. CMOS

Several authors have attempted to assess bipolar (BJT and SiGe-HBT) and CMOS device performance for rf applications [8, 19]. Figures 5.28 and 5.29 show for HBT and CMOS devices, respectively, the variation of f_T and f_{max} for reported devices and device simulations vs. BV_{ceo} and V_{DS}. It is seen that the trends of f_T and f_{max} are quite different for the two types of device. The frequencies vary for the BJT as a function of collector and base dopings and widths with opposite trends (figure 5.28), i.e., very high values can be achieved for f_T [11] or f_{max} at the expense of the other figure. A balanced SiGe transistor design provides f_T and f_{max} values of about 60–80 GHz at a BV_{ceo} of about 3 V [26, 75–79]. For the n-MOSFETs in figure 5.29, both frequency limits increase as the device is scaled down. A channel length of about 0.15 μm is required to match the f_T and f_{max} of the SiGe bipolar transistors, but at a supply voltage of only 1.5 V. Although low supply bias is of advantage for low power operation, it may cause difficulty in power amplifier design. Very high f_T and f_{max} numbers at low bias have been demonstrated for SiGe-MODFETs, indicating one way to improve MOS technology for rf application through process innovation [80].

RF performance of a MOS transistor may be improved through process modification to reduce the gate resistance, as indicated in figure 5.29. f_{max} can be increased significantly, or the same f_{max} may be achieved for a longer channel and higher supply voltage, if the FET is fabricated with a T-gate structure, which can reduce the gate resistance; also multi-finger FETs are used [81, 82].

5.6.4   Noise

Besides the device speed, the noise which is generated by a transistor relative to the associated gain is an important figure of merit. In figure 5.30 the gain is plotted against the minimum noise figure at 2 GHz for n-MOSFETs with different channel lengths, for SiGe-HBTs, and for a self-aligned BJT. It is obvious that both the noise figure and the gain improve towards smaller channel length. Best results were published for n-MOSFETs with channel lengths below 0.2 μm. If one compares at the same lithographic rule, the SiGe-HBT is superior to the MOS device, as well as to the BJT.

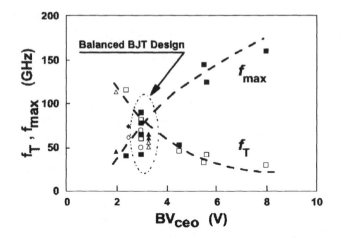

Figure 5.28.   Cutoff frequency (open symbols) and maximum oscillation frequency (solid symbols) vs. collector–emitter breakdown voltage (BV_ceo) for published Si and SiGe bipolar transistor data. After J. N. Burghartz, Proc. ESSDERC, 143–153 (1997).

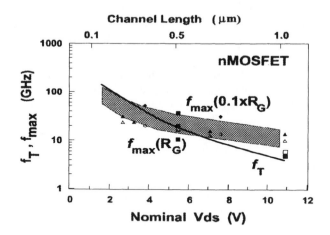

Figure 5.29.   Cutoff frequency (open symbols) and maximum oscillation frequency (solid symbols) vs. drain–source voltage (V_DS) and channel length for published and simulated MOSFETs. After J. N. Burghartz, Proc. ESSDERC, 143–153 (1997).

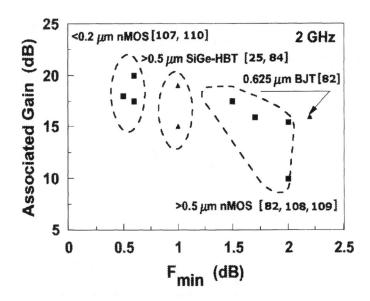

Figure 5.30. Associated gain vs. minimum noise figure of various published MOSFETs, a BJT, and SiGe-HBTs. After J. N. Burghartz, Proc. ESSDERC, 143–153 (1997).

5.6.5 Technology Choice

The ideal rf fabrication process would provide both CMOS and bipolar transistors on the same chip. A comparison of state-of-the-art BiCMOS technologies has shown that several companies offer today fabrication processes which provide transistors that are comparable in performance to those fabricated using CMOS or bipolar stand-alone processes [83]. It is now believed that the BiCMOS or SiGe-BiCMOS [84] will be the preferred technology for future single-chip high performance personal communication systems (PCS), while CMOS may lead in low performance systems [85].

Table 5.9 summarizes the performance of competing bipolar technologies for RFIC applications. Silicon bipolar technology compares extremely well with GaAs in terms of performance with the additional advantages of low cost and high volume production. Recently, f_T and f_{max} in excess of 10 GHz, less than 2 dB noise figure at 2 GHz and excellent linearity have been demonstrated in CMOS technology. Thus low-voltage submicron CMOS technologies are becoming extremely competitive for rf applications [86]. However, it should

Table 5.9. Performance comparison of competing technologies in the frequency range 1–10 GHz for RFIC applications. After C. Kermarrec et al., IEEE RFIC Symp. Dig., 65–68 (1997).

Parameter	SiGe HBT	Si BJT	GaAs HBT	GaAs MESFET	Si-BJT BiCMOS
Minimum size (μm)	0.5×1.0	0.5×1.0	2×5	0.5×5	1.2×1.5
BV_{ceo}/BV_{DS} (V)	4	4	15	8	6
f_T (GHz)	50	32	50	30	13
f_{max} (GHz)	55	35	70	60	11
G_{max} (dB) at 2 GHz	28	24	19	20	17
G_{max} (dB) at 10 GHz	16	11	13	13	1
F_{min} (dB) at 2 GHz	0.5	–	1.5	0.3	–
F_{min} (dB) at 10 GHz	0.9	–	–	0.9	–
IP_3/P_{1dB} (dB)	9	9	16	12	9
PAE (%) at 3 V	70	–	60 at 5 V	70	40
$1/f$ frequency (kHz)	0.1–1.0	0.1–1	1–10	10000	0.1–1

be pointed out that, for specific applications, one device type may dominate more clearly over the other. For instance, rf power switches are preferably built by using MOS devices due to the low standby power level. On the other hand, bipolar transistors are desirable in power amplifier design as they offer high speed at high voltage, while the MOSFETs have to be scaled down in order to provide adequate ac performance. Higher PAE values can therefore be expected for BJTs and SiGe-HBTs.

5.7   Applications of SiGe-HBTs

The major application of the SiGe technology has been in wireless communication systems in the 1–6 GHz range. Mixers, GSM power modules, dual band front-ends for GSM and DCS 1800 and DECT front-ends are the focus of attention. Additionally, LNA circuits with a noise figure of 1.6 dB and an associated gain of up to 26 dB at 5.8 GHz have been reported. The relatively small bandwidth of the 5.8 GHz LNA is an advantage for mobile communication systems, reducing the expense for the input filter. The first commercial product is a DECT front-end including an LNA with 1.6 dB noise figure and 20 dB gain combined with a 27 dBm power amplifier with 41% PAE over the whole packaged device as shown in figure 5.31. A microphotograph of a SiGe-based 5.2 GHz transceiver is shown in figure 5.32. The chip size is 2.07 mm × 2.77 mm. Table 5.10 presents a wide variety of circuits that have been demonstrated in SiGe technology showing its versatility.

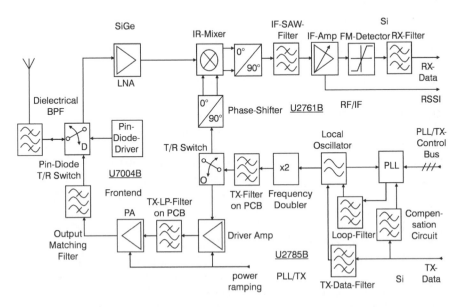

Figure 5.31. RF communication block diagram. After Temic Semiconductors, Germany.

5.8   Summary

The work described in this chapter has shown that SiGe-HBT is perhaps the simplest heterostructure transistor compatible with Si bipolar IC technology and yet offers remarkable advantages in speed and high frequency performance. The emergence of SiGe-BiCMOS combining the best in bipolar with mainstream Si IC as a trustworthy technology has been an important step for the next generation ULSI. The credentials of SiGe-HBT as a building block in rf and MMIC applications has been fully proved. Detailed studies must however be conducted on actual circuits in rf wireless applications with special reference to the problems encountered in scaling, and the power consumed and long-term reliability must be evaluated. At this time, many manufacturers continue to use a mix of GaAs and Si technology for transceivers. Work

Table 5.10. Summary of circuits using SiGe-HBT technology.

Ref.	Circuit Type	Results
[87]	Dynamic/static freq. div.	82 GHz/60 GHz, 396/689 mW
[88]	Dynamic freq. div.	79 GHz, 143 mA, 7.5 V
[89]	Static freq. div.	53 GHz, 122 mA, 6.3 V
[90]	Preamplifier	45 GHz BW, 32 dB gain
[91]	Receiver IC	10 Gb/s SONET 4.5 W, –5 V
[92]	CDR ICs	9.95 Gb/s and 12.5 Gb/s, 320 mW, 3.3 V
[93]	Transceiver chip	
[94]	Limiting amplifier	60 dB gain
		55 dB dynamic range 10 Gb/s
[95]	Optical receiver	40 Gb/s analog IC
[96]	Mixer	Conversion loss 6.5 dB,
		LO power 10 dBm
		$1/f$ noise corner freq. 3 kHz, 1 mA
[97]	2-stage K-band lumped-element amplifier	
[98]	Radio transceiver	900/1900 MHz, 2.7 V
[99]	6.25 GHz LNA	Min. NF 2.2 dB, gain 20.4 dB
		Dissipation 9.4 mW, 2.5 V supply
[100]	1.88 GHz Power amplifier	Power gain 216 dB, PAE: 54%
[101]	ECL inverter chain	16 ps/stage, 660 μA at 3.3 V
[101]	2.4 GHz Down converter (LNA + Mixer)	LNA: gain 10.5 dB
		NF 0.95 dB
		Mixer: +4 dBM
		input intercept 5 mA at 1 V (total)
[101]	Broadband amp	gain 8 dB
		Bandwidth 17
		16.8 mA at 2.5 V
[102]	12 GHz VCO	19 dBm, 5% tuning range,
		-80 dBc/Hz phase noise
[102]	12 GHz active mixer	> 0 dB gain at +3 dBm LO,
		100 kHz IF BW, 30 db isol. 96
[102]	12 GHz Power amp.	> 6 dB gain, 19 dBm output
[103]	1/128 Freq. divider	6.4–23 GHz, 1.5 W
[104]	RZ comparator	5 GHz, 1.5 V, 89 mW
[105]	Gilbert mixer	Bandwidth 12 GHz GBW > 22 GHz
[106]	12 bit DAC	1.2 Gs/s, 750 mW

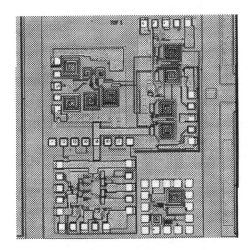

Figure 5.32. Microphotograph of the I/Q SiGe 5.2 GHz transceiver. The chip size is 2.07 mm × 2.77 mm. Source: IBM website.

should continue to establish that an all-silicon solution including SiGe-HBTs is commercially acceptable.

If one decides for the conservative way of integrating rf systems on a Si chip, one has to cope with the considerable substrate losses, resulting in a relatively high power consumption. Choosing the innovative development path will enable the rf designer to work with rf loss levels that are similar to those in a hybrid system, by using FZ-Si substrates with SOI device isolation, SOS, or micromachining techniques. This gives the advantage of a much higher integration density at the price of modifying an established Si fabrication process. In conclusion, a manufacturable SiGe-BiCMOS technology with high quality passive components will help in improving the rf/mixed-signal design system, as well as reducing cost and component count in communication products.

5.9 References

1 H. Kroemer, "Theory of a wide-gap emitter for transistors," *Proc. IRE*, vol. 45, pp. 1535–1537, 1957.

2 S. Iyer, G. L. Patton, S. S. Delage, S. Tiwari and J. M. C. Stork, "Silicon-germanium base heterojunction bipolar transistors by molecular beam

epitaxy," in *IEEE IEDM Tech. Dig.*, pp. 874–876, 1987.

3 H. Kroemer, "Heterojunction Bipolar Transistors and Integrated Circuits," *Proc. IEEE*, vol. 70, pp. 13–25, 1982.

4 D. L. Harame, E. F. Crabbe, J. D. Cressler, J. H. Comfort, J. Y.-C. Sun, S. R. Stiffler, E. Kobeda, J. N. Burghartz, J. Warnock, M. M. Gilbert, J. Malinowski and A. J. Dally, "A High-performance Epitaxial SiGe base ECL BiCMOS Technology," in *IEEE IEDM Tech. Dig.*, pp. 19–22, 1992.

5 D. Ahlgren, G. Freeman, S. Subbanna, R. Groves, D. Greenberg, J. Malinowski, D. Nguyen-Ngoc, S. Jeng, K. Stein, K. Schonenberg, D. Kiesling, B. Martin, S. Wu, D. Harame and B. Meyerson, "A SiGe HBT BiCMOS Technology for Mixed-signal RF Applications," in *IEEE BCTM Proc.*, pp. 51–56, 1997.

6 J. D. Cressler, "SiGe HBT technology: a new contender for Si-based RF and microwave circuit applications," *IEEE Trans. Microwave Th. Tech.*, vol. 46, pp. 572–589, 1998.

7 B. S. Meyerson, "Low temperature silicon epitaxy by ultrahigh vacuum/chemical vapor deposition," *Appl. Phys. Lett.*, vol. 48, pp. 797–799, 1986.

8 L. E. Larson, "Integrated circuit technology options for RFICs - present status and future directions," *IEEE J. Solid-State Circuits*, vol. 33, pp. 387–399, 1998.

9 G. L. Patton, J. H. Comfort, B. S. Meyerson, E. F. Crabbe, G. J. Scilla, E. De Fresart, J. M. C. Stork, J. Y.-C. Sun, D. L. Harame and J. N. Burghartz, "75-GHz f_T SiGe base heterojunction bipolar transistors," *IEEE Electron Dev. Lett.*, vol. EDL-11, pp. 171–173, 1990.

10 E. Kasper, H. Kibbel, J.-H. Herzog and A. Gruhle, "Growth of 100 GHz SiGe-heterobipolar transistor (HBT) structures," *Jap. J. Appl. Phys.*, vol. 33, pp. 2415–2418, 1994.

11 E. F. Crabbe, B. Meyerson, J. Stork and D. Harame, "Vertical Profile Optimization of Very High Frequency Epitaxial Si- and SiGe-Base Bipolar Transistors," in *IEEE IEDM Tech. Dig.*, pp. 83–86, 1993.

12 A. Gruhle, "SiGe Heterojunction Bipolar Transistors," in *Silicon-Based Millimeter-Wave Devices*, pp. 149–192, Springer-Verlag, Berlin, 1994.

13 A. Schuppen, A. Gruhle, H. Kibbel and U. Konig, "Mesa and planar SiGe-HBTs on MBE-wafers," *J. Mater. Sci.: Mater. Electron.*, vol. 6, pp. 298–305, 1995.

14 D. Behammer, J. N. Albers, U. Konig, D. Temmler and D. Knoll, "Si/SiGe HBTs for Application in Low Power ICs," *Solid-State Electron.*, vol. 39, pp. 471–480, 1996.

15 *IBM Tech. Disc. Bull., September*, 1992.

16 D. L. Harame, "High-performance BiCMOS process integration: Trends, issues and future directions," in *IEEE BCTM Proc.*, pp. 36–43, 1997.

17 Y.-F. Chyan, M. S. Carroll, T. G. Ivanov, A. S. Chen, W. J. Nagy, S. Chaudhry, R. W. Dail, V. D. Archer, K. K. Ng, S. Martin, M. Oh, M. R. Frei, I. C. Kizilyalli, R. Y. Huang, M. J. Thoma and K. H. Lee, "A 50 GHz 0.25 μm Implanted Base High-Energy Implanted-Collector Complementary Modular BiCMOS (HEICBiC) Technology for Low-Power Wireless Communication VLSIs," in *IEEE BCTM Proc.*, pp. 128–131, 1998.

18 J. Bock, T. F. Meister, H. Knapp, K. Aufinger, M. Wurzer, R. Gabl, M. Pohl, S. Boguth, M. Franosch and L. Treitinger, "0.5 μm/60 GHz f_{max} Implanted Base I Bipolar Technology," in *IEEE BCTM Proc.*, pp. 160–163, 1998.

19 L. E. Larson, "High-speed Si/SiGe technology for next generation wireless system applications," *J. Vac. Sci. Technol. B*, vol. 16, pp. 1541–1548, 1998.

20 R. A. Johnson, M. J. Zierak, K. B. Outama, T. C. Bahn, A. J. Joseph, C. N. Cordero, J. Malinowski, K. A. Bard, T. W. Weeks, R. A. Milliken, T. J. Medve, G. A. May, W. Chong, K. M. Walter, S. L. Tempest, B. B. Chau, M. Boenke, M. W. Nelson and D. L. Harame, "1.8 million transistor CMOS ASIC fabricated in a SiGe BiCMOS technology," in *IEEE IEDM Tech. Dig.*, pp. 217–220, 1998.

21 S. Subbanna, D. Ahlgren, D. Harame and B. Meyerson, "How SiGe Evolved into a Manufacturable Semiconductor Production Process," in *IEEE ISSCC Tech. Dig.*, pp. 66–67, 1999.

22 D. C. Ahlgren, G. Freeman, S. Subbanna, R. Groves, D. Greenberg, J. Malinowski, D. Nguyen-Ngoc, S. J. Jeng, K. Stein, K. Schonenberg, D. Kiesling, B. Martin, S. Wu, D. L. Harame and B. Meyerson, "SiGe HBT BiCMOS technology for mixed signal RF applications," in *IEEE BCTM Proc.*, pp. 195–197, 1997.

23 *IBM Tech. Disc. Bull., November*, 1992.

24 H. Presting and U. Konig, "State and Applications of Si/SiGe High Frequency and Optoelectronic Devices," in *Proc. Current Developments of Microelectronics*, pp. 139–150, 1999.

25 A. Schuppen, H. Dietrich, S. Gerlach, H. Hohnemann, J. Arndt, U. Seller, R. Gotzfried, U. Erben and H. Schumacher, "SiGe technology and components for mobile communication systems," in *IEEE BCTM Proc.*, pp. 130–133, 1996.

26 T. F. Meister, H. Schafer, M. Franosch, W. Molzer, K. Aufinger, U. Scheler, C. Walz, M. Stolz, S. Boguth and J. Bock, "SiGe Base Bipolar Technology with 74 GHz f_{max} and 11 ps Gate Delay," in *IEEE IEDM Tech. Dig.*, pp. 739–742, 1995.

27 F. H. Gaensslen, V. L. Rideout, E. J. Walker and J. J. Walker, "Very small MOSFETs for low temperature operation," *IEEE Trans. Electron Dev.*, vol. ED-24, pp. 218–229, 1977.

28 J. M. C. Stork, D. L. Harame, B. S. Meyerson and T. N. Nguyen, "Base profile design for high-performance operation of bipolar transistors at liquid nitrogen temperature," *IEEE Trans. Electron Dev.*, vol. ED-36, pp. 1503–1509, 1989.

29 E. F. Crabbe G. L. Patton, J. M. C. Stork, J. H. Comfort, B. S. Meyerson and J.-Y. C. Sun, "Low temperature operation of Si and SiGe bipolar transistors," in *IEEE IEDM Tech. Dig.*, pp. 17–20, 1990.

30 J. D. Cressler, D. L. Harame, J. H. Comfort, J. M. C. Stork, B. S. Meyerson and T. E. Tice, "Silicon–germanium heterojunction bipolar technology: The next leap in silicon," in *Tech. Dig. 1994 Int'l. Solid-State Circuits Conf.*, pp. 24–27, 1994.

31 J. D. Cressler, E. F. Crabbe, J. H. Comfort, J. Y.-C. Sun and J. M. C. Stork, "An Epitaxial Emitter-Cap SiGe-Base Bipolar Technology

Optimized for Liquid-Nitrogen Temperature Operation," *IEEE Electron Dev. Lett.*, vol. 15, pp. 472–474, 1994.

32 A. J. Joseph, J. D. Cressler and D. M. Richey, "Operation of SiGe Heterojunction Bipolar Transistors in the Liquid-Helium Temperature Regime," *IEEE Electron Dev. Lett.*, vol. 16, pp. 268–270, 1995.

33 J. D. Cressler, J. H. Comfort, E. F. Crabbe, G. L. Patton, J. M. C. Stork, J. Y.-C. Sun and B. S. Meyerson, "On the profile design and optimization of epitaxial Si- and SiGe base bipolar technology for 77 K applications - Part I : Transistor DC design considerations," *IEEE Trans. Electron Dev.*, vol. 40, pp. 525–541, 1993.

34 A. Gruhle, H. Kibbel, U. Konig, U. Erben and E. Kasper, "MBE-grown Si/SiGe HBTs with high β, f_T, and f_{max}," *IEEE Electron Dev. Lett.*, vol. 13, pp. 206–208, 1992.

35 J. D. Cressler, J. H. Comfort, E. F. Crabbe, G. L. Patton, J. M. C. Stork, J. Y.-C. Sun and B. S. Meyerson, "On the profile design and optimization of epitaxial Si- and SiGe base bipolar technology for 77 K applications - Part II: Circuit performance issues," *IEEE Trans. Electron Dev.*, vol. 40, pp. 542–556, 1993.

36 A. J. Joseph, J. D. Cressler, D. M. Richey, R. C. Jaeger and D. L. Harame, "Neutral Base Recombination and its Influence on the Temperature Dependence of Early Voltage and Current Gain–Early Voltage Product in UHV/CVD SiGe Heterojunction Bipolar Transistors," *IEEE Trans. Electron Dev.*, vol. 44, pp. 404–413, 1997.

37 K. Imai, T. Yamazaki, T. Tashiro, T. Tatsumi, T. Niino, N. Aizaki and M. Nakamae, "A CRYO-BiCMOS technology with Si/SiGe heterojunction bipolar transistors," in *IEEE BCTM Proc.*, pp. 90–93, 1990.

38 G. L. Patton, B. L. Harame, J. M. C. Stork, B. S. Meyerson, G. J. Scilla and E. Ganin, "Graded-SiGe base, poly-emitter heterojunction bipolar transistors," *IEEE Electron Dev. Lett.*, vol. 10, pp. 534–536, 1989.

39 J. D. Cressler, J. H. Comfort, E. F. Crabbe, G. L. Patton, W. Lee, J. Y.-C. Sun, J. M. C. Stork and B. S. Meyerson, "Sub-30-ps ECL circuit operation at liquid-nitrogen temperature using self-aligned epitaxial SiGe base bipolar transistors," *IEEE Electron Dev. Lett.*, vol. 12, pp. 166–168, 1991.

40 J. H. Comfort, G. L. Patton, J. D. Cressler, W. Lee, E. F. Crabbe, B. S. Meyerson, J. Y.-C. Sun, J. M. C. Stork, P.-F. Lu, J. N. Burghartz, J. Warnock, G. Scilla, K.-Y. Toh, M. D'Agostino, C. Stanis and K. Jenkins, "Profile leverage in self-aligned epitaxial Si or SiGe base bipolar technology," in *IEEE IEDM Tech. Dig.*, pp. 21–24, 1990.

41 J. S. Yuan, "Modeling Si/Si$_{1-x}$Ge$_x$ Heterojunction Bipolar Transistors," *Solid-State Electron.*, vol. 35, pp. 921–926, 1992.

42 J. C. Sturm, E. J. Prinz and C. W. Magee, "Graded-base Si/Si$_{1-x}$Ge$_x$/Si heterojunction bipolar transistors grown by rapid thermal chemical vapor deposition with near-ideal electrical characteristics," *IEEE Electron Dev. Lett.*, vol. EDL-12, pp. 303–305, 1991.

43 G. L. Patton, J. M. C. Stork, J. H. Comfort, E. F. Crabbe, B. S. Meyerson, D. L. Harame and J. Y.-C. Sun, "SiGe base heterojunction bipolar transistors: physics and design issues," in *IEEE IEDM Tech. Dig.*, pp. 13–16, 1990.

44 S. C. Jain, P. Balk, M. S. Goorsky and S. S. Iyer, "Strain relaxation in GeSi layers with uniform and graded composition," *Microelectronic Engineering*, vol. 15, pp. 131–134, 1991.

45 H. Schumacher, U. Erben and A. Gruhle, "Noise characterization of Si/SiGe heterojunction bipolar transistors at microwave frequencies," *Electronics Lett.*, vol. 28, pp. 1167–1168, 1992.

46 D. L. Harame, J. M. C. Stork, B. S. Meyerson, E. F. Crabbe, G. J. Scilla, E. de Fresart, A. C. Megdanis, C. L. Stanis, G. L. Patton, J. H. Comfort, A. A. Bright, J. B. Johnson and S. S. Furkay, "30 GHz polysilicon-emitter and single-crystal-emitter graded SiGe base PNP transistors," in *IEEE IEDM Tech. Dig.*, pp. 33–36, 1990.

47 E. J. Prinz, P. M. Garone, P. V. Schwartz, X. Xiao and J. C. Sturm, "The effect of base–emitter spacers and strain-dependent densities of states in Si/Si$_{1-x}$Ge$_x$/Si heterojunction bipolar transistors," in *IEEE IEDM Tech. Dig.*, pp. 639–642, 1989.

48 E. J. Prinz, P. Garone, P. Schwartz, X. Xiao and J. Sturm, "The effects of base dopant outdiffusion and undoped Si$_{1-x}$Ge$_x$ junction space layers in Si/Si$_{1-x}$Ge$_x$/Si heterojunction bipolar transistors," *IEEE Electron Dev. Lett.*, vol. EDL-12, pp. 42–44, 1991.

49 A. Pruijmboom, J. W. Slotboom, D. J. Gravesteijn, C. W. Fredriksz, A. A. van Gorkum, R. A. van de Heuvel, J. M. L. van Rooij-Mulder, G. Streutker and G. F. A. van de Walle, "Heterojunction bipolar transistors with SiGe base grown by molecular beam epitaxy," *IEEE Electron Dev. Lett.*, vol. 12, pp. 357–359, 1991.

50 J. W. Slotboom, G. Streutker, A. Pruijmboom and D. J. Gravesteijn, "Parasitic energy barriers in SiGe HBTs," *IEEE Electron Dev. Lett.*, vol. 12, pp. 486–488, 1991.

51 Z. A. Shafi, C. J. Gibbings, P. Ashburn, I. R. C. Post, C. G. Tuppen and D. J. Godfrey, "The Importance of Neutral Base Recombination in Compromising the Gain of Si/SiGe Heterojunction Bipolar Transistors," *IEEE Trans. Electron Dev.*, vol. 38, pp. 1973–1976, 1991.

52 S. C. Jain, H. J. Osten, B. Dietrich and H. Rucker, "Growth and properties of strained $Si_{1-x-y}Ge_xC_y$ layers," *Semicond. Sci. Technol.*, vol. 10, pp. 1289–1302, 1995.

53 A. Gruhle, H. Kibbel and U. Konig, "The reduction of base dopant outdiffusion in SiGe heterojunction bipolar transistors by carbon doping," *Appl. Phys. Lett.*, vol. 75, pp. 1311–1313, 1999.

54 H. J. Osten, D. Knoll, B. Heinemann and P. Schley, "Increasing Process Margin in SiGe Heterojunction Bipolar Technology by Adding Carbon," *IEEE Trans. Elec. Dev.*, vol. 46, pp. 1910–1912, 1999.

55 B. Heinemann, D. Knoll, G. Fischer, D. Kruger, G. Lippert, H. J. Osten, H, H. Rucker, W. Ropke, P. Schley and B. Tillack, "Control of steep boron profiles in Si/SiGe heterojunction bipolar transistors," in *Proc. ESSDERC'97*, pp. 544–547, 1997.

56 P. A. Potyraj, K. J. Petrosky, K. D. Hobart, F. J. Kub and P. E. Thompson, "A 230 watt S-band SiGe HBT," in *IEEE MTT-S Dig.*, pp. 673–676, 1996.

57 U. Erben, M. Wahl, A. Schuppen and H. Schumacher, "Class-A SiGe HBT Power Amplifiers at C-Band Frequencies," *IEEE Microwave and Guided Wave Lett.*, vol. 5, pp. 435–436, 1995.

58 A. Schuppen, S. Gerlach, H. Dietrich, D. Wandrei, U. Seiler and U. Konig, "1 W SiGe Power HBT's for Mobile Communication," *IEEE Microwave and Guided Wave Lett.*, vol. 6, pp. 341–343, 1996.

59 A. Schuppen, H. Dietrich, D. Zerrweck, H. van den Ropp, K. Burger, N. Gellrich, J. Arndt, M. Lentmaier, B. Jehl, J. Imschweiler, W. Kraus, F. Voswinkel, T. Asbeck, H. Conzelmann, W. Arndt, R. Kirchmann, A. Voigt and K. Worner, "Silicon Germanium ICs on the RF Market," in *Proc. Current Developments of Microelectronics*, pp. 151–162, 1999.

60 J. N. Burghartz, M. Soyuer and K. A. Jenkins, "Microwave Inductors and Capacitors in Standard Multilevel Interconnect Silicon Technology," *IEEE Trans. Microwave Th. Tech.*, vol. 44, pp. 100–104, 1996.

61 A. K. Agrawal, M. C. Driver, M. H. Hanes, H. M. Hobgood, P. G. McMullin, H. C. Nathanson, T. W. O'Keefe, T. J. Smith, J. R. Szendon and R. N. Thomas, "MICROX - An Advanced Silicon Technology for Microwave Circuits up to X-Band," in *IEEE IEDM Tech. Dig.*, pp. 687–690, 1991.

62 J. Y.-C. Chang, A. A. Abidi, M. Gaitan, "Large Suspended Inductors on Silicon and their Use in a 2-μm CMOS RF Amplifier," *IEEE Elec. Dev. Lett.*, vol. 14, pp. 246–248, 1993.

63 V. Milanovic, M. Gaitan, E. D. Bowen and M. E. Zaghloul, "Micromachined Coplanar Waveguides in CMOS Technology," *IEEE Microwave and Guided Wave Lett.*, vol. 6, pp. 380–381, 1996.

64 B. Senapati, C. K. Maiti and N. B. Chakrabarti, "Silicon heterostructure devices for RF wireless communication," in *IEEE VLSI Design Conf. Proc.*, pp. 488–491, 2000.

65 Y.-H. Xie, M. R. Frei, A. J. Becker, C. A. King, D. Kossives, L. T. Gomez and S. K. Theiss, "An Approach for Fabricating High-Performance Inductors on Low-Resistivity Substrates," *IEEE J. Solid-State Circuits*, vol. 33, pp. 1433–1438, 1998.

66 N. M. Nguyen and R. G. Meyer, "Si IC-Compatible Inductors and LC Passive Filters," *IEEE J. Solid-State Circuits*, vol. 25, pp. 1028–1031, 1990.

67 J. Long and M. Copeland, "Modeling, Characterization and Design of Monolithic Inductors for Silicon RF ICs," in *IEEE CICC Proc.*, pp. 185–188, 1996.

68 J. N. Burghartz, M. Soyuer and K. A. Jenkins, "Integrated RF and Microwave Components in BiCMOS Technology," *IEEE Trans. Elec. Dev.*, vol. 43, pp. 1559–1570, 1996.

69 J. N. Burghartz, D. C. Edelstein, K. A. Jenkins, C. Jahnes, C. Uzoh, E. J. O'Sullivan, K. K. Chan, M. Soyuer, P. Roper and S. Cordes, "Monolithic Spiral Inductors Using a VLSI Cu-Damascene Interconnect Technology and Low-Loss Substrates," in *IEEE IEDM Tech. Dig.*, pp. 99–102, 1996.

70 J. N. Burghartz, D. C. Edelstein, M. Soyuer, H. A. Ainspan and K. A. Jenkins, "RF Circuit Design Aspects of Spiral Inductors on Silicon," *IEEE J. Solid-State Circuits*, vol. 33, pp. 2028–2034, 1998.

71 J. N. Burghartz, K. A. Jenkins and M. Soyuer, "Multilevel-Spiral Inductors Using VLSI Interconnect Technology," *IEEE Elec. Dev. Lett.*, vol. 17, pp. 428–430, 1996.

72 K. B. Ashby, W. C. Finley, J. J. Bastek and S. Moinian, "High Q Inductors for Wireless Applications in a Complementary Silicon Bipolar Process," in *IEEE BCTM Proc.*, pp. 179–182, 1994.

73 J. N. Burghartz, M. Soyuer, K. A. Jenkins, M. Kies, P. Dolan, K. Stein, J. Malinowski and D. L. Harame, "RF Components Implemented in an Analog SiGe Bipolar Technology," in *IEEE BCTM Proc.*, pp. 138–141, 1996.

74 A. A. Abidi, "CMOS-only RF and baseband circuits for a monolithic 900 MHz wireless transceiver," in *IEEE BCTM Proc.*, pp. 35–42, 1996.

75 A. Schuppen, U. Erben, A. Gruhle, H. Kibbel, H. Schumacher and U. Konig, "Enhanced SiGe Heterojunction Bipolar Transistors with 160 GHz-f_{max}," in *IEEE IEDM Tech. Dig.*, pp. 743–746, 1995.

76 D. L. Harame, J. H. Comfort, J. D. Cressler, E. F. Crabbe, J. Y.-C. Sun, B. S. Meyerson and T. Tice, "Si/SiGe Epitaxial-Base Transistors - Part I: Materials, Physics, and Circuits," *IEEE Trans. Electron Dev.*, vol. 42, pp. 455–468, 1995.

77 F. Sato, T. Hashimoto, T. Tatsumi, M. Soda, H. Tezuka, T. Suzaki and T. Tashiro, "A Self-Aligned SiGe Base Bipolar Technology Using Cold Wall UHV/CVD and Its Application to Optical Communication IC's," in *IEEE BCTM Proc.*, pp. 82–88, 1995.

78 M. Kondo, K. Oda, E. Ohue, H. Shimamoto, M. Tanabe, T. Onai and K. Washio, "Sub-10 fJ ECL/68 A 4.7 GHz divider ultra-low-power SiGe base bipolar transistors with a wedge-shaped CVD-SiO$_2$ isolation structure and a BPSG-refilled trench," in *IEEE IEDM Tech. Dig.*, pp. 245–248, 1996.

79 A. Pruijmboom, D. Terpstra, C. E. Timmering, W. B. de Boer, M. J. J. Theunissen, J. W. Slotboom, R. J. E. Hueting and J. J. E. M. Hageraats, "Selective-Epitaxial Base Technology with 14 ps ECL-Gate Delay, for Low Power Wide-band Communication Systems," in *IEEE IEDM Tech. Dig.*, pp. 747–750, 1995.

80 K. Ismail, "Si/SiGe High-Speed Field-Effect Transistors," in *IEEE IEDM Tech. Dig.*, pp. 509–512, 1995.

81 A. E. Schmitz, R. H. Walden, L. E. Larsen, S. E. Rosenbaum, R. A. Metzger, J. R. Behnke and P. A. Macdonald, "A Deep-Submicrometer Microwave/Digital CMOS/SOS Technology," *IEEE Elec. Dev. Lett.*, vol. 12, pp. 16–17, 1991.

82 S. Voinigescu, S. Tarasewicz, T. Macalwee and J. Ilowski, "An assessment of state-of-the-art 0.5 μm bulk CMOS technology for RF applications," in *IEEE IEDM Tech. Dig.*, pp. 721–724, 1995.

83 J. N. Burghartz, "BiCMOS Process Integration and Device Optimization: Basic Concepts and New Trends," *Electrical Eng.*, vol. 79, pp. 313–327, 1996.

84 D. Harame, D. Nguyen-Ngoc, K. Stern, L. Larson, M. Case, S. Kovacic, S. Voinigescu, J. Cressler, T. Tewksburg, R. Gorves, E. Eld, D. Sunderland, D. Rensch, S. Jeng, J. Malinowski, M. Gilbert, K. Schonenberg, D. Ahlgren and B. Meyerson, "SiGe HBT Technology: Device and Application Issues," in *IEEE IEDM Tech. Dig.*, pp. 731–734, 1995.

85 L. E. Larson, "Integrated Circuit Technology Options for RFICs - Present Status and Future Directions," in *IEEE CICC Proc.*, pp. 169–176, 1997.

86 A. A. Abidi, "Low-Power Radio-Frequency ICs for Portable Communications," *Proc. IEEE*, vol. 83, pp. 544–569, 1995.

87 K. Washio, E. Ohue, K. Oda, R. Hayami, M. Tanabe, H. Shinamoto, T. Harad and M. Kondo, "82 GHz Dynamic Frequency Divider in 5.5ps ECL SiGe HBTs," in *IEEE ISSCC Tech. Dig.*, pp. 210–211, 2000.

88 H. Knapp, T. Meister, M. Wurzer, D. Zvschg, K. Aufinger and L. Treitinger, "A 79 GHz Dynamic Frequency Divider in SiGe Bipolar Technology," in *IEEE ISSCC Tech. Dig.*, pp. 208–209, 2000.

89 M. Wurzer, T. Meister, H. Knapp, K. Aufinger, R. Schreiter, S. Boguth and L. Treitinger, "53 GHz Static Frequency Divider in a SiGe Bipolar Technology," in *IEEE ISSCC Tech. Dig.*, pp. 206–207, 2000.

90 T. Masuda, K.-I. Ohhata, F. Arakawa, N. Shiramizu, E. Ohue, K. Oda, R. Hayami, M. Tanabe, H. Shimamoto, M. Kondo, T. Harada and K. Washio, "45GHz Transimpedance 32dB Limiting Amplifier and 40Gb/s 1:4 High-Sensitivity Demultiplexer with Decision Circuit using SiGe HBTs for 40Gb/s Optical Receiver," in *IEEE ISSCC Tech. Dig.*, pp. 60–61, 2000.

91 Y. Greshishchev, P. Schvan, J. Showell, M-L. Xu, J. Ojha and J. Rogers, "A Fully-Integrated SiGe Receiver IC for 10 Gb/s Data Rate," in *IEEE ISSCC Tech. Dig.*, pp. 53–53,447, 2000.

92 M. Meghelli, B. Parker, H. Ainspan and M. Soyuer, "SiGe BiCMOS 3.3 V Clock and Data Recovery Circuits for 10Gb/s Serial Transmission Systems," in *IEEE ISSCC Tech. Dig.*, pp. 56–57, 2000.

93 M. Bopp, M. Alles, D. Eichel, S. Gerlach, R. Gotzfried, F. Gruson, M. Kocks, G. Krimmer, R. Reimann, B. Roos, M. Siegle and J. Zieschang, "A DECT transceiver chip set using SiGe technology," in *IEEE ISSCC Tech. Dig.*, pp. 68–69, 1999.

94 Y. M. Greshishchev and P. Schvan, "A 60 dB Gain 55 dB Dynamic Range 10 Gb/s Broadband SiGe HBT Limiting Amplifier," in *IEEE ISSCC Tech. Dig.*, pp. 382–383, 1999.

95 T. Masuda, K. Ohhata, K. Oda, M. Tanabe, H. Shimamoto, T. Onai and K. Washio, "40 Gb/s analog IC chipset for optical receiver using SiGe HBTs," in *IEEE ISSCC Tech. Dig.*, pp. 314–315, 1998.

96 K. M. Strohm, J.-F. Luy, T. Hackbarth and S. Kosslowski, "MOTT SiGe SIMMWICs," in *IEEE MTT-S Dig.*, pp. 1691–1694, 1998.

97 L.-H. Lu, J. S. Rieh, P. Bhattacharya, L. P. B. Katehi, E. T. Croke, G. E. Ponchak and S. A. Alterovitz, "K-band Si/SiGe HBT MMIC amplifiers using lumped passive components with a micromachined structure," in *IEEE Radio Frequency Integrated Circuits (RFIC) Symp.*, pp. 17–20, 1998.

98 J. Sevenhans, B. Verstraeten, G. Fletcher, H. Dietrich, W. Rabe, J. L. Bacq, J. Varin and J. Dulongpont, "Silicon germanium and silicon bipolar RF circuits for 2.7 V single chip radio transceiver integration," in *IEEE Custom Integrated Circuits Conf.*, pp. 409–412, 1998.

99 H. Ainspan, M. Soyuer, J.-O. Plouchart and J. Burghartz, "A 6.25-GHz low DC power low noise amplifier in SiGe," in *IEEE Custom Integrated Circuits Conf. Proc.*, pp. 177–180, 1997.

100 G. N. Henderson, M. F. O'Keefe, T. E. Boles, P. Noonan, J. M. Sledziewski and B. M. Brown, "SiGe bipolar junction transistors for microwave power applications," in *IEEE MTT-S Dig.*, pp. 1299–1302, 1997.

101 J. R. Long, M. A. Copeland, S. J. Kovacic, D. S. Malhi and D. L. Harame, "RF analog and digital circuits in SiGe technology," in *IEEE ISSCC Tech. Dig.*, pp. 82–83, 1996.

102 L. Larson, M. Case, S. Rosenbaum, D. Rensch, P. Macdonald, M. Matloubian, M. Chen, D. Harame, J. Malinowski, B. Meyerson, M. Gilbert and S. Maas, "Si/SiGe HBT technology for low-cost monolithic microwave integrated circuits," in *IEEE ISSCC Tech. Dig.*, pp. 80–81, 1996.

103 M. Case, S. Knorr, L. Larson, D. Rensch, D. Harame, B. Meyerson and S. Rosenbaum, "A 23 GHz static 1/128 frequency divider implemented in a manufacturable Si/SiGe HBT process," in *IEEE BCTM Proc.*, pp. 121–124, 1995.

104 W. Gao, W. M. Snelgrove, T. Varelas, S. J. Kovacic and D. L. Harame, "A 5-GHz SiGe HBT return-to-zero comparator," in *IEEE BCTM Proc.*, pp. 166–169, 1995.

105 J. Glenn, M. Case, D. Harame and B. Meyerson, "12-GHz Gilbert Mixers Using a Manufacturable Si/SiGe Epitaxial-base Bipolar Technology," in *IEEE BCTM Proc.*, pp. 186–189, 1995.

106 D. L. Harame, K. Schonenberg, M. Gilbert, D. Nguyen-Ngoc, J. Malinowski, S.-J. Jeng, B. S. Meyerson, J. D. Cressler, R. Groves, G. Berg, K. Tallman, K. Stein, G. Hueckel, C. Kermarrec, T. Tice, G. Fitzgibbons, K. Walter, D. Colavito and D. Houghton, "A 200 mm SiGe HBT Technology for Wireless and Mixed-Signal Applications," in *IEEE IEDM Tech. Dig.*, pp. 437–440, 1994.

107 H. S. Momose, E. Morifuji, T. Yoshitomi, T. Ohguro, I. Saito, T. Morimoto, Y. Katsumata and H. Iwai, "High-Frequency AC Characteristics of 1.5 nm Gate Oxide MOSFETs," in *IEEE IEDM Tech. Dig.*, pp. 105–108, 1996.

108 P. R. de la Houssaye, C. E. Chang, B. Offord, G. Imthurn, R. Johnson, P. M. Asbeck, G. A. Garcia and I. Lagnado, "Microwave Performance of Optically Fabricated T-Gate Thin Film Silicon-on-Sapphire Based MOSFETs," *IEEE Elec. Dev. Lett.*, vol. 16, pp. 289–292, 1995.

109 J. P. Colinge, J. Chen, D. Flandre, J. P. Raskin, R. Gillon and D. Vanhoenacker, "A Low-Voltage, Low-Power Microwave SOI MOSFET," in *Proc. IEEE Int. SOI Conf.*, pp. 128–129, 1996.

110 D. Hisamoto, S. Tanaka, T. Tanimoto, Y. Nakamura and S. Kimura, "Silicon RF Devices Fabricated by ULSI Process Featuring 0.1-μm SOI-CMOS and Suspended Inductors," in *Dig. VLSI Technol. Symp.*, pp. 104–105, 1996.

Chapter 6

Heterostructure Field Effect Transistors

Over the past 20 years, the channel length of MOS transistors has been halved at intervals of approximately three or four years. The continuous shrinking of the size of MOS transistors has led to increasing performance (clock speed) in electronic systems (e.g., computers, mobile phones) and increasing packing density in Si chips. As a result, increasingly sophisticated and powerful electronic products have appeared at prices similar to earlier generations of the product. The question that arises is how long this trend can continue.

A number of factors are posing a threat to this evolution of CMOS scaling technology. The reasons are manifold. First, the channel length of the MOS transistor is defined using optical lithography, which is limited by the wavelength of light. It therefore becomes increasingly difficult to design new generations of optical lithography tools. The current thinking is that optical lithography can reach channel lengths of around 0.15 μm, but it is not clear whether it can meet the challenge of smaller geometries. Instead of defining the channel length of a lateral MOS transistor using lithography, an effort is being made to define the channel length of a vertical MOS transistor using epitaxy. Extremely thin layers can be produced using epitaxy which would allow MOS transistors to be produced with channel lengths below 0.025 μm.

Secondly, short channel effects such as drain induced barrier lowering (DIBL) and punch-through problems become more difficult to manage. It is becoming increasingly difficult to obtain shallow enough source and drain junctions. Highly abrupt, vertically and laterally non-uniform SUPER-HALO doping profiles will be required for control of short-channel effects in the 0.05 μm channel-length regime. Gate oxide thickness is predicted to be tunneling-current limited below 20 Å, or roughly 7 atomic layers. Therefore, it is necessary to explore other methods for further improvement of the speed of silicon-based ULSI devices beyond the device scaling limit.

In this chapter, a review of the present status of the fabrication and characterization of Si HFETs in the Si/SiGe and SiGeC material systems is presented. Heterojunction MOSFETs may use a strained SiGe/SiGeC channel or strained Si or strained Ge channel. The substrate for a compressively (tensilely) strained channel would have a lower (larger) lattice constant. The channels may lie on the surface or be buried. The other freedom is to use a vertical channel. The choice of the cap layer for a buried SiGe channel is an important issue having bearing on the performance of the device. A brief discussion of design considerations is given.

6.1   SiGe Quantum Well HFETs

Several factors are available for enhancing the drive current in a short-channel (of length L and width W) drain current (I_{DS}). Consider the short-channel MOSFET drive current equation [1],

$$I_{DS} = \frac{\mu_{eff} C_{ox} W/L}{1 + V_{DS}/E_{sat}L} \left[(V_{GS} - V_T) V_{DS} - \frac{V_{DS}^2}{2} \right] \qquad (6.1)$$

where μ_{eff} is the effective mobility, C_{ox} is the oxide capacitance per unit area, E_{sat} is the saturated effective field, V_T is the threshold voltage, and V_{DS} and V_{GS} are the drain and gate biases, respectively. However, all the above variables in the equation 6.1 are not independent.

The carrier mobility in equation 6.1 can be adjusted by modifying the channel material to increase the drive current. As discussed in Chapter 3, strained Si alloy heterostructures are useful for reducing the scattering rate with carriers in the most populated band having a lower effective mass in the transport direction. This contributes to improvement in mobility. The low hole mobility of Si compared to its electron counterpart necessitates a much higher area for the p-channel MOSFETs to match the current drives in a CMOS technology. Therefore attention has been directed to improving the hole mobility, though gain in electron mobility will be an added advantage. Figure 6.1 shows the relative performance gain in terms of delay as predicted [2] for CMOS devices below 0.1 μm. It is clear that the initial mobility improvement factor is most beneficial, allowing 25%–50% gains in relative CMOS performance before being slowed by parasitic resistance and carrier velocity limitations. Taking advantage of the increased hole mobility in strained $Si_{1-x}Ge_x$, a high performance p-channel FET can be fabricated to achieve properly balanced CMOS devices. Excellent reviews are available in the literature on the application of SiGe heterostructures for different kinds of FET devices [3, 4].

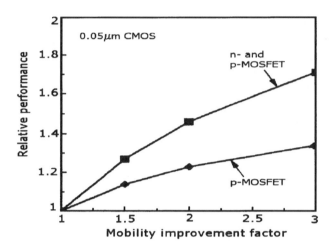

Figure 6.1.   Projected relative performance gain of CMOS circuits with improvements in either the hole mobility alone or both the electron and the hole mobilities. After Y. Taur and E. J. Nowak, IEEE IEDM Tech. Dig., 215–218 (1997), copyright © 1997 IEEE.

An interesting variation is a device structure making use of carbon alloys, namely SiC/SiGeC channel p-MOSFETs. Ge channel MOSFETs deserve special attention as Ge has the highest mobility among group-IV semiconductors. But there are technological issues involved. The fabrication and performance of Ge channel MOSFETs are therefore examined. Strained Si MOSFETs with surface or buried channels are currently considered to be the most attractive candidate for replacing conventional Si devices in VLSI technology because both hole and electron mobilities can be enhanced. Both p- and n-channel HFETs are discussed in some detail. There has been considerable work in recent years on building isolated devices on silicon-on-insulator. The performance of HFETs on SOI/SOS is considered. In order that an HFET provides the enhanced performance expected one must fabricate the gate, drain and source structures with care and attention. In order that the contemplated device structures may be integrated with Si IC technology, it is necessary to postulate and examine a compatible integrated SiGe-CMOS process. The development in this area is also discussed.

The first reported SiGe-FET device was a p-channel MODFET with a TiSi$_2$ Schottky gate [5]. A Si$_{0.8}$Ge$_{0.2}$ strained layer was sandwiched between two 1000 Å thick p$^+$ Si layers. The holes were injected from the p$^+$ layer to the Si$_{0.8}$Ge$_{0.2}$ layer where they formed a two-dimensional high-mobility hole gas constituting the p-channel. However, the device showed large gate leakage at temperatures above 77 K or gate voltages above 0.8 V, primarily due to the Schottky gate. A MOSFET using an oxide between the channel and the gate is useful to circumvent the problem by increasing the barrier to hole injection to 4.0 eV. Though functionally preferred, it is not feasible to place the higher mobility structure directly below the insulator. It has been found by several research groups that oxidation of Si$_{1-x}$Ge$_x$ results in Ge segregation to the underlying substrate, leading to SiO$_2$ with a Ge-rich underlayer. The oxides are of poor quality with a large number of interface states accompanied by poor turn-on characteristics of the fabricated devices. Spacing the SiGe channel away from the gate with a thin Si cap alleviates the problem. The, choice of Si cap thickness plays an important role in the characteristics of the transistor. Leaving an unconsumed Si cap on top of the channel layer thus forming a buried SiGe p-MOS structure has the following advantages:

(i) The type-I band structure of the resulting device is useful to confine the holes in the valence band quantum well leading to carrier transport in high mobility channel layer.

(ii) There is reduced scattering by surface roughness and oxide charges.

(iii) The mobility degradation factor caused by scattering from the metal/SiO$_2$ interface roughness in the case of thin oxides, which are needed for short-channel applications, is reduced.

However, it is preferable to keep the unconsumed Si cap layer as thin as possible for the following reasons:

(i) to prevent carrier transport through the parasitic Si surface channel;

(ii) to maximize the gate-to-channel capacitance and hence to increase the SiGe MOSFET transconductance as the unconsumed Si cap in series with the oxide capacitance reduces the effective gate capacitance.

Another virtue of the SiGe p-MOS structure is that, because the channel is buried, hot carrier degradation of the oxide is expected to be reduced. In addition, the increased probability of impact ionization in the drain region in the lower bandgap alloy might provide an energy dissipation mechanism, which would reduce the hot carrier population. The smaller bandgap, however, leads to lowering of the breakdown field and increased drain leakage. So it is desirable to fabricate buried channel SiGe devices with an optimized Si cap.

There has been considerable interest in SiGe channel p-MOS devices, where the carriers (holes) are confined in a narrow SiGe quantum well channel (see figure 6.2), rather than at the usual Si/SiO_2 interface. These Si/SiGe devices exhibited higher carrier mobility and transconductance for two reasons: the Si/SiGe interface is an improvement over the standard Si/oxide interface and the carrier mobility in the channel is higher because the in-plane effective mass in the lowest-lying (first) quantum well subband is considerably smaller than in bulk Si.

The inversion layer at a Si/SiO_2 interface represents the most widely used quasi two-dimensional system. The carriers at the interface are confined in a potential well created by gate bias induced band bending. The potential well is ordinarily assumed to be triangular, but its actual shape is determined by the distribution of the dopants and the carriers. The potential wells that occur in Si/SiGe/Si heterostructures may be assumed to be rectangular with barriers determined by the band offsets on two sides. Wells can also be created by doping; an example is provided by a p-i-n-i. δ-doping is an important application of such a well. Expressions for wave function and energy for square wells are well known and are given by

$$\Psi(x) = A \sin\left(\frac{n\pi x}{w}\right) \qquad (6.2)$$

where w is the width of the barrier, n is the quantum number and A is the normalization constant. The energy for an infinitely deep well is given by

$$E_n = \frac{\pi^2 \hbar^2 n^2}{2m^* w^2} \qquad (6.3)$$

where m^* is the effective mass of the carrier in the well.

The ground state energy for a quantized inversion layer is therefore smaller when the transverse mass is larger. It is important to note the variation of the energy with mass and barrier width. For finite barrier heights such as occur in Si/SiGe heterostructures, the wave function will be nonzero at the interface and will decay into the barrier region.

The eigenfunctions for a triangular well defined by a constant electric field (ε) are

$$\psi(x) = A_i\left(\frac{x}{w} - \frac{E_n}{q\varepsilon}\right) \qquad (6.4)$$

where A_i is the Airy function and

$$w = \left(\frac{\hbar^2}{2m^* q\varepsilon}\right)^{1/3} \qquad (6.5)$$

and the energies are

$$E_n = \frac{1}{2}\frac{\hbar^2}{m^* w^2}\left[\frac{3\pi}{2}\left(n + \frac{3}{4}\right)\right]^{2/3} \qquad (6.6)$$

The ground state energy is seen to depend on mass as $(m^*)^{-1/3}$ and with field as $\varepsilon^{2/3}$.

When two quantum wells are coupled by a barrier region, the eigenfunctions combine to form doublets corresponding to symmetric and asymmetric modes. Resonant tunneling diodes and transistors make use of transitions between states in coupled wells. In the weak coupling limit where the doublet splitting is small compared to the subband separation, one can extend the concept to superlattices and compute the energy dispersion relation as in the tight binding approximation. Multiple quantum well structures are widely used in optoelectronics. The energy states in MQM can be computed by the transfer matrix method. The variation of the energy states with electric field is the basis of the function of modulators.

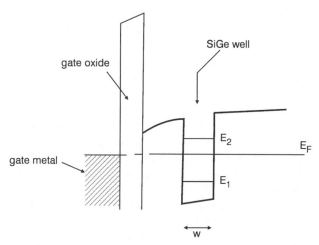

Figure 6.2. Band diagram of a SiGe p-MOS quantum well. After IBM Tech. Disc. Bull., 127–130 (1993).

In quantum well devices, optimal performance requires that the carrier density in the first subband be maximized, but at the same time the Fermi level should not approach the higher-lying (second) subband, where the in-plane mass is higher and the mobility is consequently degraded (see figure 6.2 for a schematic band diagram of a standard device). Evidently, increasing the energy separation (E_2-E_1) can contribute to a better performance by increasing the maximum achievable density in the first subband.

One possibility is to increase the depth of the SiGe well by using higher Ge concentrations. This has the disadvantage of reducing the critical thickness of the well before dislocations occur [6] and a narrower well pushes E_1 up towards the Si band edge, reducing the maximum carrier density in the well. Proposed is an alternative technique, the step-grading of the SiGe channel profile, that relies on the difference in the subband wavefunctions to increase the subband energy separation without increasing the Ge content. In addition, the step-graded channel profile reduces the lattice mismatch at the Si/SiGe interfaces. Finally, although all the device work in SiGe has focused on p-MOS devices because the Si/SiGe band offset occurs almost entirely in the valence band if unstrained Si substrates are used, the same step-grading technique can be applied to increase the electron subband separation in n-MOS devices grown on relaxed SiGe substrates.

Since the implementation of a step-graded SiGe quantum well profile is fully compatible with the standard growth procedure of Si/SiGe heterostructures, the advantage of the step-graded profile is not offset by more complex fabrication. The depth of the SiGe well grown on Si substrates is roughly proportional to the Ge content. Consequently, the implementation of the step-graded SiGe profile simply replaces the standard $Si_{1-x}Ge_x$ layer of thickness w with a sequence of three layers. Since the width and Ge content of the three SiGe layers can be adjusted independently, an optimum configuration can be calculated for a device under operating conditions. The three-layer growth sequence confers the added benefit of reducing the lattice mismatch at the Si/SiGe interfaces, which may itself improve the channel mobility.

The proposed bandgap modification of SiGe channels should increase the maximum carrier density in Si/SiGe heterostructure MOSFETs. Although the above description pertains largely to p-MOSFETs, which have already been demonstrated, an analogous technique can be used for n-MOS strained Si devices grown on relaxed SiGe substrates.

The p-type MOSFET has attracted much more attention than its MODFET counterpart (for a review see Reference 7). Three main approaches have been investigated and demonstrated: devices with a surface SiGe channel, a strained Si surface channel and a buried modulation doped heterostructure [8–10]. The surface SiGe channel approach is the simplest and the easiest to integrate with n-type MOSFETs toward the fabrication of complementary circuits. When integrated with conventional Si n-MOSFETs, a surface SiGe channel should improve the hole transport and help reduce the performance discrepancy between p- and n-type devices. As a result, the size of the p-type devices in conventional CMOS can be reduced, leading to a higher packing density and lower load capacitance for digital circuits. This direction offers the easiest and most affordable way to introduce SiGe to Si FET devices.

6.2 Design of SiGe channel p-HFETs

The critical design parameters for SiGe channel FETs are the choice of gate material, the method of threshold voltage adjustment, the SiGe profile in the channel, the channel and cap doping, and the gate oxide thickness. The impact of each of these design parameters on device performance was studied by Verdonckt-Vandebroek *et al.* [11]. The key design parameter of a MOSFET is the device transconductance. This can be enhanced by maximizing the number of high mobility holes confined to the SiGe channel while minimizing the density of low mobility holes, which flow at the Si/SiO_2 interface. An important figure of merit is the cross-over gate voltage at which the number of holes in the SiGe well equals that of holes in the parasitic Si cap channel.

6.2.1 Effect of SiGe Layer Thickness and Ge Mole Fraction

The thickness and Ge mole fraction of the SiGe layer play a critical role in determining the hole confinement gate voltage range. A larger band offset at the top Si/SiGe interface results in better hole confinement [12]. Strain relaxation issues set the upper limit on the amount of band offset that can be achieved by the incorporation of Ge. For a fixed amount of strain, it is advantageous to minimize the thickness of the SiGe layer and increase the Ge mole fraction.

Several workers have reported simulation results on various p-HFET structures for different Ge mole fractions and profiles [11, 12]. Quantum effects were not taken into account; they are expected to affect carrier transport when the SiGe channel thickness is comparable to the hole de Broglie wavelength. The simulation results show that the thickness of the SiGe channel has very little effect on the hole densities in the SiGe and the surface channel. Thinning the SiGe layer improves the stability of the strained film, thus allowing higher Ge contents to be used at the same stability. Figure 6.3 shows the simulated hole densities for 10%, 20%, 30% and 40% Ge mole fractions. With increasing Ge mole fraction, the hole confinement window widens. Both the cross-over voltage and the hole density in the SiGe channel increase with increasing Ge concentration. It is apparent that the Ge concentration should be more than 30% to ensure that most of the holes flow through the SiGe channel.

It is possible to obtain significant improvement in performance if channels with very high Ge concentration (40%–50%) are used. Grading the Ge profile in the channel in this case can satisfy the Matthews–Blakeslee criterion for critical layer thickness. Since most of the carriers flow through the surface of the inversion layer, the valence band offset should be maximized at the

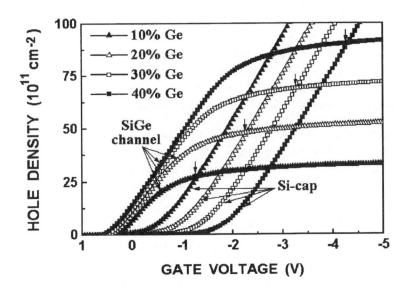

Figure 6.3. One-dimensional Poisson simulations of the Si cap and SiGe channel charges in an n^+-gate SiGe p-HFET, comparing various Ge mole fractions in the SiGe channel. The SiGe channel is 31 nm wide, the Si cap layer is 2 nm wide and the oxide thickness is 9 nm.

top of the SiGe channel. Locating the peak Ge concentration at the top of the channel moves the cross-over point to higher voltages. The increased hole density in the higher mobility channel will lead to a larger drive current in the graded profile device. The creation of a built-in electric field due to the graded Ge profile forces the holes towards the top of the inversion layer resulting in an increase of gate to channel capacitance. This leads to steeper turn-on at low gate bias.

Voinigescu *et al.* [13] proposed a triangular profile by grading down the top Ge concentration to 0% over 20–30 Å to reduce interface scattering. Figure 6.4 shows the vertical cross-section showing a schematic Ge profile with triangular and rectangular shape. Hole mobilities in excess of 400 cm^2/Vs were obtained experimentally for transistors with 0%–50% triangular Ge profiles. Figure 6.5 shows the comparison of experimental transfer characteristics for n$^+$-gate SiGe p-MOSFETs with triangular (0%–50% Ge) and rectangular (25% Ge) channel profiles. Compared with devices having rectangular Ge profiles, the

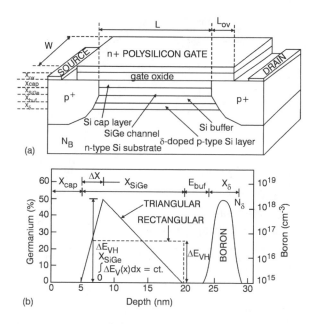

Figure 6.4. (a) Layer structure of a Si/SiGe/Si p-MOSFET. (b) Vertical cross-section showing schematic Ge and B profiles. After S. P. Voinigescu et al., IEEE IEDM Tech. Dig., 369–372 (1994), copyright ©1994 IEEE.

MOSFETs with triangular profiles demonstrated 30%–40% improvement in mobility, transconductance and cutoff frequency.

6.2.2   Effect of Oxide and Si cap Thickness

In a SiGe p-HFET, in addition to the oxide thickness t_{ox}, the ratio of oxide layer thickness to Si cap thickness t_{cap} plays an important role in determining the hole confinement gate voltage range V_{crit}, which is defined as the point at which the parasitic surface channel turns on in addition to the SiGe channel [14]. It has been shown that V_{crit} increases as the ratio t_{ox}/t_{cap} increases. Therefore, it is sufficient to use a 70 Å gate oxide while minimizing the cap thickness, in order to maximize the ratio. The thinner the Si cap, the better the hole confinement, the charge control capacitance, and thus, the transconductance. A cap thickness of 50 Å is found to provide a 40% improvement in hole confinement.

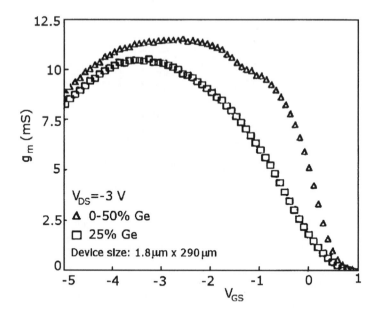

Figure 6.5. Comparison of experimental transconductance characteristics in saturation for SiGe p-MOSFETs with triangular and rectangular Ge profiles. Boron dose in the δ-doping is 1.5×10^{12} cm^{-3}, peak hole mobility is 400 cm^2/Vs for V_{DS} of –0.05 V. After S. P. Voinigescu et al., IEEE IEDM Tech. Dig., 369–372 (1994), copyright ©1994 IEEE.

Reduction of gate oxide thickness decreases the cross-over voltage. Decrease in the Si cap thickness increases both the SiGe transconductance and the cross-over voltage. However, since the current flows very close to the gate oxide, interface scattering will degrade the hole mobility for thinner cap layers and optimization of Si cap thickness should be determined by a mobility/capacitance trade-off along with the gate voltage window.

6.2.3 Channel Doping and Threshold Adjustment

The threshold voltage of the SiGe channel can be expressed as [11]

$$V_{\mathrm{T}}(\mathrm{SiGe}) = V_{\mathrm{T}}(\mathrm{Si}) - \Delta E_{\mathrm{v}}/q - Q_{\mathrm{channel}}/C_{\mathrm{ox}} + Q'_{\mathrm{channel}}/C_{\mathrm{eq}} \qquad (6.7)$$

with $V_{\mathrm{T}}(\mathrm{Si})$ defined as the onset of strong inversion in a Si MOSFET and C_{eq} the equivalent gate capacitance in series combination of the oxide and the Si cap capacitance. $Q_{channel}$ is the charge in the depletion region of a Si MOSFET and $Q'_{channel}$ is the background charge in the SiGe channel and in the depleted region underneath the SiGe channel.

For n^+-gate p-MOSFET devices, a positive V_T shift (–1.4 V to –1.2 V) is desirable for lowering the power supply voltage. However, threshold adjustment is necessary to move V_T to even lower values. For the p^+-gate, the introduction of the SiGe channel shifts the threshold voltage from –0.2 to 0 V. This is unacceptable for digital CMOS applications requiring additional n-type dopants in the channel to increase V_T. When the threshold is adjusted using uniform doping such that both the n^+- and p^+-poly gate MOSFETs have V_T of –0.6 V, the hole confinement in the SiGe well for the n^+-gate is significantly better than that for p^+-gate design. In the case of the p^+-poly gate, the parasitic Si channel turns on shortly after threshold and the cross-over voltage is as low as –1.1 V compared to –3.0 V for the n^+-poly design. Additionally, the p-type dopants used for V_T adjustment in the n^+-gate design give rise to a significant increase in the hole density in the SiGe channel. The band discontinuity is sufficiently large for the n^+-gate to confine most of the holes in the SiGe channel, while most of them flow in the Si cap for the p^+-gate due to poor confinement in the SiGe channel. Therefore, the n^+-poly-Si gate SiGe p-MOSFET is preferred to the p^+-gate for operating at higher power supply voltages.

Since most of the carriers flow through the surface of the inversion layer, the valence band offset should be maximized at the top of the SiGe channel. For the n^+-gate SiGe p-HFET, where a modulation doped boron layer is used underneath the SiGe channel for threshold adjustment, the valence band offset is required at the bottom of the channel as well. For a p^+-gate HFET, where the threshold is adjusted by increasing the channel doping, no valence band discontinuity is needed at the bottom of the channel. Therefore, for identical SiGe thickness and identical integrated Ge doses, the peak Ge mole fraction for the p^+-gate is higher than that of the n^+-gate design. A SiGe p-MOSFET with modulation doped B layer (termed MODMOS) has been experimentally demonstrated by IBM researchers [11]. The device used an n^+-polySi gate to illustrate the design concepts discussed above with a graded Ge profile to optimize the hole confinement and p^+ modulation doping underneath the SiGe channel to adjust the threshold voltage. Locating the modulation doped hole injector layer underneath the channel rather than above it has the advantage of working with lower Si cap thickness and avoiding process sensitive thinning of the injector cap layer.

6.2.4   Choice of gate material: n^+ vs. p^+-poly

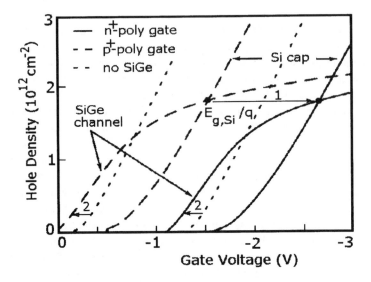

*Figure 6.6. Simulated hole densities in SiGe and Si channel as a function of gate voltage for both n^+- and p^+-gate SiGe p-HFETs with 50 Å Si cap and a 90 Å wide 30% SiGe channel with n-type doping of 5×10^{16} cm^{-3}. 70 Å thick gate oxide and an interface state density of 5×10^{10} cm^{-2}/eV were assumed in computation. After S. Verdonckt-Vandebroek et al., IEEE Trans. Electron Dev., **41**, 90–102 (1994), copyright ©1994 IEEE.*

The type of gate material strongly influences the degree of hole confinement to the SiGe channel. The submicron Si p-MOSFET using an n^+-poly gate suffers from severe short-channel effects often leading to use of dual-work function poly-Si gates in modern Si-CMOS technology. However, the introduction of a SiGe channel can extend the p-channel MOSFET operation into the deep submicron regime. Figure 6.6 shows the simulated hole densities in the SiGe and parasitic Si channel at the Si/SiO$_2$ interface as a function of gate voltage for both n^+- and p^+-gate SiGe p-HFETs with 50 Å Si cap and a 90 Å wide 30% SiGe channel with n-type doping of 5×10^{16} cm^{-3} and 70 Å thick gate oxide. An interface state density of 5×10^{10} cm^{-2}/eV was assumed in computation.

For both the gates, the SiGe channel shows a lower turn-on voltage compared to the parasitic Si channel. At higher gate voltage the holes in the parasitic Si layer screen the gate potential and fewer holes are added to the SiGe channel with increasing gate voltage, limiting the maximum concentration of high mobility holes. The gate voltage window is about 1.4 V, after which the parasitic charge in the Si surface channel exceeds the SiGe channel charge. The p$^+$- and n$^+$-gate SiGe p-HFETs show identical hole confinement, but the difference in work function of the gates results in a horizontal shift equal to the Si bandgap.

6.3 Experimental HFETs

To prevent three-dimensional growth, strained SiGe layers are grown at a relatively low temperature. Hence a low thermal budget process should be adopted for the fabrication of devices. If the epitaxial growth of the SiGe film is performed first, then a low thermal budget process is required for the fabrication of the devices. This may not result in the optimum performance for devices. To alleviate these problems, techniques for selective epitaxial growth can be utilized.

First, n-type MOSFETs are fully fabricated using conventional technology, then they are covered fully with CVD oxide. Windows are opened in this oxide where SiGe is selectively grown. Finally, the p-type MOSFETs are then fabricated in these areas. The advantages of this technique are twofold. First, the thermal budget constraints are limited only for the fabrication of p-type devices. Second, higher degrees of strain can be accommodated in small areas (islands) of SiGe with a much better quality due to the lower density of threading dislocations that can be achieved.

The growth of good quality oxide on SiGe layers is also a technological challenge. It has been demonstrated that the thermal oxidation rate of SiGe is enhanced by a factor of 3 compared to Si. This technique requires heating the sample to elevated temperatures (700–800 °C) leading to Ge segregation or "snow plowing" at the SiGe/SiO$_2$ interface [15]. This results in a higher trap density (D_{it}) at the surface (about 10^{12} cm^{-2}/eV), which plays an important role in degrading the hole transport properties at this interface and results in poor turn-on characteristics of the devices.

One way to solve the Ge "snow plowing" problem is to grow a sacrificial Si cap layer that can be consumed during the growth of a good quality oxide [8, 16]. However, the choice of Si cap thickness plays an important role in the characteristics of the transistor. For the case of a surface SiGe channel device, the Si layer needs to be totally consumed. This places stringent requirements on the control of the oxidation process.

Microwave and electron cyclotron resonance (ECR) oxidation [17–20] have been demonstrated to relax the requirement for a high processing temperature (< 500 ^{o}C). Dielectric films fabricated using this technique have been shown to be stoichiometric and composed of both SiO_2 and GeO_2. Other techniques like wet rapid thermal oxidation (WRTO), chemical vapor deposition, wet thermal oxidation, and plasma enhanced chemical vapor deposition of oxides have been used to fabricate p-type SiGe channel MOSFETs [21–23]. Several design and simulation studies on the effect of the integrated Ge concentration, cap thickness, gate oxide and profile on the performance of MOSFET devices have been conducted [11, 13, 24]. The best reported results are summarized in table 6.1.

Table 6.1. Some of the reported results for SiGe p-MOSFETs on Si, SOI and SOS substrates.

L_{eff} μm	Mode	$g_{m,ext}$(mS/mm) 300 K	77 K	Gate Oxide	Mat.	Sub.	Ref.
1.0	Enh.	14	5	70 Å Ther.	n^+poly	SOS	[25]
1.0	Enh.	48	60	100 Å ECR	n^+poly	Si	[18]
0.9	Enh.	–	–	70 Å PECVD	n^+poly	Si	[11]
0.7	Enh.	64	–	50 Å Ther.	poly-Si	Si	[8]
4.0	Enh.	–	50	500 Å CVD	Al	Ge	[22]
1.0	Dep.	80	–	65 Å WRTO	p^+poly	SIMOX	[26]
0.25	Enh.	167	201	71 Å Ther.	$TiSi_2$	Si	[16]

6.3.1 $Si_{1-x}Ge_x$ p-MOSFETs

Theoretically, the mobility of holes in SiGe p-MOSFETs should show enhancement over that in bulk Si. The enhancement in Hall mobility does not give the true picture since the performance should be compared on completion of the device processing after the structure has been subjected to source–drain implantation, gate oxidation and annealing. One of the first experimental devices showing the enhancement of hole mobility in MBE grown SiGe over bulk Si was demonstrated by Nayak *et al.* [8] by comparing the room temperature hole mobility based on a long-channel device. The device structure is shown in figure 6.7. The mobility, obtained from the slope of the saturation transconductance vs. gate voltage, of the buried SiGe channel device was 155 cm^2/Vs while the control Si devices yielded a mobility of

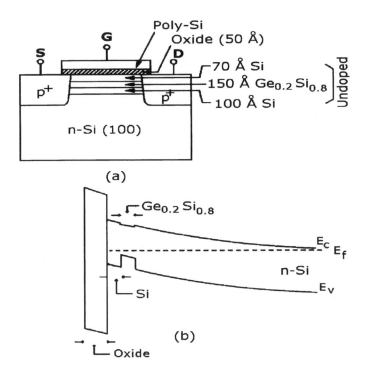

*Figure 6.7. Schematic diagram of a SiGe p-MOSFET structure. After D. K. Nayak et al., IEEE Electron Dev. Lett., **EDL-12**, 154–156 (1991), copyright ©1991 IEEE.*

122 cm^2/Vs. Garone *et al.* [23, 27] have also reported the hole mobility enhancement in an Al-gate buried Si$_{1-x}$Ge$_x$ p-MOSFET using PECVD gate oxide (125 Å). In addition to the optimization of hole density using a one-dimensional Poisson solver, they used quantum mechanical modeling to find the effect of cap layer thickness on the inversion layer mobility. It is important to distinguish between the effective mobility (used by circuit designers) and inversion layer mobility in SiGe devices. The effective mobility assumes that the hole density under the gate is

$$Q_{\mathrm{inv}} = C_{\mathrm{ox}}(V_{\mathrm{GS}} - V_{\mathrm{T}}) \qquad (6.8)$$

where Q_{inv} is the inversion layer charge per unit area, C_{ox} is the oxide capacitance per unit area and V_{GS} and V_{T} are the gate and threshold voltages of the device, respectively. The effective mobility is generally plotted as a function of effective transverse field defined by [28]

$$E_{\text{eff}} = (Q_{\text{dep}} + Q_{\text{inv}}/\eta)/\epsilon_{\text{Si}} \qquad (6.9)$$

where Q_{dep} is depletion charge per unit area and ϵ_{Si} is the relative permittivity of Si. η is the empirical factor for universal mobility behavior for all doping and biases and is chosen as $1/2$ and $1/3$ for electrons and holes, respectively.

In the case of buried channel SiGe transistors, the gate capacitance is lower than C_{ox} because of the series capacitance of the Si cap layer. Thus the number of carriers is overestimated using C_{ox}. So modeling and calculating the inversion layer mobility from the simulated hole density is more meaningful for SiGe devices. Figure 6.8 shows the peak effective mobility as a function of temperature for control Si and $Si_{0.7}Ge_{0.3}$ devices with a Si cap of 105 Å. The disproportionately large increase of the SiGe mobility relative to Si is due to the reduced surface scattering in the former. Surface scattering dominates due to the reduction of phonon scattering at low temperature. Therefore, the variation of surface scattering mobility (μ_{sr}) as a function of average separation of the carriers from the surface (Z_{avg}) needs to be known for designing the cap layer. Surface scattering mobility is calculated from the measurement of bulk (μ_{bulk}) and inversion layer (μ_{inv}) mobilities using Matthiessen's rule:

$$\frac{1}{\mu_{\text{sr}}} = \frac{1}{\mu_{\text{inv}}} - \frac{1}{\mu_{\text{bulk}}} \qquad (6.10)$$

The surface scattering follows the inverse square law giving the mobility in terms of average separation distance as [29]

$$\mu_{\text{sr}} \propto (Z_{\text{avg}})^2 \qquad (6.11)$$

Taking into account the presence of carriers in both the SiGe well and Si/SiO_2 channel, the relation of the effective fields and average spacing to the hole potential in the devices is depicted in figure 6.9, where

$$E_{\text{eff},Si/SiO_2} = (Q_{\text{dep}} + Q_{\text{SiGe}} + \frac{1}{3}Q_{\text{Si}/\text{SiO}_2})/\epsilon_{\text{Si}} \qquad (6.12)$$

and

$$E_{\text{eff,SiGe}} = (Q_{\text{dep}} + \frac{1}{3}Q_{\text{SiGe}})/\epsilon_{\text{Si}} \qquad (6.13)$$

Hole densities in the SiGe well and at the Si/SiO_2 interface obtained from a one-dimensional Poisson solver can be used [27] to take a weighted average of the SiGe surface mobility term and the Si/SiO_2 surface mobility

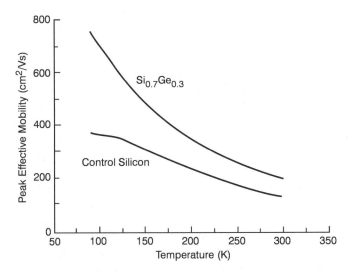

Figure 6.8. *Peak effective mobility vs. temperature for $Si_{0.7}Ge_{0.3}$ channel p-MOSFET with a cap thickness of 105 Å. After P. M. Garone et al., IEEE IEDM Tech. Dig., 29–32 (1991), copyright ©1991 IEEE.*

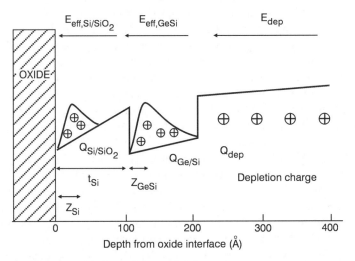

Figure 6.9. *Relation of the effective fields and average spacings to the hole potential in the devices. After P. M. Garone et al., IEEE IEDM Tech. Dig., 29–32 (1991), copyright ©1991 IEEE.*

term. The calculated surface mobility in combination with bulk mobility gives the inversion layer mobility for a given SiGe device at a given effective field. The model fits very well across the whole range of effective field showing the inversion mobility much higher than the underestimated effective mobility. At low temperature, the model overestimates the mobility over the measured value indicating the existence of additional Coulombic scattering potential, which is field-independent and predominant at low temperature. An additional constant mobility term of 1470 cm^2/Vs attributed to alloy scattering in the SiGe channel gives much better fit with experimental data at low temperature.

All the devices described above were processed in university laboratories without much emphasis on compatibility with state-of-the-art CMOS processing, namely submicron (0.25 μm) LDD devices, standard isolation, silicided gate and shallow source–drain junction. Researchers at IBM [16] have demonstrated 0.25 μm $Si_{1-x}Ge_x$ CMOS devices using an integrable process module, i.e. LOCOS isolation, threshold and deep well implants, titanium silicided gates compatible with conventional Si processing. Identical phosphorus threshold-adjust and deep-well implants and anneal were carried out for both the Si and SiGe structures. The undoped $Si_{1-x}Ge_x$ ($0 < x < 0.25$) channel and Si cap (70 Å/105 Å) layers were grown by selective UHVCVD at 530 °C. Gate oxides (70 Å) were either thermally grown or deposited using a PECVD process. For devices with thermal oxides grown at 700 °C (without Ge redistribution), the Si cap thickness was 105 Å to account for Si consumption during oxidation. A p^+-poly-Si gate with pre-amorphized source–drain junction of depth 1200–1500 Å followed by self-aligned titanium silicide gates was used in the process. The long-channel linear-region field effect mobility at room temperature was found to increase from 95 cm^2/Vs for Si to 150 cm^2/Vs for $Si_{0.8}Ge_{0.2}$. The mobility enhancement was slightly more at 82 K, from 250 cm^2/Vs (Si) to 400 cm^2/Vs ($Si_{0.8}Ge_{0.2}$). The transconductance improvement was found to decrease as the channel length was reduced due to onset of carrier velocity saturation.

A SiGe MOSFET called a modulation doped SiGe-MOSFET or p-MODMOS device has been experimentally demonstrated by an IBM group [11, 24]. The device used an n^+-polysilicon gate with a graded Ge profile to optimize the hole confinement and p^+-modulation doping underneath the SiGe channel for adjusting the threshold voltage. Locating the modulation doped hole injector layer underneath the channel rather than above it has the advantage of working with lower Si cap thickness and avoiding process-sensitive thinning of the injector cap layer. The p-MODMOS device was fabricated with three different SiGe channel profiles as shown in figure 6.10. The Ge concentration was graded over 15 nm from 25% near the gate to

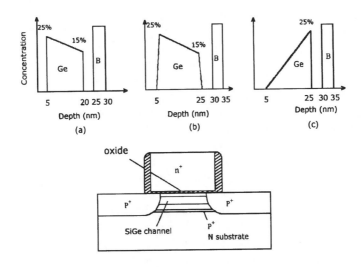

Figure 6.10. Schematic cross-section of SiGe p-MODMOS devices with three different channel gradings: (a) the abrupt, (b) the graded and (c) the retrograded profile. After S. Verdonckt-Vandebroek et al., IEEE Trans. Electron. Dev., **41**, *90–102 (1994), copyright ©1994 IEEE.*

15% at the bottom of the channel. The first set of devices with abrupt SiGe heterointerfaces (profile a) exhibited poor electrical characteristics due to high interface roughness. In the second set of devices, a linearly graded Ge profile (profile b) at the top and the bottom graded over 2.5 nm and at the channel graded over 15 nm was employed. A third set of experimental devices were fabricated with a retrograde Ge profile (c) with the maximum Ge concentration 25% at the bottom and 0% at the top of a 20 nm SiGe channel. The devices were processed using an As-doped n^+-poly-Si gate, 7 nm PECVD oxide and an 0.14 μm deep source–drain junction with $TiSi_2$ contact. In addition to reduced ionized impurity scattering, reported advantages of p-MODMOS over an implanted buried channel p-MOSFET are:

(i) the p-MODMOS does not suffer from short-channel effects since the p^+-doped layer used for V_T adjustment is very narrow and very close to the gate, and

(ii) since the conducting holes are physically separated from the impurity atoms, no threshold voltage change occurs due to impurity freezeout at low temperature. This can be verified by the absence of abnormal subthreshold characteristics at low temperature.

Linear transconductance as a function of gate voltage of the fabricated 10 μm × 10 μm devices [11] at room and low (82 K) temperature is shown in figure 6.11. The transconductance at 300 K is flat over a broader range of gate voltages than at 82 K. This is because even when the low-mobility Si surface channel turns on at high gate voltages along with the SiGe channel at 300 K, the average gate-to-channel capacitance increases due to closer proximity of the holes to the gate. The transconductance drops when the Si channel fully turns on because the degradation of mobility due to surface scattering becomes larger than the increase in capacitance. At 82 K, the transconductance exhibits a pronounced peak when the SiGe channel turns on with fewer holes flowing to the cap thus indicating a better confinement.

The maximum g_m in the retrograded profile is lower than in the graded channel since the built-in electric field due to the Ge profile confines the holes to the bottom of the SiGe channel at low gate voltages, thus resulting in a lower gate-to-channel capacitance. In addition the holes are drawn closer to the ionized impurities in the case of the retrograded channel design. The hole mobility was extracted from the device data for abrupt and graded profiles. As expected, the mobility is very much degraded in an abrupt profile due to the rough Si/SiGe interface at the top of the channel. Therefore, experimental observations suggest that a graded channel design is preferable to an abrupt or retrograde channel design.

All the above SiGe p-MOSFETs have shown improved device performances compared to that of bulk Si. Ge concentration in the alloy in principle is the most significant factor for comparing the enhancement ratio. It is encouraging to note that Goto *et al.* [30] have reported a mobility enhancement of about 70% at 300 K in a strained $Si_{0.5}Ge_{0.5}$ channel MOS structure. Cornell University research team has achieved a transit frequency of 23 GHz and a maximum oscillation frequency of 35 GHz in a 0.2 μm $Si_{0.6}Ge_{0.4}$ p-MOSFET, which is comparable to state-of-the-art Si n-MOS of similar geometry. These indicate the possibility [31] that SiGe p-MOS device technology could mature to have an impact on deep submicron, e.g., 0.18 μm, Si-CMOS.

6.3.2 Ge-channel p-MOSFETs

The fabrication of p-MOSFETs incorporating a bulk Ge channel should increase the device drive current significantly because of higher hole mobility in Ge compared to Si or $Si_{1-x}Ge_x$. As distinguished from the strained layer devices discussed earlier, p-MOSFET devices using bulk unstrained Ge have

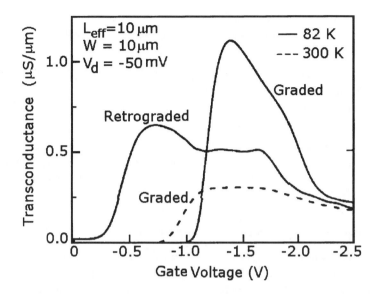

*Figure 6.11. Linear conductance of p-MOSFET with graded-Ge channel design, indicating enhanced confinement at 82 K relative to 300 K. The 82 K transconductance of the MOSFET with a retrograded channel design is also plotted. After S. Verdonckt-Vandebroek et al., IEEE Trans. Electron. Dev., **41**, 90–102 (1994), copyright ©1994 IEEE.*

also been reported. Growth of high quality gate oxide on bulk Ge is the key problem due to the hygroscopic nature of GeO_2.

A p-channel MOSFET using a native germanium oxynitride gate was fabricated by Martin *et al.* [32]. Measured low field mobility was $770 \ cm^2/Vs$ in 7 μm channel length devices. In spite of the superior hole mobility in the Ge channel, it is unlikely that devices made on Ge substrates will ever be attractive for large-scale device production. Therefore, Ge channel devices made on a Si substrate appear more promising. However, the critical layer thickness of bulk Ge on Si is only about 4 nm, too small as an inversion layer thickness without quantum confinement effects. The use of a compositionally graded relaxed $Si_{1-x}Ge_x$ substrate similar to that used for growing tensile strained Si can be a fruitful solution.

Wang *et al.* [33] have reported a high mobility Ge channel heterostructure with a hole mobility as high as $55,000 \ cm^2/Vs$ at 4 K. Shubnikov–de-Haas measurements done at MIT have revealed a significant reduction in hole effective mass m_h^* ($0.1m_0$ in the strained Ge channel compared to the bulk Ge heavy-hole effective mass of $0.28m_0$). However, the mobility in the bulk Ge channel may be limited by following factors:

(i) interface roughness between the $Si_{1-x}Ge_x$ spacer layer and the Ge layer grown without a proper surfactant, and
(ii) very small channel width which may cause the hole wave function to be pushed into the relaxed SiGe layer.

Therefore, in spite of the highest reported mobility, manufacturing issues like low critical layer thickness, growth of the relaxed buffer layer and the problem of a gate dielectric on Ge may complicate its implementation in current CMOS design.

6.3.3   SiGe- and Ge-channel n-MOSFETs

Most studies have concentrated on p-MOSFET devices. A few studies have been reported on the fabrication of compressive strained SiGe or Ge-channel n-MOSFETs on Si. From theoretical considerations, the in-plane mobility of electrons in compressive strained SiGe should be lower than in bulk Si because of the higher in-plane effective mass of electrons in the Δ_4 band (see Chapter 3). However, the only result reported on SiGe n-MOSFETs [34] with Ge implanted channel showed a higher drain current at any given V_{GS}–V_T compared to control Si devices. A channel with 16% Ge concentration was formed by implanting Ge^+ at an energy of 80 keV with a dose of 6×10^{16} cm^{-2}. Assuming the same effective channel length in both cases, the electron mobility is found to be higher in SiGe devices in contradiction to the theoretical prediction. More experiments on the effect of compressive strain on electron mobility are necessary to explain the above observation. n-MOSFETs using an unstrained Ge channel have been reported [32, 35] to have high electron mobilities in the range 940–1200 cm^2/Vs. However, n-channel devices with a strained Ge layer have not yet been reported.

6.4   SiC/SiGeC-channel p-HFETs

Due to the approximately 4.17% lattice mismatch of Si and Ge, there is an equilibrium critical thickness for pseudomorphic growth of SiGe layers on Si, which depends on the desired strain in the layer. SiGe technology, developed for over a decade, has been plagued by this problem for device applications requiring high Ge mole fraction. The thermal stability of strained Si and compressively strained $Si_{1-x}Ge_x$ layers is a major concern in many device structures. Consequently, the design flexibility is limited for applications involving low Ge concentration (in the buffer layer in the case of strained Si), thinner active layer and relatively low process temperature windows.

Incorporating smaller size carbon atoms substitutionally into the SiGe system enables one to compensate the strain, which leads to increase in thermal stability and critical layer thickness [36, 37]. Growth of strain-compensated ternary SiGeC layers and relaxed buffer layers (as a template for growing strained Si) have been reported by many researchers [36–39]. This has paved the way to extending SiGe-based heterostructures allowing more flexibility in strain and bandgap engineering. Carbon-containing alloys promise to expand the range of device applications of silicon-based heterostructures.

It has been shown that carbon reduces the strain in SiGe at a faster rate than it increases the bandgap [40]. Thus for a given bandgap, a larger critical thickness can be obtained for $Si_{1-x-y}Ge_xC_y$ films compared to those without carbon. The extracted valence band offset in $Si/Si_{1-x-y}Ge_xC_y$ heterostructures also decreases much more slowly than predicted with increasing carbon in the alloy, so that for a given lattice mismatch to Si, the valence band offset is larger for SiGeC than for Si. Therefore, both the band alignment and the valence band offset in ternary alloys are favorable for various device applications, as they reduce the possibility of process-induced strain relaxation, while confining the holes in the valence band quantum well. The conduction band offset is small, at least for Ge mole fractions less than 0.7 in strained SiGe, limiting the application to only p-HFET/MODFET devices.

p-HFET devices have been fabricated with partially strain-compensated $Si_{1-x-y}Ge_xC_y$ films with Ge-to-C ratio of 30:1 while maintaining a large valence band offset. Bulk Si, epi-Si, $Si_{1-x}Ge_x$ and completely strain compensated $Si_{1-x-y}Ge_xC_y$ channel devices have also been fabricated for comparison [41, 42]. The dc characteristics of a partially strain-compensated $Si_{1-x-y}Ge_xC_y$ channel p-HFET have been reported. The devices were evaluated between room temperature and 77 K and were compared to those of $Si_{1-x}Ge_x$ and control Si devices. The room temperature low field effective mobility in $Si_{1-x-y}Ge_xC_y$ devices is found to be higher than that of $Si_{1-x}Ge_x$ grown in the metastable regime and Si devices at low gate bias. However, with increasing transverse fields and with decreasing temperatures, $Si_{1-x-y}Ge_xC_y$ devices show degraded performance. The enhancement at low gate bias is attributed to the strain stabilization effect of C. This application of $Si_{1-x-y}Ge_xC_y$ in p-HFETs demonstrated the potential benefits in the use of C for strain stabilization of the binary alloy.

Figure 6.12 shows the room temperature I_{DS}–V_{DS} normalized for oxide thickness differences between the samples and subthreshold I_{DS}–V_{GS}

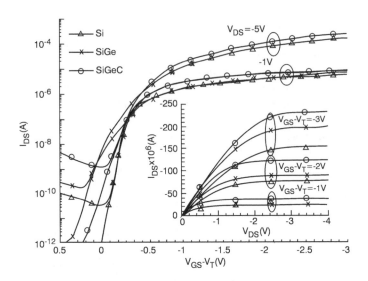

Figure 6.12. I_{DS} *vs.* V_{GS}–V_T *for epitaxial Si, $Si_{0.8}Ge_{0.2}$ (SiGe) and $Si_{0.793}Ge_{0.2}C_{0.007}$ (SiGeC) p-MOSFETs for linear and saturation values of V_{DS} for 10 μm × 10 μm devices. Inset shows I_{DS} vs. V_{DS} for increasing values of V_{GS}–V_T. The curves have been normalized for oxide thickness variations between the samples. After S. John et al., Appl. Phys. Lett.,* **74**, *847–849 (1999).*

characteristics of $Si_{0.8}Ge_{0.2}$, $Si_{0.793}Ge_{0.2}C_{0.007}$ and control Si transistors with the same doping with a channel length of 10 μm. The subthreshold slopes are 101, 90 and 75 mV/decade for $Si_{0.8}Ge_{0.2}$, $Si_{0.793}Ge_{0.2}C_{0.007}$ and control Si devices, respectively. All devices exhibit good saturation and turn-off characteristics. However, the $Si_{0.793}Ge_{0.2}C_{0.007}$ transistor exhibits a high current at the same V_{GS}–V_T than $Si_{0.8}Ge_{0.2}$ devices and control Si devices as shown in the inset.

A comparison of the field effect mobilities for epitaxial Si, $Si_{0.8}Ge_{0.2}$, $Si_{0.793}Ge_{0.2}C_{0.007}$, and lightly doped bulk Si p-MOSFETs at room and liquid nitrogen temperatures is shown in figure 6.13. The peak mobility at 300 K is enhanced to 190 cm^2/Vs for $Si_{0.793}Ge_{0.2}C_{0.007}$ in comparison to 140 cm^2/Vs for the $Si_{0.8}Ge_{0.2}$ devices. The ternary alloy sample shows the highest peak mobility whereas the effective mobility for $Si_{0.8}Ge_{0.2}$ devices is only slightly higher than that of epitaxial Si and lower than bulk Si devices. It is known

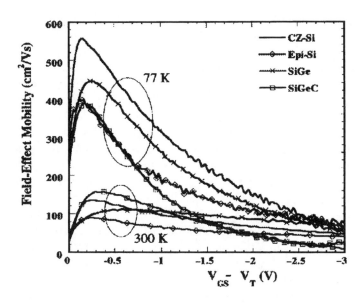

*Figure 6.13. Linear field effect mobility (μ_{fe}) for 1×10^{15} cm^{-3} doped bulk Si, 2×10^{17} cm^{-3} doped epitaxial Si (Si), Si$_{0.8}$Ge$_{0.2}$ (SiGe) and Si$_{0.793}$Ge$_{0.2}$C$_{0.007}$ (SiGeC) p-MOSFETs as a function of V_{GS}–V_T for 10 μm $\times$ 10 μm devices at room and liquid nitrogen temperatures. After S. John et al., Appl. Phys. Lett., **74**, 847–849 (1999).*

that the in-plane hole mobility in compressively strained Si$_{1-x}$Ge$_x$ is enhanced due to the lifting of valence band degeneracy and modification of the band structure.

6.5 Si$_{1-y}$C$_y$ p-MOSFETs

As discussed earlier, tensile strained Si layers have attracted considerable attention for advanced CMOS devices because of the enhancement of in-plane mobility of both the electrons and holes compared to bulk Si [43]. Despite much work, however, relaxed Si$_{1-x}$Ge$_x$ buffer layers still suffer from threading dislocation densities of the order of 10^4–10^5 cm^{-2}, making future large-scale applications uncertain. Secondly, the method of growing several microns thick relaxed Si$_{1-x}$Ge$_x$ buffer layers with continuous or step grading (to minimize the propagation of threading and misfit dislocations) possesses an inherent technological complexity, in addition to the difficulty of integrating mixed devices using strained Si, unstrained Si and $Si_{1-x}Ge_x$, because of the problems of lithography on complex surface topographies. Due to the much smaller lattice constant of C (mismatch $\sim 52\%$), the incorporation of C into a Si epitaxial layer results in a strained $Si_{1-y}C_y$ alloy with a band structure akin to that of a tensile strained Si layer on a SiGe buffer. Since the conduction band offset in Si/SiGe structures is mainly induced by the strain, one expects that a corresponding strain in $Si_{1-y}C_y$/Si should lead to a similar confinement by producing type-II band alignment without the need of a thick relaxed SiGe buffer layer. The tensile strained $Si_{1-y}C_y$ layer thus appears promising for both n-MOSFET and p-MOSFET devices. Quinones *et al.* [44] demonstrated $Si_{1-y}C_y$ MOSFETs using UHVCVD grown films. Some of the key results are discussed below.

Epitaxial layers of $Si_{1-y}C_y$ were grown on lightly doped, 10^{15} cm^{-3}, n-type Si(100) substrates by UHVCVD at 500 °C using Si_2H_6 and CH_3SiH_3. PH$_3$ was used as an *in situ* source for n-type doping. XRD rocking curves of the Si(004) plane demonstrated that $Si_{1-y}C_y$ alloys do result in tensile strain. AFM analysis also showed that the addition of small amounts of carbon, up to 1.5%, does not result in increased surface roughness over Si epitaxial layers. p-MOSFET devices were fabricated with films containing 0.5% and 1.0% C. The strain compensation ratio of Ge to carbon is generally accepted to be 8.5 to 10:1. Therefore, the deposited $Si_{1-y}C_y$ layers are effectively strain-equivalent to SiGe buffer layers of 4.25%–5.0% and 8.5%–10% Ge content, respectively. Based on device simulation using MEDICI [45], a heavily doped (10^{18} cm^{-3}), 100 Å thick Si cap layer (60–70 Å consumed during processing) was grown to prevent C segregation into the gate oxide without the formation of any parasitic Si surface channel. The channel layer was doped to 5×10^{16} cm^{-3}. The epi-Si sample consisted of the same structure without C in the channel. A low doped substrate, 10^{15} cm^{-3}, was used as the bulk sample. An LDD, p$^+$-polySi gate process was used for device fabrication. An LTO oxide was used to define the active region. A 30 min, 750 °C wet oxidation was then used to grow the gate oxide (≈ 70 Å), followed by polysilicon deposition (≈ 1600 Å) at 650 °C. The LDD region was formed via a 5 keV, 5×10^{15} cm^{-2}, BF$_2$ implant followed by a polysilicon edge reoxidation step (750 °C, 100% O$_2$, 15 min). Source–drain contacts were formed by a 15 keV, 5×10^{15} cm^{-2}, BF$_2$ implant and annealing in 100% O$_2$ for 30 min at 750 °C, followed by Al metallization. Such a low temperature process prevented the loss of substitutional C that may occur in a conventional high thermal budget process.

The threshold voltage (V_T) and oxide thickness (t_{ox}) for all the devices were independently measured and normalized to the Si epi sample to prevent any discrepancies due to process-induced variations. Typical output characteristics

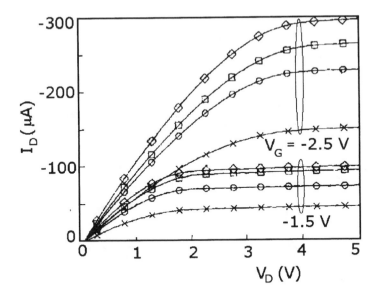

Figure 6.14. I_{DS}–V_{DS} *characteristics for Si bulk (squares), Si epi (circles),* $Si_{0.995}C_{0.005}$ *and* $Si_{0.99}C_{0.01}$ *p-MOSFET 10 μm × 10 μm devices.*

are shown in figure 6.14 for Si bulk, Si epi, $Si_{0.99}C_{0.01}$ and $Si_{0.995}C_{0.005}$ devices (10 μm × 10 μm). The 0.5% C device exhibits higher drive current than either the Si epi or Si bulk devices. Additionally, the 0.5% C device also shows (figure 6.15) the highest linear transconductance (g_m) for both long-channel and short-channel devices. Subthreshold characteristics are good for all the devices, with leakage current below 10^{-12} A. The increase in subthreshold slope from 86 mV/decade (Si-epi, bulk, and 0.5% C) to 127 mV/decade (1.0% C) may be due to increased heterointerface states in the alloy layer.

It is evident from the figure 6.15 that the mobility degrades as C content in the film is increased from 0.5% $(g_m = 5.49\ \mu S)$ to 1.0% $(g_m = 2.06\ \mu S)$. The 34% hole mobility enhancement over epi in the 0.5% C device is attributed to the band structure modifications in the $Si_{1-y}C_y$ layer resulting in lifting of the degeneracy in the valence band similar to that seen in strained Si on a SiGe buffer. On the other hand, 1.0% C in Si may represent very strong perturbation (due to the 52% lattice mismatch) in the host crystal, leading to the generation of localized states/charge centers. This, along with an increased alloy scattering factor, may lead to the degradation of mobility in 1.0% C samples compared to that of 0.5% C samples. Therefore, in fabricating dilute

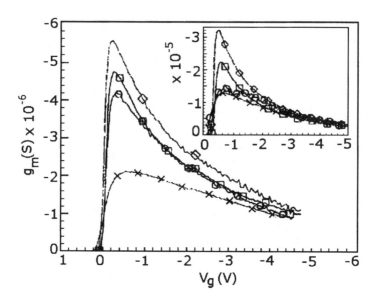

Figure 6.15. Linear transconductance (g_m) for $W \times L = 10\ \mu m \times 10\ \mu m$ and $10\ \mu m \times 0.5\ \mu m$ (inset) p-MOSFET devices. Si bulk (squares), Si-epi (circles), $Si_{0.995}C_{0.005}$ (diamonds) and $Si_{0.99}C_{0.01}$ (hatches).

C containing devices, an optimization must be made to minimize C induced interface charges and alloy scattering rate.

6.6   Vertical MOSFETs

Device down scaling is the current trend in commercial integrated circuit production. During the last 20 years, functionality of the integrated circuits has increased while the cost per function has decreased. Innovations in technology and device design have made it possible to overcome the barriers of device scaling. However, when the semiconductor industry takes a leap into the deep submicron regime, designers are faced with a number of fundamental limitations. The first challenge is the lithography which is reaching its limits and at present there are no alternative options that can assure a high throughput, reproducibility and cost effectiveness in production.

Vertical MOS technology offers the perspective to combine standard optical lithography and the possibility of fabricating channel lengths with nanometer resolution [46–51]. One of the important advantages of vertical

MOS transistor technology is that the channel length scaling is not limited by the minimum lithographic resolution. The transistor channel length is determined by shallow trench etching and epitaxial layer growth techniques. The epitaxial growth of the different layers, either by MBE or CVD, decouples the channel length definition from the requirements of the lithography, making it feasible to define channel lengths accurately in the nanometer regime.

Vertical MOSFETs were first reported by Rodgers and Meindl in 1974 [52]. Due to the wet etch and diffusion technologies available at that time, development of the vertical MOSFET was not successful. In the 1990s, many developments took place and vertical MOS transistors with channel length as short as 70 nm [53] have been demonstrated. These devices showed a good current drive and high transconductance and have the potential for extremely high packing density. Pein and Plummer have implemented a three-dimensional sidewall flash EPROM cell which provides higher packing density and higher read currents, and supports industry standard programs and erase voltages with controllable erase characteristics [54]. Auth and Plummer [55] have presented a scaling theory for fully-depleted cylindrical surrounding-gate MOSFETs. However, vertical MOSFET devices suffer from drain induced barrier lowering (DIBL), causing threshold voltage roll-off and roll-up of the subthreshold slope [56].

6.6.1   Vertical SiGe p-HFETs

A vertical SiGe p-HFET using a strained SiGe channel has been reported by Liu *et al.* [51, 57]. The structure combines the merits of a very short channel device without a critical lithography process and a higher hole mobility in the channel region. Like the in-plane transport, the hole mobility in SiGe normal to the growth plane has been predicted to be significantly larger than for the unstrained counterpart. This study gave the experimental evidence of the enhancement of out-of-plane hole mobility in strained SiGe using the vertical structure.

The starting substrate was 3–5 Ω cm n-type $< 100 >$ Si 100 mm diameter wafers. Subsequently an RIE process with HBr and Cl_2 was employed to etch the mesa region for the vertical channel. After RIE, Si polish etch and a sacrificial oxidation were used to remove RIE-induced damage. A low temperature oxide (LTO) field oxide was then deposited for isolation. This was followed by 100 Å rapid thermal oxide for the gate oxide and polysilicon deposition for gate electrode formation. The gate was formed by an anisotropic etch to form a self-aligned sidewall polysilicon gate. Two-stage P implantation (150 keV, 5×10^{13} cm^{-2} and 200 keV, 7×10^{13} cm^{-2}) was done for channel doping. Ge at 200 keV with a dose of 7×10^{16} cm^{-2} was implanted into the central 60 mm diameter of the wafer; the remaining area of the substrate was used for the fabrication of control Si devices. This eliminated the possibility of channel length variations and thermal budget in Si and SiGe layers from wafer to wafer in a single wafer RIE process. The peak value of Ge mole fraction was 15% at a depth of 1500 Å below the surface, as shown in the SIMS profile of the device in figure 6.16. As evident, a graded SiGe channel and built-in electric field is obtained by Ge mole fraction variation along the vertical direction. The recrystallization anneal of the SiGe films was done by rapid thermal anneal at 1000 °C, for 3 min. Low energy (10 keV) BF$_2$ implant was used to form the source and drain regions. The rest of the process was a standard MOS process, based on low temperature oxide (LTO) deposition for isolation and contact hole etching, metal deposition and patterning followed by a forming gas anneal.

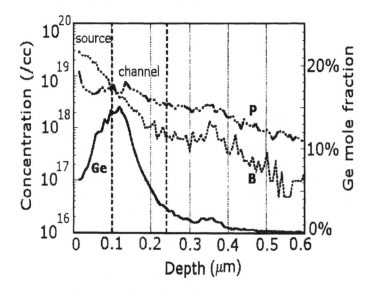

Figure 6.16. SIMS profiles along the vertical channel of the p-MOSFET. After K. C. Liu et al., IEEE DRC Dig., 128–129 (1997), copyright ©1997 IEEE.

The main features of the process are: (i) the vertical SiGe channel formed using high-dose Ge implantation followed by solid phase epitaxial regrowth, in which a graded SiGe profile provides a built-in electric field, (ii) circumvention of the problems associated with the thermal oxidation of a SiGe film by utilizing the vertical structure with gate oxide grown on the sidewall and Ge implanted after oxide formation and (iii) a process fully compatible with the current CMOS techniques and a device with channel length below 0.2 μm. The process flow and structure of the vertical SiGe p-HFET [51] are shown in figure 6.17.

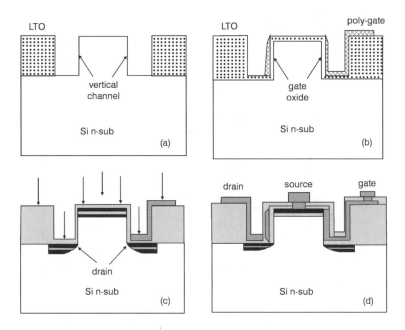

*Figure 6.17. The process flow of the vertical SiGe p-MOS transistor and structure: (a) RIE pillar etch, sacrificial oxide growth and removal, isolation oxide deposition and patterning; (b) growth of gate oxide, self-aligned sidewall polysilicon gate formation; (c) channel, Ge, and S/D implantation; and (d) isolation oxide deposition, contact hole etch and metal patterning. After K. C. Liu et al., IEEE Electron Dev. Lett., **19**, 13–15 (1998), copyright ©1998 IEEE.*

Figure 6.18a shows the I_{DS}–V_{DS} characteristics of 0.2 μm length vertical MOSFETs using Si and SiGe channels. As seen in the characteristics, the drain currents obtained for the SiGe channel p-HFET are about 100% higher than those for the control Si devices fabricated on the same wafer. Threshold voltages of both the devices were quite high because of unoptimized channel doping. The enhancement in drive current in SiGe was attributed to the

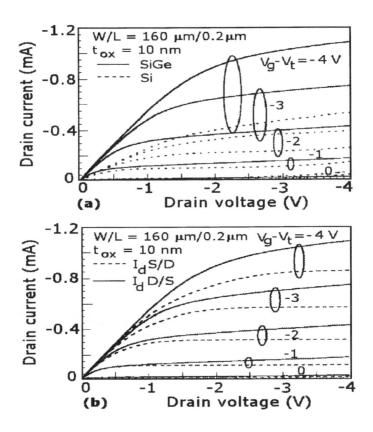

Figure 6.18. (a) I − V characteristics of SiGe (solid line) and Si (dashed line) vertical p-MOSFETs with 0.2 μm channel length, and (b) I − V characteristics of SiGe vertical p-MOS transistor with source contact on the top and drain on the bottom (dashed line) and in reverse mode (solid line). After K. C. Liu et al., IEEE Electron Dev. Lett., 19, 13–15 (1998), copyright ©1998 IEEE.

combined effect of higher hole mobility in SiGe and a built-in electric field along the channel.

Interchanging source and drain connections yielded a lower value of drive current for the same MOSFET. Figure 6.18b shows the corresponding I_{DS}–V_{DS} characteristicswith drain contact on the top and source contact at the bottom (referred to as reverse mode). The lower value of drain current compared to that of figure 6.18a indicated the role of the built-in electric field in enhancing the drain current in SiGe vertical HFETs. Channel mobility of holes could not be extracted from the device structure because of very short-channel lengths in the vertical direction. However, the linear peak transconductance of the SiGe device was found to be 200 mS in comparison to 75 mS for the control Si device. This is direct evidence of the enhancement of out-of-plane hole mobility in a strained SiGe layer. The transconductance of the devices, which is very low compared to state-of-the-art Si devices, is limited by the high source and drain series resistances.

At high transverse fields, the linear g_m of SiGe devices is comparable to that of Si as the holes are drawn near to the oxide interface and hence the g_m becomes surface roughness scattering limited. The I_{DS}–V_{GS} characteristics of both the Si and SiGe vertical devices showed a higher subthreshold slope compared to planar bulk Si devices. The presence of end-of-range defects along with gate oxide damage due to Ge implantation leading to higher gate interface state and off-state leakage current may contribute to higher subthreshold slopes in the vertical structures. The use of an epitaxially grown channel rather than an implanted one may be beneficial for practical device applications.

Vertical n-MOSFETs using an SiGe channel are also attractive because of the theoretical prediction of higher out-of-plane electron mobility compared to bulk Si. In compressively strained SiGe, electrons populate perpendicularly in the Δ_4 band where the longitudinal effective mass of the electron is much lower than the transverse effective mass. In the case of a vertical n-MOSFET the transport direction is the longitudinal one. The improved transport properties of electrons perpendicular to the growth plane in SiGe has also been experimentally observed [58] in vertical n-MOSFET devices. This is in contrast to the degradation of in-plane electron mobility reported in the case of strained SiGe planar HFETs. The alloy scattering due to Ge has been proposed to be small relative to impurity scattering due to higher doping in the case of 0.2 μm vertical MOSFETs. The experimental observations indicate the potential of high performance SiGe vertical CMOS devices exploiting the enhancement of mobility of both the carriers in the out-of-plane direction compared to planar devices with degraded in-plane electron mobility.

6.6.2   Scaling of Vertical FETs

By introducing a thin SiGeC diffusion barrier layer in the source and drain regions, boron diffusion has been suppressed, and this has enabled the scaling of vertical p-channel MOSFETs to under 1000 Å in channel length [59]. The output current–voltage and the subthreshold drain current vs. gate voltage characteristics for devices with a channel length of 700 Å are shown in figure

6.19 and one notes that well-behaved drain currents and subthreshold curves are obtained. A transconductance of 100 mS/mm at $V_{DS} = -1.1$ V was reported for a 60 Å gate oxide in planar control Si devices.

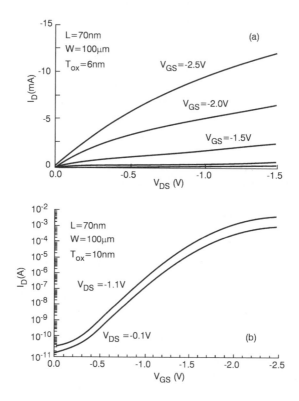

*Figure 6.19. (a) Output current–voltage and (b) subthreshold drain current vs. gate voltage characteristic for $L = 70$ nm. Gate oxide was 6 or 10 nm measured on the planar Si surface. After M. Yang et al., IEEE Electron. Dev. Lett., **20**, 301–303 (1999), copyright ©1999 IEEE.*

Due to the high channel doping (2.5×10^{18} cm^{-3}), the subthreshold slopes are relatively large, 190 mV/decade. Longer channel length devices (0.1 μm) with lower doping (0.9×10^{18} cm^{-3}) and the same gate oxide thickness had better slopes (88 mV/decade). However, when further scaling of the device was attempted with a 25 nm channel length, the devices suffered from the onset of punchthrough, but the gate still can control the drain current in the linear region as shown in figure 6.20. Note that the strained SiGeC has a bandgap

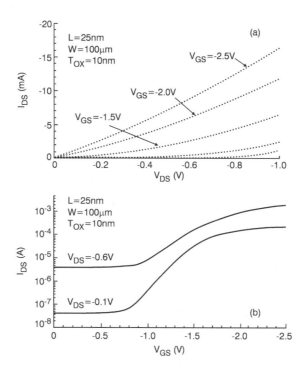

Figure 6.20. (a) Output current–voltage and (b) subthreshold drain current vs. gate voltage for devices with L = 25 nm. Gate oxide was 100 Å measured on the planar Si surface. After M. Yang et al., IEEE Electron. Dev. Lett., **20,** *301–303 (1999), copyright ©1999 IEEE.*

0.15 eV lower than Si, with most of the offset in the valence band [60]. It has been predicted that a narrower bandgap in the source region would help reduce the floating body effects [61], but their effect in scaled vertical p-channel devices is uncertain.

6.7 Strained Si p-HFETs

While biaxially compressed $Si_{1-x}Ge_x$ offers many desirable properties, most of the advantages are encountered in the valence band causing an enhancement in hole mobility. To realize improvements in electron mobility and a usable conduction band offset, it is necessary for the material to be in biaxial tension.

As discussed in Chapter 3, a smaller lattice constant Si epi-layer will be in biaxial tension when grown on a larger lattice constant relaxed $Si_{1-x}Ge_x$.

In this case, type-II band offset occurs and the structure has several advantages over the more common type-I band alignment, as a large band offset (on the order of 100 meV or more) is obtained in both the conduction and valence bands, relative to the relaxed $Si_{1-x}Ge_x$ layer [62]. Strained Si both provides larger conduction and valence band offsets and does not suffer from alloy scattering (hence mobility degradation) [63]. The significant improvement in both electron and hole mobility shows the possibility of both n- and p-type FET devices for strained-Si/SiGe-based heterostructure CMOS (HCMOS) technology.

As mentioned earlier, strained Si is more difficult to grow compared to strained $Si_{1-x}Ge_x$, since bulk $Si_{1-x}Ge_x$ substrate is currently not available and, until recently, growth of relaxed $Si_{1-x}Ge_x$ without forming a large concentration of defects due to dislocations was difficult.   However, the ability to achieve both n-MOS and p-MOS devices using strained Si provides a promising alternative for next generation high performance SiGe-CMOS technology [7, 64].

It has been predicted [63] and demonstrated experimentally that the hole mobility is improved in strained Si [9, 10, 65]. The enhancement in the hole mobility was found to be 40% at room temperature and 200% at 77 K. To eliminate the parasitic hole well that forms in the relaxed buffer, a modified structure that has a pseudomorphic SiGe layer has also been proposed [65]. These structures offer marginal improvement in the hole mobility, but require considerable effort in the growth of the relaxed buffer. As a direct consequence of this thick relaxed buffer, the integration with n-type Si MOSFETs will result in a nonplanar topography.

The tensile strain in Si grown on relaxed SiGe buffer raises the light-hole band and lowers the heavy-hole band leading to a significant increase in the low field hole mobility. Observation of hole mobility enhancement in strained Si p-MOSFETs was first demonstrated by Nayak *et al.* [66]. The initial devices were fabricated on a 1 μm thick uniform-composition partially-relaxed SiGe buffer, which is known to have a very high defect density [67], and this resulted in a limited performance (subthreshold slope 111 mV/decade).

Recently, an improved device structure and process to fabricate high performance strained Si p-MOSFETs with a step-graded completely-relaxed thick (3 μm) SiGe buffer layer (defect density $< 10^5$ cm^{-2}), a low thermal budget (maximum temperature 700 ^{o}C) and a high-quality (100 Å) gate oxide has been reported The device structure used is shown in figure 6.21a. It was shown that the high field channel mobility of a device with Ge concentration of

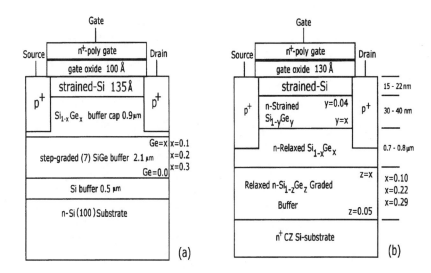

Figure 6.21. Schematic diagram of a strained Si p-MOSFET: (a) strained Si grown on a fully relaxed SiGe buffer layer (abrupt) and (b) strained Si grown on a grade-back $Si_{1-y}Ge_y$ layer (graded).

0.18 in the SiGe buffer is 40% and 200% higher at 300 K and 77 K, respectively, than that of a similarly processed bulk Si p-MOSFET. This is a consequence of the biaxial tensile strain in Si which improves in-plane hole mobility. Rim *et al.* [65] have also reported enhanced hole mobilities in surface-channel p-MOSFETs (see figure 6.21b) employing strained Si on pseudomorphic $Si_{1-y}Ge_y$ on fully relaxed $Si_{1-x}Ge_x$ buffer layers.

Figure 6.22 shows the variation of low V_{DS} (–0.1 and –0.3 V) transconductances of strained Si and control Si p-MOSFETs at 300 K for the device structure shown in figure 6.21a. For low V_{GS} the output current is small and remains nearly constant up to –0.65 V. The gate voltage at which peak transconductance occurs depends on the value of V_{DS} and the device type, i.e., control Si or strained Si. The control Si device shows one large peak at –1.7 V. But two peaks are perceptible at –1.5 V and –1.9 V, respectively, for the strained Si devices at 300 K. The peak at –1.5 V corresponds to hole confinement at the strained-Si/SiGe-buffer interface.

At higher gate voltage, however, the holes at the SiO_2/strained-Si interface dominate the channel conduction and the device becomes a surface channel device. The transition from buried channel to surface channel is clearly seen

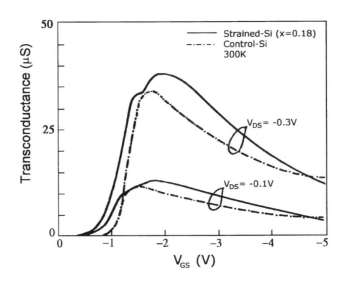

*Figure 6.22. Linear transconductances of a long-channel (L × W = 100 μm × 300 μm) strained Si (on an 18% Ge buffer layer) and a control Si p-MOSFET at 300 K. After C. K. Maiti et al., Solid-State Electron., **41**, 1863–1869 (1997).*

from the transconductance plot at 77 K (see figure 6.23). The two peaks are clearly seen at –1.55 V and –2.7 V. The I_{DS}–V_{GS} characteristics at 77 K for both the strained Si and control Si devices are also shown in this figure. It will be noticed that there is substantial current at V_{GS} close to zero, particularly for the control Si device. For the strained Si device the characteristics indicate an accumulation current threshold of about –1 V. When the temperature is reduced to 77 K, the mobility improves in both Si and strained Si, the factor of improvement depending on the scattering mechanisms operating at the applied gate voltage.

The transverse field dependence of MOS device parameters has assumed a greater importance because the thinner gate dielectrics and higher doping levels used in submicron MOSFETs lead to very high transverse electric fields well above 0.5 MV/cm. It is well known that such high fields cause a degradation in device performance. The variation of the effective mobility with electric field is often used as a basis of comparison of developed MOS devices and for computer aided design. The transconductance factor, field effect mobility and effective mobility computed from I_{DS}–V_{GS} characteristics at

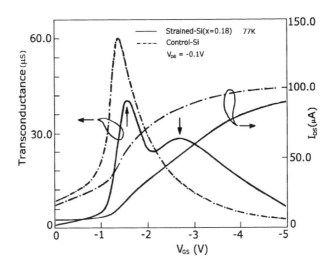

Figure 6.23. Linear transconductances of a long-channel (L × W = 100 μm × 300 μm) strained Si and a control Si p-MOSFET at 77 K. Drain currents for the devices are also shown (right scale). After C. K. Maiti et al., Solid-State Electron., 41, 1863–1869 (1997).

room and liquid nitrogen temperature (LNT) have been compared for strained Si and control Si accumulation p-MOSFET devices [10].

Figure 6.24 shows the variation of computed field effect mobility and effective mobility for strained Si and control Si at 77 K. The effective field values corresponding to gate voltage (assuming $V_{FB} = -1$ V) are indicated. The presence of the surface and parasitic channel at the strained-Si/SiO$_2$ and SiGe/Si interface is indicated by the transconductance (see figure 6.23). Above $V_{GS} = -2.5$ V, the strained Si device shows an improvement in both field effect mobility and effective mobility.

6.8 Strained Si n-MOSFETs

The mobility of the Si inversion layer is different from that in bulk Si. Therefore the two-dimensional characteristics of carriers in the inversion layer should be considered to explain the physical origin of mobility enhancement in strained Si MOSFETS. In the inversion layer on a Si(100) surface, six valleys in

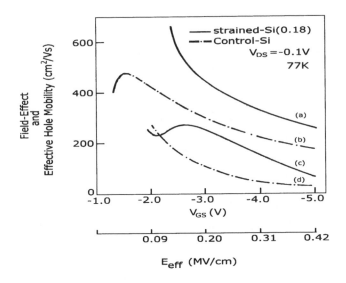

Figure 6.24. Comparison of the field effect and effective hole mobility of long-channel strained Si and control Si p-MOSFETs at 77 K: (a) μ_{fe} of strained Si, (b) μ_{fe} of control Si, (c) μ_{eff} of strained Si and (d) μ_{eff} of control Si. The effective electric field values applicable at 77 K for a current threshold value of −1.0 V are also indicated. After C. K. Maiti et al., Solid-State Electron., 41, 1863–1869 (1997).

the conduction band are classified into twofold degenerate valleys with m_l perpendicular to the Si/SiO_2 interface and fourfold degenerate valleys with m_t parallel to the interface. The electronic states in the inversion layer are quantized into subbands. Even in an unstrained Si MOS structure, there is band splitting between the subband energies in the Δ_2 and Δ_4 valleys, originating from the difference in the effective mass perpendicular to the Si/SiO_2 interface. In the subband structure of a strained Si MOS, the band splitting of the conduction band $\Delta E_{\mathrm{strain}}$ is superposed on the subband energies in an unstrained Si MOS. Therefore in a strained Si n-MOSFET, although mobility limited by intravalley acoustic phonon scattering is affected very little by the strain, the mobility limited by intervalley phonon scattering rapidly increases with the strain or Ge content in the relaxed SiGe buffer. The reasons for the enhancement of inversion mobility in a strained Si n-MOS are twofold: the reduction of intervalley phonon scattering as a function of increasing strain and the reduced occupancy of the electrons in Δ_4 valleys, which have a lower mobility due to the stronger interaction with intervalley phonons.

Very high electron mobilities demonstrated in strained Si layers suggest a great potential for this material in high transconductance n-MOSFETs. To date, in-plane electron mobilities approaching 3000 cm^2/Vs have been reported in long-channel MOSFETs with both surface and buried channel [43]. Figure 6.25 shows the schematic diagrams of several possible configurations of strained Si channel MOSFETs. All the structures have thick, relaxed Si$_{1-x}$Ge$_x$ buffer layers, consisting of a layer with linearly-graded Ge content, followed by a constant Ge content layer. The surface channel device (figure 6.25a) has a single layer of thin strained Si grown on top of the relaxed buffer layer. This layer is oxidized to form the gate oxide. The buried strained Si channel device (figure 6.25b) has a layer of strained Si buried beneath a thin layer of relaxed Si$_{1-x}$Ge$_x$. An additional layer of strained Si is necessary to form the gate oxide on top of the Si$_{1-x}$Ge$_x$, but ideally this additional Si layer (sacrificial layer) should be consumed during oxidation. If this sacrificial layer is not consumed fully, then a very thin layer of Si left between the gate oxide and the Si$_{1-x}$Ge$_x$ barrier layer (figure 6.25c), which can act as a parallel conducting channel, strongly affects the device performance. Depending on the dopant type in the layers, these structures can be used for n- or p-MOSFETs.

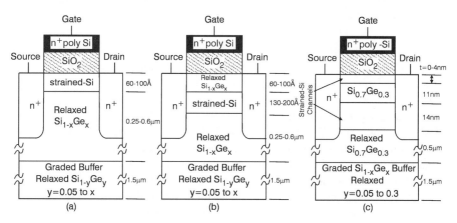

Figure 6.25. Device structures for strained Si MOSFETs with (a) Si on the surface, (b) Si buried and (c) dual strained Si channels.

Welser *et al.* [43, 68] have fabricated both p- and n-MOSFETs using all these device structures and some of their results on n-MOSFETs are presented

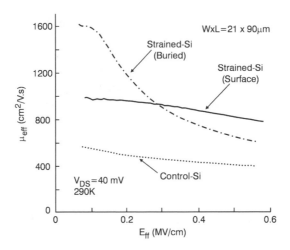

*Figure 6.26.   Effective low field mobility vs. effective field for different n-MOSFETs. The surface channel strained Si mobility shows a fairly constant mobility enhancement compared to that of control Si device, while the buried strained Si mobility peaks at low fields, but decreases rapidly at higher fields. After J. Welser et al., IEEE Electron Dev. Lett., **15**, 100–102 (1994), copyright © 1994 IEEE.*

below. Long-channel ($L \times W = 10$ μm $\times$ 168 μm) surface and buried n-MOSFET devices fabricated on relaxed $Si_{0.7}Ge_{0.3}$ buffer layers showed well-behaved output characteristics. The effective low field mobilities for these device structures are shown in figure 6.26. For the surface channel strained Si device μ_{eff} is enhanced compared to the control Si device and has a similar dependence on the effective electric field. The peak mobility is 1000 cm^2/Vs, which shows an 80% enhancement over control Si devices (550 cm^2/Vs). The peak mobility value for buried channel devices is over 1600 cm^2/Vs, which is almost 3 times that of control Si device. Room temperature effective mobility vs. electric field curves of surface-channel, strained Si n-MOSFETs with different Ge content in the buffer layer are shown in figure 6.27 along with the mobility extracted from a control Si device. Strained Si mobility increases with increasing strain (more Ge content in the relaxed buffer layer) and has little dependence on the effective electric field.

Presently, high quality relaxed SiGe substrates are not available and thick relaxed SiGe buffer layers are difficult to integrate in conventional Si

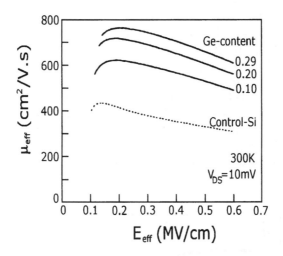

Figure 6.27. Effective mobility of surface-channel, strained Si n-MOSFETs at room temperature. Strained Si mobility increases with increasing strain (more Ge content in the relaxed buffer layer). After J. Welser et al., IEEE IEDM Tech. Dig., 373–376 (1994), copyright ©1994 IEEE.

technology. Although several methods have been used to deposit relaxed SiGe buffer layers, such as by grading the Ge in the layer, by chemical-mechanical polishing of a graded SiGe layer or by using a stepwise SiGe buffer layer, tensile strained Si device performance is still limited because of (i) residual strain in the buffer layer which may cause further relaxation in the active layer, (ii) a poor understanding of how dislocations affect dopant diffusion and (iii) the fact that, in a p-HFET device, due to the valence band offset, the holes are confined in the relaxed SiGe layer where the hole mobility is lower [10, 65].

For the growth of device quality strained Si, it is necessary to grow a graded SiGe layer several microns thick in which most of the dislocations are trapped, followed by a relaxed $Si_{1-x}Ge_x$ buffer layer of constant concentration. However, the growth of a several microns thick buffer layer with continuous or step grading possesses inherent technological complexity in addition to the difficulty of integrating mixed devices using strained Si, unstrained Si and $Si_{1-x}Ge_x$ because of the problems of lithography on complex surface topography. An alternative technique of realizing strained Si n-MOS devices on a relaxed $Si_{1-x}Ge_x$ layer formed by a single step implant followed by solid phase epitaxy and growth of a constant Ge buffer layer has been demonstrated by John *et al.* [69].

In this case, ^{74}Ge with a dose of 5×10^{16} cm^{-2} and energy of 184 keV was implanted on Si wafers followed by recrystallization anneal for 70 min at 1050 °C in a N$_2$ ambient. The high temperature anneal resulted in a relaxed SiGe film with graded Ge profile over 1500 Å with a peak Ge concentration of 12%. A thin (2000 Å) Si$_{0.85}$Ge$_{0.15}$ buffer layer of constant Ge concentration followed by a tensile strained Si layer of thickness 180–200 Å was grown using UHVCVD on implanted wafers. Figure 6.28 shows the SIMS depth profile of the resultant heterostructure. Using the (004) spectra in conjunction with a symmetric (224) reflection, the relaxation in the buffer layer of the Ge implanted sample was found to be 75%.

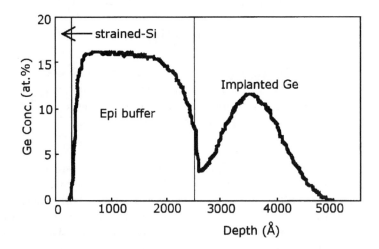

Figure 6.28. SIMS spectra of Ge implanted and annealed sample after growth of 2000 Å Si$_{0.85}$Ge$_{0.15}$ buffer layer and 200 Å strained Si layer.

Strained Si n-MOSFET devices grown on implanted relaxed SiGe layers were fabricated using the self-aligned n$^+$-poly-gate process with channel doping of 2×10^{17} cm^{-3}. Results were compared with control Si devices on an epitaxial Si layer and Si grown on a 2000 Å Si$_{0.85}$Ge$_{0.15}$ buffer (referred to as Ge unimplanted device). Figure 6.29 shows the transconductance (g_m) as a function of V_{GS}–V_T for epitaxial control Si, Ge implanted and Ge unimplanted devices with $W \times L = 10$ μm $\times$ 10 μm. The peak mobility is found to be the highest for Ge implanted transistors ($\approx$ 20% increase), while the mobility of

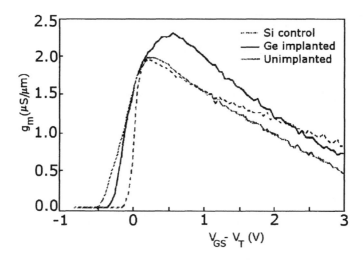

Figure 6.29. Transconductance g_m as a function of V_{GS}–V_T for epitaxial control Si, Ge implanted and unimplanted 10 µm × 10 µm n-MOSFETs.

unimplanted samples is slightly lower than that of the control Si devices. The drive current for Ge implanted samples is found (figure 6.30) to be about 20% higher than that of epitaxial control Si devices, while that of unimplanted Ge devices is found to be slightly lower than that of the control Si devices.

The enhancement of mobility in the implanted device was attributed to the reduced in-plane effective mass of electrons in the strained Si channel layer. The degradation of mobility [70] in the unimplanted sample indicated the propagation of threading dislocations into the channel layer and the presence of misfit dislocations in the non-pseudomorphic buffer layer of constant Ge concentration. On the other hand, the presence of a graded $Si_{1-y}Ge_y$ ($y = 0$ to 0.12) layer underneath the buffer $Si_{0.85}Ge_{0.15}$ in the case of the Ge implanted sample prevented dislocation propagation into the strained Si channel. The electron mobility enhancement ratio in implanted strained Si samples was much lower than that reported by Welser *et al.* [43, 71]. A much lower fraction of Ge content, 15% compared to 30% in the References 43 and 71 above, in the buffer layers and an unoptimized recrystallization anneal cycle for the formation of relaxed SiGe layers are the probable reasons. The subthreshold slopes were found to be 74 mV/decade, 76

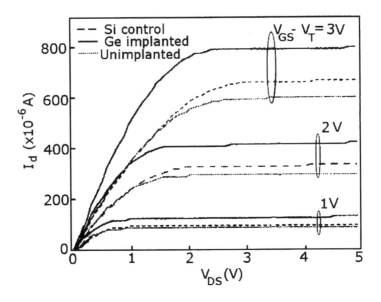

Figure 6.30. I_{DS} *vs.* V_{DS} *as a function of* $V_{GS}-V_T$ *for 10 μm × 10 μm n-MOSFETs, epitaxial control Si, Ge implanted and Ge unimplanted devices.*

mV/decade and 90 mV/decade for epitaxial control Si, Ge implanted and Ge unimplanted MOSFETs, respectively. The study demonstrated the potential of an implanted relaxed $Si_{1-x}Ge_x$ layer as a template for growing strained Si. An optimized recrystallization anneal and increase of the Ge fraction in the relaxed layer may be useful for further enhancement of the electron mobility in strained Si layers.

6.9   HFETs on SOI/SOS

Due to the low power consumption and high-speed performance, SOI MOSFETs are attractive candidates for future ULSI devices. However, the floating body effect is a major obstacle which must be overcome before they can be used in practical applications. This effect causes a lowering of the drain breakdown voltage, an anomalous decrease in subthreshold swing and kinks in the dc characteristics. These phenomena are caused by the isolation of the channel region from the substrates by buried oxide layers and the accumulation of holes generated by impact ionization near the drain region.

FET devices realized on thin SOI/SiGe/Si substrates, as shown in figure

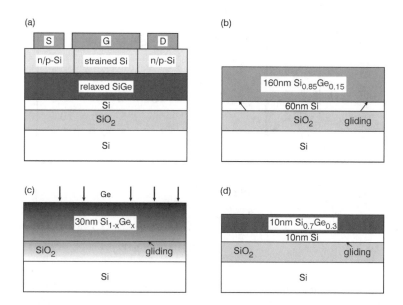

*Figure 6.31. (a) Schematic representation of a strained Si n- or p-MOSFET on a virtual $SOI/Si_{1-x}Ge_x$ substrate and different concepts for fabrication of such substrates which are based on gliding of the top Si/SiGe layers on the oxide layer at high substrate temperature T_s. (b) Relaxation of a thick SiGe layer on a thinner SOI layer during thermal annealing ($d_{SiGe} > d_{SOI} > d_{MB,crit}$). T-shaped lines illustrate dislocations within the SOI layer. (c) Ge diffusion into an SOI substrate during deposition at high T_s. (d) Relaxation (and interdiffusion) of very thin SiGe and SOI layers without dislocation formation during annealing ($d_{SiGe}, d_{SOI} \leq d_{MB,crit}$). After K. Brunner et al., Thin Solid Films, **321**, 245–250 (1998).*

6.31, may overcome these problems and may additionally offer some progress in performance concerning, for example, improved velocity overshoot, reduced operation voltage, perfect dielectric isolation and small junction capacitances. Furthermore, improved transport properties of the p-channel in such strained Si layers may lift the bottleneck caused by the poor performance of p-FETs in CMOS circuits [72].

Reported room temperature hole mobility enhancement using the SiGe

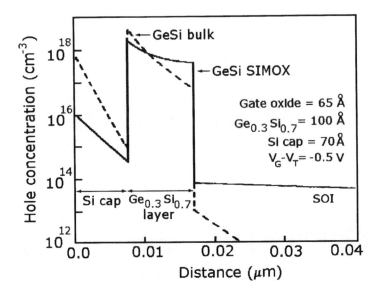

*Figure 6.32. Computed hole distribution in SiGe bulk and SiGe SIMOX devices. After D. K. Nayak et al., IEEE Electron Dev. Lett., **14**, 520–522 (1993), copyright ©1993 IEEE.*

channel is seen to vary from 30% to 70% (depending upon Ge concentration) over its Si counterpart [30]. Additional improvement in channel mobility has been observed in SiGe p-HFETs fabricated on silicon-on-insulator substrates [26]. In a fully depleted SOI device, the vertical electric field and band bending at the Si surface are significantly reduced. This property gives rise to improved hole confinement in a buried SiGe channel.

Figure 6.32 shows the simulated hole distribution as a function of band bending in SiGe SIMOX and SiGe bulk devices [26]. The device structure used a substrate doping of 5×10^{16} cm^{-3}, 1500 Å SOI layer, 100 Å Si$_{0.7}$Ge$_{0.3}$ layer and a 100 Å Si cap layer. In simulation, the top and bottom interfaces of SOI were assumed to be free of traps.

In figure 6.32, the gate overdrive, V_{GS}–V_T, is taken as –0.5 V, where V_{GS} is the gate voltage and V_T is the threshold voltage. The hole concentration at the Si surface for the SiGe SIMOX is about 2 orders of magnitude smaller than that for the SiGe bulk device. This means that the channel conduction through the parasitic surface channel (at the Si/SiO$_2$ interface) is significantly diminished in SiGe SIMOX due to reduced band bending at the surface. Also,

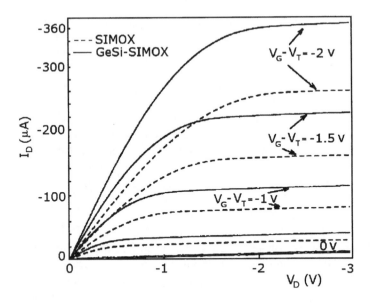

Figure 6.33. Comparison of $I - V$ characteristics of long-channel (25 μm × 10 μm) p-MOSFETs on SiGe and SIMOX. After D. K. Nayak et al., IEEE Electron Dev. Lett., 14, 520–522 (1993), copyright ©1993 IEEE.

this reduced band bending results in a more uniform hole concentration in the quantum well for the SiGe SIMOX device. The centroid of the hole distribution in the $Si_{0.7}Ge_{0.3}$ quantum well of the SiGe SIMOX device is located farther away from the Si/SiO_2 interface than that in the SiGe bulk device. For the same Si cap layer thickness, this reduces Si/SiO_2 surface scattering for the SiGe SIMOX device and results in a further improvement in channel mobility.

The $I - V$ characteristics of a long-channel device are shown in figure 6.33 for SIMOX and SiGe SIMOX devices. The drain current of the SiGe SIMOX device is 45%–65% higher in linear and saturation regions compared to that for the SIMOX device. This improvement in drain current of the SiGe SIMOX device over that of the SIMOX device is significant, considering the fact that the effective gate capacitance of the SiGe SIMOX device is smaller than that of the SIMOX device.

Among the various SOI technologies, silicon-on-sapphire is still the best candidate for microwave integrated circuits. Some of the desirable properties of sapphire include high dielectric constant, low dielectric loss, reduced self-heating due to good thermal conductivity, reduced substrate capacitance, and full dielectric isolation due to the high resistivity of sapphire. The challenge of fabrication of SiGe-HFETs on SOS comes from the natural difficulty of depositing strained SiGe on an already strained thin film SOS in which the lattice parameter is smaller than that of bulk Si [73].

Mathew *et al.* [25] have reported on the various issues involved in the realization of SiGe p-HFETs and the dc and low frequency noise; they used two-dimensional simulation to examine the profile optimization issues for a 0.25 μm p-HFET design. Several devices were fabricated on 100 nm and 50 nm SOS films. The 100 nm SOS films have three different SiGe profiles (flat 20% Ge, graded 20% Ge and flat 15% Ge) and a control Si device on SOS for comparison purposes. The graded 20% Ge in the 100 nm SOS film had a triangular distribution with the peak mole fraction of 20% oriented toward the top surface. The 50 nm SOS films had a flat 15% Ge and a true control Si device on them. The results show that the SiGe p-HFETs have higher low field mobility, transconductance, and cutoff frequency than a comparable Si p-MOSFET. All the devices were fabricated (with minor alterations) in a standard CMOS process using an n-polysilicon gate and an 80 Å gate oxide.

The typical linear $I - V$ characteristic of the SiGe p-HFETs and Si p-MOSFETs on both the 50-nm and 100-nm SOS films for a 1.0 μm gate length is shown in figure 6.34. The SiGe p-HFETs show higher current drive than the Si p-MOSFETs. The turn-off behavior of the device is excellent and the SiGe p-HFETs show the expected reduction in threshold voltage due to Ge-induced band offset [12]. The low field mobility in the 100 nm and 50 nm SiGe on SOS devices at 300 K is shown in figure 6.35. All the SiGe profiles show improvement compared to the Si profiles. The effect of thinning the SOS layer from 100 nm to 50 nm in the bulk Si p-MOSFET is seen as a reduction in μ_{eff}. The addition of Ge enhances the mobility in all the SiGe p-HFETs, however, irrespective of the thickness of the starting SOS.

Simulated cutoff frequency for 0.25 μm p-HFETs with 20% and 50% Ge profiles are shown as a function of gate bias in figure 6.36. Since the cutoff frequency at high V_{GS} is of practical interest, a good hole confinement is very important to obtain high g_{m} with a high Ge mole fraction (0.5) and a thin Si cap layer (50 Å). The spectrum of the input-referred gate voltage noise (S_{VD}) vs. frequency for a 10 μm long device is shown in figure 6.37. All devices exhibited a $1/f$ dependence. It is observed that the input-referred noise level in SiGe p-HFETs is considerably lower than in Si p-MOSFETs.

The tunneling of carriers in the inversion layer to the oxide traps is one of the major $1/f$ noise mechanisms in MOSFETs. From theory, the largest noise contribution comes from the traps nearest the quasi-Fermi level of the conducting carriers [74]. One of the reasons for the observed improvement of

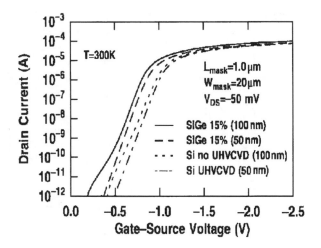

*Figure 6.34. Typical linear I_{DS}–V_{GS} characteristics of 1.0 μm gate length SiGe p-FETs for both 50 nm and 100 nm SOS films. After S. J. Mathew et al., IEEE Trans. Electron Dev., **46**, 2323–2332 (1999), copyright ©1999 IEEE.*

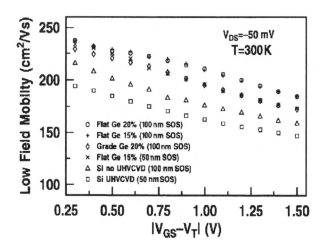

*Figure 6.35. Low field mobility extracted for the SiGe p-HFETs and Si p-MOSFETs on both 100 nm and 50 nm SOS films shown as a function of gate overdrive at 300 K. After S. J. Mathew et al., IEEE Trans. Electron Dev., **46**, 2323–2332 (1999), copyright ©1999 IEEE.*

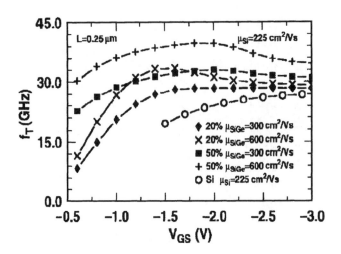

*Figure 6.36. Computed cutoff frequency over the gate bias of a 0.25 μm design for various SiGe profiles and SiGe mobility estimations. After S. J. Mathew et al., IEEE Trans. Electron Dev., **46**, 2323–2332 (1999), copyright ©1999 IEEE.*

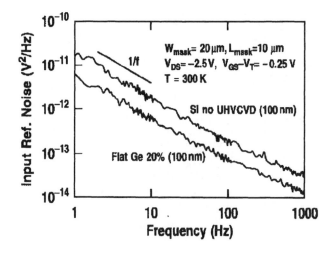

*Figure 6.37. Spectral density of input-referred gate noise voltage for a 20% SiGe p-HFET and a Si p-MOSFET on 100 nm SOS. After S. J. Mathew et al., IEEE Trans. Electron Dev., **46**, 2323–2332 (1999), copyright ©1999 IEEE.*

$1/f$ noise in SiGe p-HFETs is that the Ge-induced valence band offset increases the separation between the quasi-Fermi level of holes and the valence band edge at the surface. Due to this band offset, the separation of the quasi-Fermi level from the valence band edge at the surface is larger in the SiGe p-HFETs than in the Si p-MOSFETs, by roughly ΔE_v. From a device design point of view, use of higher Ge content and thinner Si cap should make the hole confinement gate voltage range wider [14]. As a result, the hole quasi-Fermi level can be kept well above the valence band edge over a wider gate voltage range, further lowering the $1/f$ noise over the entire bias range.

When the devices were cooled to 85 K using liquid nitrogen to study the effect of temperature on hole confinement, the SiGe p-HFETs showed a secondary peak which was not observed at room temperature. Figure 6.38 shows the linear transconductance as a function of gate overdrive for the SiGe and Si p-HFETs at 300 and 85 K. The Si p-MOSFETs continue to show only a single peak with cooling. The secondary peak in the linear transconductance is a signature of hole confinement occurring in the SiGe channel [75].

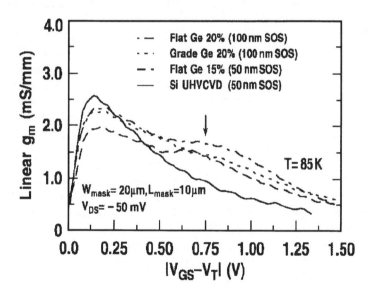

*Figure 6.38. Linear transconductance for the various SiGe and Si p-MOSFETs at 85 K showing the secondary peak, which is an indication of hole confinement occurring in the SiGe channel. After S. J. Mathew et al., IEEE Trans. Electron Dev., **46**, 2323–2332 (1999), copyright © 1999 IEEE.*

6.10   Source/Drain Engineering

For process simplicity, the poly-Si gate electrodes in a CMOS process are typically doped after patterning by the same ion implantation steps used to form the source–drain regions. However, this doping method does not yield sufficiently high active dopant concentration at the critical gate–dielectric interface, leading to problems associated with depletion of the gate under strong inversion bias [76]. In order to suppress the poly-Si gate depletion effect (PDE), higher gate implant dosage or annealing temperature is needed, and boron penetration through the thin gate oxide is inevitably enhanced.

Scaling of vertical p-channel MOSFETs with the source and drain doped with boron during low temperature epitaxy is limited by the diffusion of boron during subsequent sidewall gate oxidation. By introducing thin SiGeC layers in the source and drain regions, this diffusion has been suppressed, enabling the scaling of vertical p-channel MOSFETs down to 250 Å in channel length [59]. As the strained SiGe and SiGeC have a lower bandgap than Si, with most of the offset in the valence band, use of these semiconductors in the source and drain region would help reduce the floating body effects.

Nishiyama *et al.* [61] have proposed the formation of a narrow bandgap material   such as SiGe in the source region and have shown that this structure is effective in preventing the floating body effect of the fully depleted MOSFETs. In addition, the increased probability of impact ionization in the drain region in a lower bandgap alloy might provide an energy dissipation mechanism, which would reduce the hot carrier population. However, the lower bandgap may lead to lowering of the breakdown field and increased drain leakage. Implantation of Ge species has been employed to realize the structure due to its simplicity and compatibility with conventional VLSI processes. Poly-SiGe has also been reported as a promising alternative gate material for single-gate CMOS technology. Poly-SiGe-gated devices with lower gate sheet resistance, higher current drive and less PDE compared to poly-Si-gated devices have been demonstrated [77–79].

In SOI technology, several technology challenges exist.   First, series resistances associated with the source–drain regions limit the device performance by reducing the output current, especially when the SOI films are thin. Second, the floating body effect causes early source-to-drain breakdown and complicates design considerations when a reduced voltage supply is used. In order to overcome these technical impediments, source–drain engineering with large angle Ge implantation and a pre-amorphization technique to improve the current drive as well as breakdown voltage for fully depleted SOI MOSFETs has been proposed [79].

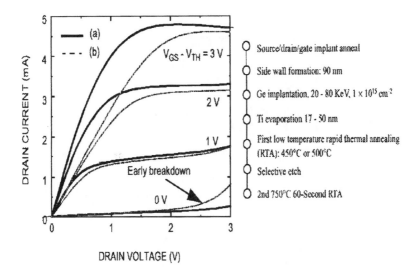

Figure 6.39. Comparison of output characteristics of two different salicide processes: (a) Ge-implanted ($L_{\text{eff}} = 0.32 \ \mu m$, Ge: $1 \times 10^{15} \ cm^{-2}$ at 40 keV, Ti = 32 nm, 450 °C 1 min RTA, selective etch, 750 °C RTA 1 min), and (b) conventional two-step RTA without Ge implantation ($L_{\text{eff}} = 0.29 \ \mu m$, Ti = 22 nm, 600 °C 1 min RTA, selective etch, 800 °C RTA 1 min). $W = 10 \ \mu m$, $t_{\text{ox}} = 7 \ nm$, $t_{\text{SOI}} = 90 \ nm$, and $N_A = 1 \times 10^{15} \ cm^{-3}$ for both devices. The silicidation process flow for Ge pre-amorphization is also included. After T. C. Hsiao et al., Proc. ESSDERC'97, 516–519 (1997).

Ge large angle tilt implantation is applied to amorphize the Si films prior to silicidation. The $I - V$ characteristics of two different salicide processes with 0.3 μm effective channel lengths are shown in figure 6.39. Process flow for Ge pre-amorphization is also included. In addition to reducing source–drain resistance, Ge implantation improves the breakdown characteristics. In devices with Ge implantation the breakdown voltage is increased by 0.3 V compared to devices without Ge implantation.

A large angle Ge implantation was used to amorphize the Si films, thereby reducing source–drain series resistances and suppressing floating body effects simultaneously. This tilt implant makes a wider amorphous layer that reaches the SOI device body close to the buried oxide interface (see figure 6.40). Together with the SiGe pocket, the minority carrier (holes in the source)

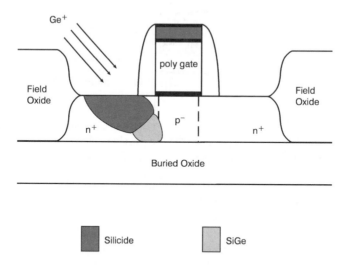

Figure 6.40.   Schematic representation of large angle Ge implant and the formation of silicide layer and SiGe pocket. After T. C. Hsiao et al., Proc. ESSDERC'97, 516–519 (1997).

recombination velocity increases and hole injection can potentially be further improved. Source/drain parasitic resistances are substantially reduced and due to the formation of a metal–semiconductor barrier near the source/channel junction, the floating body effects are reduced.

Figure 6.41 shows the output characteristics of two identical SOI n-MOS 0.25 μm devices with Ge implant angles of 70° and 45°, respectively. Large angle implantation further improves the breakdown voltage by 0.6 V. Large angle Ge implantation enhances the hole flow into the source by creating a SiGe pocket and a metal–semiconductor barrier. As a result, further reduction of the emitter injection efficiency can be accomplished.

p-channel MOS transistors have been fabricated with raised SiGe sources and drains to demonstrate the concept of latch-up suppression using SiGe. The concept relies on the reduced bandgap of SiGe to suppress the gain of the parasitic bipolar transistors formed by the source (emitter), n-well (base) and substrate (collector) and by the source (emitter), n-well (base) and drain (collector). A reduction in parasitic gain by a factor of two was obtained for 16% Ge and by a factor of eighteen for 20% Ge.

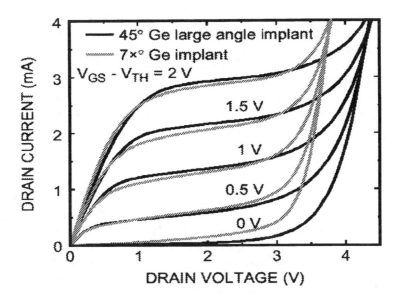

Figure 6.41. Output characteristics of two 0.25 μm SOI n-MOS devices with and without Ge large angle implant. W = 10 μm, t_{ox} = 7 nm, t_{SOI} = 140 nm and N_A = 1 × 10^{16} cm^{-3} for both devices. After T. C. Hsiao et al., Proc. ESSDERC'97, 516–519 (1997).

6.11 SiGe-CMOS Process

A process for fabricating both a p-HFET and an n-MOSFET in a single SiGe channel region has been proposed [80]. The fully integrated SiGe CMOS fabrication process solves the thermal budget and multi-deposition problems and, furthermore, results in a planar structure at the expense of only one additional mask.

A typical process sequence is as follows: first, a boron-doped layer is grown with a typical concentration and thickness of 4 × 10^{17} cm^{-3} and 350 Å. Next, an undoped SiGe channel is grown and followed by a thin intrinsic Si cap. This cap separates the gate oxide from the SiGe channel to minimize surface scattering. After processing, a quasi-uniform doping profile is obtained with a peak doping level of typically 3.5 × 10^{17} cm^{-3} underneath the channel and a surface–junction depth of 400 Å. At the top of the channel where most of the holes flow, the doping concentration is below 1 × 10^{16} cm^{-3} and impurity scattering is hence minimized.

The SiGe is passivated with a thin (100 Å) undoped Si epi layer. The n-well is self-passivated as positive charge in the oxide on the sidewalls will lead to accumulation. The p-well is passivated by a boron implant. An additional mask follows to adjust the n-MOSFET threshold voltage. Low temperature processing is then used to complete the devices, including a single n^+-poly-Si deposition for the gates of both the n-MOSFET and p-HFET. The resulting n-MOSFET is a surface-channel device and the p-HFET a sub-surface-channel device with holes confined in the SiGe channel.

6.12   Summary

Enhancement of the performance of FETs was established in the early stage of development of SiGe heterostructure technology. Band structure modifications responsible for the improvement were also identified. Significant advances that have taken place over the last fifteen years include extension of the alloy composition to include carbon, use of vertical structures and innovations to include bipolar and MOS devices and complementary MOS devices in the same structure within the framework of Si IC technology. Results obtained as presented in this chapter are quite impressive.

Various types of heterostructure MOSFETs realized in Si technology have been considered. As a buried channel SiGe p-HFET structure is expected to reduce hot carrier degradation of the oxide, it is desirable to fabricate buried channel SiGe devices with an optimized Si cap layer.

It is worth considering some general features of the structures studied. The simplest HFET is a p-SiGeC quantum well on Si. The corresponding n-MOS with electron wells uses strained Si on relaxed silicon–germanium. The structure next in order of complexity for p-MOS satisfying the requirement of a top Si layer to allow growth of good oxynitride consists of Si cap/SiGe/Si. The results reported in this chapter refer to many instances of this p-MOS buried channel structure. The vertical MOSFET offers higher packing densities, is easier to use for channel engineering, involves a simpler fabrication process and offers lower DIBL and substrate bias effect and higher punchthrough voltage.

The results of studies by different groups have shown that incorporation of carbon in silicon–germanium relaxes the rigid relation between strain and band offset and stabilizes the alloy. This is because the linear relationship between Ge and carbon fractions and the resulting strain has parameters different from that for valence band offsets.

Performance improvement of strain compensated $Si_{1-x-y}Ge_xC_y$ channel p-MOSFETs over $Si_{1-x}Ge_x$ channel devices has been demonstrated as a result of less process induced relaxation in the ternary layer. The incorporation of controlled amounts of carbon can provide a wider process window for device fabrication with strained $Si_{1-x}Ge_x$ films with thickness near the metastable critical limit. The complete strain compensation of the film by incorporating higher amounts of C (Ge-to-C ratio = 10:1), however, results in a degradation of the device characteristics compared to the binary one.

A word of caution is warranted regarding theoretical estimates of improvement principally because of the uncertainty about the magnitude of the alloy scattering potential. While assessing experimental results displaying higher current drive capability, one must note that a surface channel is quite often present contributing its share of the current. It is also necessary to ascertain that an improvement is reflected in the transconductance. To be fair, comparison of p-HFETs should be made with bulk type Si MOSFETs using an n^+-poly-Si gate.

Unfortunately, there have not been many studies on strained layer n-MOSFETs, even though it has been established that tensilely strained Si n-MOSFETs can provide an almost 2:1 improvement in effective mobility. This is because of the need for a thick relaxed SiGe pseudo-substrate and perhaps the n-MOSFET performance to start with is satisfactory. Buried channel tensilely strained n-MOSFETs would have a minimal configuration of Si/SiGe/Si/SiGe. This implies that a surface channel will always be present because of the oxide requirement. A tensilely strained buried channel p-MOSFET appears difficult to realize because its interface with SiGeC has a valence band offset impeding the flow of carriers from the bulk to the channel; the surface channel is destined to dominate.

Current activity apart from consolidating results obtained in research laboratories appears to be directed towards development of the SiGe CMOS process in an SOI environment. SOI-CMOS technology, in general, is of great interest at short gate lengths due to its improved short-channel effects, reduced parasitics and reduced processing complexity compared to traditional CMOS technology on bulk Si. This is considered to be the technology of the next generation. From a design point of view, the SiGe on SOI/SOS technology provides a number of tunable design parameters such as Ge content, Si cap thickness and modulation doping, in addition to oxide thickness and doping levels, which can be used to tailor the performance. Detailed investigation is warranted into appropriate SOI growth technology for SiGeC film incorporation and performance of fabricated devices. Many innovations are being proposed and studied. A layer structure of Si/SiGe/Si/SiGe has been proposed to integrate HBT and the so-called charge injection transistor (CHINT) [81, 82]. A CMOS using tensilely strained n-channel Si- and compressively strained p-channel Ge-MOS deserves fuller study.

6.13   References

1 S. Wolf, *Silicon processing for the VLSI era – The Submicron MOSFET.* Lattice Press, Calif., 1995.

2 Y. Taur and E. J. Nowak, "CMOS devices below 0.1 μm: How will Performance Go?," in *IEEE IEDM Tech. Dig.*, pp. 215–218, 1997.

3 S. C. Jain, *Germanium–Silicon Strained Layers and Heterostructures.* Academic Press Inc., New York, 1994.

4 S. C. Jain and W. Hayes, "Structure, properties and applications of Ge_xSi_{1-x} strained layers and superlattices," *Semicond. Sci. Technol.*, vol. 6, pp. 547–576, 1991.

5 T. P. Pearsall and J. C. Bean, "Enhancement and Depletion-Mode p-Channel Ge_xSi_{1-x} Modulation-Doped FETs," *IEEE Electron Dev. Lett.*, pp. 308–310, 1986.

6 R. People and J. C. Bean, "Band alignments of coherently strained Ge_xSi_{1-x}/Si heterostructures on $< 001 >$ Ge_ySi_{1-y} substrates," *Appl. Phys. Lett.*, vol. 48, pp. 538–540, 1986.

7 C. K. Maiti, L. K. Bera and S. Chattopadhyay, "Strained Si Heterostructure Field Effect Transistors," *Semicond. Sci. Technol.*, vol. 13, pp. 1225–1246, 1998.

8 D. K. Nayak, J. C. S. Woo, J. S. Park, K. L. Wang and K. P. MacWilliams, "Enhancement-Mode Quantum-Well Ge_xSi_{1-x} PMOS," *IEEE Electron Dev. Lett.*, vol. EDL-12, pp. 154–156, 1991.

9 D. K. Nayak, K. Goto, A. Yutani, J. Murota and Y. Shiraki, "High-Mobility Strained Si PMOSFETs," *IEEE Trans. Electron Dev.*, vol. 43, pp. 1709–1715, 1996.

10 C. K. Maiti, L. K. Bera, S. S. Dey, D. K. Nayak and N. B. Chakrabarti, "Hole mobility enhancement in strained Si p-MOSFETs under high vertical fields," *Solid-State Electron.*, vol. 41, pp. 1863–1869, 1997.

11 S. Verdonckt-Vandebroek, F. Crabbe, B. S. Meyerson, D. L. Harame, P. J. Restle, J. M. C. Stork and J. B. Johnson, "SiGe-Channel Heterojunction p-MOSFETs," *IEEE Trans. Electron Dev.*, vol. 41, pp. 90–102, 1994.

12 G. F. Niu and G. Ruan, "Threshold voltage and inversion charge modeling of graded SiGe channel modulation-doped p-MOSFETs," *IEEE Trans. Electron Dev.*, vol. 42, pp. 2242–2246, 1995.

13 S. P. Voinigescu, C. A. T. Salama, J.-P. Noel and T. I. Kamins, "Optimized Ge Channel Profiles for VLSI Compatible Si/SiGe p-MOSFETs," in *IEEE IEDM Tech. Dig.*, pp. 369–372, 1994.

14 G. F. Niu, G. Ruan and D. H. Zhang, "Modeling of hole confinement gate voltage range for SiGe Channel p-MOSFETs," *Solid-State Electron.*, vol. 39, pp. 69–73, 1996.

15 F. K. LeGoues, R. Rosenberg, T. Nguyen, F. Himpsel and B. S. Meyerson, "Oxidation studies of SiGe," *J. Appl. Phys.*, vol. 65, pp. 1724–1728, 1989.

16 V. P. Kesan, S. Subbanna, P. J. Restle, M. J. Tejwani, J. M. Aitken, S. S. Iyer and J. A. Ott, "High Performance 0.25 μm p-MOSFETs with Silicon Germanium Channels for 300K and 77K Operation," in *IEEE IEDM Tech. Dig.*, pp. 25–28, 1991.

17 M. Mukhopadhyay, S. K. Ray, C. K. Maiti, D. K. Nayak and Y. Shiraki, "Electrical properties of oxides grown on strained SiGe layer at low temperatures in a microwave oxygen plasma," *Appl. Phys. Lett.*, vol. 65, pp. 895–897 (see also Vol. 66(12), p.1566, March 20, 1995 issue), 1994.

18 P. W. Li, E. S. Yang, Y. F. yang, J. O. Chu and B. S. Meyerson, "SiGe pMOSFET's with Gate Oxide Fabricated by Microwave Electron Cyclotron Resonance Plasma Processing," *IEEE Electron Dev. Lett.*, vol. 15, pp. 402–405, 1994.

19 P. W. Li, H. K. Liou, E. S. Yang, S. S. Iyer, T. P. Smith, III and Z. Lu, "Formation of stoichiometric SiGe oxide by electron cyclotron resonance plasma," *Appl. Phys. Lett.*, vol. 60, pp. 3265–3267, 1992.

20 P. W. Li and E. S. Yang, "SiGe gate oxide prepared at low-temperatures in an electron cyclotron resonance plasma," *Appl. Phys. Lett.*, vol. 63, pp. 2938–2940, 1993.

21 D. K. Nayak, J. S. Park, J. C. S. Woo, K. L. Wang, G. K Yabiku and K. P. MacWilliams, "High Performance Quantum-Well PMOS on SIMOX," in *IEEE IEDM Tech. Dig.*, pp. 777–780, 1992.

22 E. Murakami, K. Nakagawa, A. Nishida and M. Miyao, "Fabrication of strain-controlled SiGe/Ge MODFET with ultrahigh hole mobility," *IEEE Trans. Electron Dev.*, vol. 41, pp. 857–861, 1994.

23 P. M. Garone, V. Venkataraman and J. C. Sturm, "Hole mobility enhancement in MOS gated Ge_xSi_{1-x} heterostructure inversion layers," *IEEE Electron Dev. Lett.*, vol. 13, pp. 56–58, 1992.

24 S. Verdonckt-Vandebroek, E. F. Crabbe, B. S. Meyerson, D. L. Harame, P. J. Restle, J. M. C. Stork, A. C. Meydanis, C. L. Stanis, A. A. Bright, G. M. W. Kroesen and A. C. Warren, "High-mobility modulation-doped grades SiGe-Channel p-MOSFET's," *IEEE Electron Dev. Lett.*, vol. EDL-12, pp. 447–449, 1991.

25 S. J. Mathew, G. Niu, W. B. Dubbelday and J. D. Cressler, "Characterization and Profile Optimization of SiGe pFETs on Silicon-on-Sapphire," *IEEE Trans. Electron Dev.*, vol. 46, pp. 2323–2332, 1999.

26 D. K. Nayak, J. C. S. Woo, G. K. Yabiku, K. P. MacWilliams, J. S. Park and K. L. Wang, "High Mobility GeSi PMOS on SIMOX," *IEEE Electron Dev. Lett.*, vol. 14, pp. 520–522, 1993.

27 P. M. Garone, V. Venkataraman and J. C. Sturm, "Hole Confinement in MOS-Gated Ge_xSi_{1-x}/Si Heterostructures," *IEEE Electron Dev. Lett.*, vol. 12, pp. 230–232, 1991.

28 S. C. Sun and J. D. Plummer, "Electron Mobility in Inversion and Accumulation Layers on Thermally Oxidized Silicon Surfaces," *IEEE Trans. Electron Dev.*, vol. 27, pp. 1497–1508, 1980.

29 P. M. Garone, V. Venkataraman and J. C. Sturm, "Mobility Enhancement and Quantum Mechanical Modeling in Ge_xSi_{1-x} Channel MOSFETs From 90 to 300 K," in *IEEE IEDM Tech. Dig.*, pp. 29–32, 1991.

30 K. Goto, J. Murota, T. Maeda, R. Schutz, K. Aizawa, R. Kircher, K. Yokoo and S. Ono, "Fabrication of a $Si_{1-x}Ge_x$ channel metal–oxide–semiconductor field-effect transistor (MOSFET) containing high Ge fraction layer by low pressure chemical vapor deposition," *Jap. J. Appl. Phys.*, vol. 32, pp. 438–441, 1993.

31 T. E. Whall and E. H. C. Parker, "Silicon–Germanium Heterostructures - Advanced Materials and Devices for Silicon Technology - Review," *J. Mater. Sci.: Mater. Electron.*, vol. 6, pp. 249–264, 1995.

32 S. C. Martin, L. M. Hitt and J. J. Rosenberg, "p-Channel germanium MOSFETs with high channel mobility," *IEEE Electron Dev. Lett.*, vol. 10, pp. 325–326, 1989.

33 K. L. Wang, S. G. Thomas and M. O. Tann, "SiGe Band Engineering for MOS, CMOS and Quantum Effect Devices," *J. Mater. Sci.: Mater. Electron.*, vol. 6, pp. 311–324, 1995.

34 C. R. Selvakumar and B. Hecht, "SiGe channel n-MOSFET by Germanium Implantation," *IEEE Electron Dev. Lett.*, vol. EDL-12, pp. 444–446, 1991.

35 J. J. Rosenberg and S. C. Martin, "Self aligned Germanium MOSFETs using a nitrided native oxide gate insulator," *IEEE Electron Dev. Lett.*, vol. 9, pp. 639–640, 1988.

36 K. Eberl, S. S. Iyer, S. Zollner, J. C. Tsang and F. K. LeGoues, "Growth and strain compensation effects in the ternary $Si_{1-x-y}Ge_xC_y$ alloy system," *Appl. Phys. Lett.*, vol. 60, pp. 3033–3035, 1992.

37 J. L. Regolini, F. Gisbert, G. Dolino and P. Boucaud, "Growth and characterization of strain compensated $Si_{1-x-y}Ge_xC_y$ epitaxial layers," *Mater. Lett.*, vol. 18, pp. 57–60, 1993.

38 Z. Atzmon, A. E. Bair, E. J. Jaquez, J. W. Mayer, D. Chandrasekhar, D. J. Smith, R. L. Hervig and M. D. Robinson, "Chemical vapor deposition of heteroepitaxial $Si_{1-x-y}Ge_xC_y$ films on (001)Si Substrates," *Appl. Phys. Lett.*, vol. 65, pp. 2559–2561, 1994.

39 J. Mi, P. Warren, P. Letourneau, M. Judelewicz, M. Gailhanou, M. Dutoit, C. Dubois and J. C. Dupuy, "High quality $Si_{1-x-y}Ge_xC_y$ epitaxial layers grown on (100) Si by rapid thermal chemical vapor deposition using methylsilane," *Appl. Phys. Lett.*, vol. 67, pp. 259–261, 1995.

40 A. St. Amour, C. W. Lice, J. C. Sturm, Y. Lacroix and M. L. W. Thewalt, "Defect-free band-edge photo-luminescence and bandgap measurement of pseudomorphic $Si_{1-x-y}Ge_xC_y$ alloy layers on Si(100)," *Appl. Phys. Lett.*, vol. 67, pp. 3915–3917, 1995.

41 S. K. Ray, S. John, S. Oswal and S. K. Banerjee, "Novel SiGeC Channel Heterojunction PMOSFET," in *IEEE IEDM Tech. Dig.*, pp. 261–264, 1996.

42 S. John, S. K. Ray, y, E. Quinones, S. K. Oswal and S. K. Banerjee, "Heterostructure P-channel metal–oxide–semiconductor transistor utilizing a $Si_{1-x-y}Ge_xC_y$ channel," *Appl. Phys. Lett.*, vol. 74, pp. 847–849, 1999.

43 J. Welser, J. L. Hoyt and J. F. Gibbons, "Electron Mobility Enhancement in Strained Si N-type Metal-Oxide-Semiconductor Field-Effect Transistors," *IEEE Electron Dev. Lett.*, vol. 15, pp. 100–102, 1994.

44 E. Quinones, S. K. Ray, K. C. Liu and S. Banerjee, "Enhanced Mobility PMOSFETs Using Tensile-Strained $Si_{1-x}C_x$ Layers," *IEEE Electron Dev. Lett.*, vol. 20, pp. 338–340, 1999.

45 Technology Modeling Associates, *MEDICI, 2D Semiconductor Device Simulator, Ver. 4.0*, 1997.

46 D. Behammer, L. Vescan, R. Loo, J. Moers, A. Muck, H. Luth and T. Grabolla, "Selectively grown vertical Si-pMOS transistor with short-channel lengths," *Electronics Lett.*, vol. 32, pp. 406–407, 1996.

47 H. Gossner, F. Wittmann, I. Eisele, T. Grabolla and D. Behammer, "Vertical MOS technology with sub-0.1-μm channel lengths," *Electronics Lett.*, vol. 31, pp. 1394–1396, 1995.

48 H. Gossner, I. Eisele and L. Risch, "Vertical Si-metal–oxide–semiconductor field effect transistors with channel lengths of 50 nm by molecular beam epitaxy," *Jpn. J. Appl. Phys.*, vol. 33, pp. 2423–2428, 1994.

49 Z. Rav-Noy, U. Schreter, S. Mukai, E. Kapon, J. S. Smith, L. C. Chiu, S. Margalit and A. Yariv, "Vertical FETs in GaAs," *IEEE Electron Dev. Lett.*, vol. EDL-5, pp. 228–230, 1984.

50 T. Aeugle, L. Risch, W. Rosner, T. Schulz and D. Behammer, "Advanced self aligned SOI concepts for vertical MOS transistors with ultra-short-channel lengths," in *Proc. ESSDERC'97*, pp. 628–631, 1997.

51 K. C. Liu, S. K. Ray, S. K. Oswal and S. K. Banerjee, "A deep submicron $Si_{1-x}Ge_x$/Si vertical PMOSFET fabricated by Ge ion implantation," *IEEE Electron Dev. Lett.*, vol. 19, pp. 13–15, 1998.

52 T. J. Rodgers and J. D. Meindl, "VMOS: High-Speed TTL Compatible MOS Logic," *IEEE J. Solid-State Circuits*, vol. SC-9, pp. 239–250, 1974.

53 L. Risch, W. H. Krautschneider, F. Hofmann, H. Schafer, T. Aeugle and W. Rosner, "Vertical MOS transistors with 70 nm channel length," *IEEE Trans. Electron Dev.*, vol. 43, pp. 1495–1498, 1996.

54 H. B. Pein and J. D. Plummer, "A 3-D sidewall flash EPROM cell and memory array," *IEEE Electron Dev. Lett.*, vol. 14, pp. 415–417, 1993.

55 C. Auth and J. Plummer, "Scaling theory for cylindrical, fully depleted, surrounding-gate MOSFET's," *IEEE Electron Device Lett.*, vol. 18, pp. 74–76, 1997.

56 K. De Meyer, M. Caymax, N. Collaert, R. Loo and P. Verheyen, "The vertical heterojunction MOSFET," *Thin Solid Films*, vol. 336, pp. 299–305, 1998.

57 K. C. Liu, S. K. Ray, S. K. Oswal, N. B. Chakraborti, R. D. Chang, D. L. Kencke and S. K. Banerjee, "Enhancement of drain current in vertical SiGe/Si PMOS transistors using novel CMOS technology," in *Device Research Conf. Dig.*, pp. 128–129, 1997.

58 K. C. Liu, X. Wang, E. Quinones, X. Chen, X. D. Chen, D. Kencke, B. Anantharam, R. D. Chang, S. K. Ray, S. K. Oswal, C. Y. Tu and S. K. Banerjee, "A Novel Sidewall Strained Si Channel nMOSFET," in *IEEE IEDM Tech. Dig.*, pp. 63–66, 1999.

59 M. Yang, C.-L. Chang, M. Carroll and J. C. Sturm, "25-nm p-Channel Vertical MOSFET's with SiGeC Source-Drains," *IEEE Electron Dev. Lett.*, vol. 20, pp. 301–303, 1999.

60 C. L. Chang, A. St. Amour and J. C. Sturm, "The effect of carbon on the valence band offset of compressively strained $Si_{1-x-y}Ge_xC_y/(100)$ Si heterostructures," *Appl. Phys. Lett.*, vol. 70, pp. 1557–1559, 1997.

61 A. Nishiyama, O. Arisumi, M. Yoshimi, "Suppression of the floating-body effect in partially-depleted SOI MOSFETs with SiGe source structure and its mechanism," *IEEE Trans. Electron Dev.*, vol. 44, pp. 2187–2192, 1997.

62 R. People, "Physics and Applications of Ge_xSi_{1-x}/Si Strained Layer Heterostructures," *IEEE J. Quantum Elec.*, vol. QE-22, pp. 1696–1710, 1986.

63 D. K. Nayak and S. K. Chun, "Low field mobility of strained Si on (100) $Si_{1-x}Ge_x$ substrate," *Appl. Phys. Lett.*, vol. 64, pp. 2514–2516, 1994.

64 F. Schaffler, "High-mobility Si and Ge structures," *Semicond. Sci. Technol.*, vol. 12, pp. 1515–1549, 1997.

65 K. Rim, J. Welser, J. L. Hoyt and J. F. Gibbons, "Enhanced Hole Mobilities in Surface-channel Strained Si p-MOSFETs," in *IEEE IEDM Tech. Dig.*, pp. 517–520, 1995.

66 D. K. Nayak, J. C. S. Woo, J. S. Park, K. L. Wang and K. P. MacWilliams, "High-mobility p-channel metal–oxide–semiconductor field-effect transistor on strained Si," *Appl. Phys. Lett.*, vol. 62, pp. 2853–2855, 1993.

67 E. A. Fitzgerald, Y.-H. Xie, M. L. Green, D. Brasen, A. R. Kortan, J. Michel, Y.-J. Mii and B. E. Weir, "Totally relaxed Ge_xSi_{1-x} layers with low threading dislocation densities grown on Si substrates," *Appl. Phys. Lett.*, vol. 59, pp. 811–813, 1991.

68 J. Welser, J. L. Hoyt and J. F. Gibbons, "NMOS and PMOS transistors fabricated in Strained Silicon/relaxed Silicon–Germanium structures," in *IEEE IEDM Tech. Dig.*, pp. 1000–1003, 1992.

69 S. John, S. K. Ray, E. Quinones and S. K. Banerjee, "Strained Si NMOSFET on relaxed $Si_{1-x}Ge_x$ formed by ion implantation of Ge," *Appl. Phys. Lett.*, vol. 74, pp. 2076–2078, 1999.

70 K. Ismail, "Effect of dislocations in strained-Si/SiGe on electron mobility," *J. Vac. Sci. Technol. B*, vol. 14, pp. 2776–2779, 1996.

71 J. Welser, J. L. Hoyt, S. Takagi and J. F. Gibbons, "Strain Dependence of the Performance Enhancement in Strained Si n-MOSFETs," in *IEEE IEDM Tech. Dig.*, pp. 373–376, 1994.

72 K. Brunner, H. Dobler, G. Abstreiter, H. Schäfer and B. Lustig, "Molecular beam epitaxy growth and thermal stability of $Si_{1-x}Ge_x$ layers on extremely thin silicon-on-insulator substrates," *Thin Solid Films*, vol. 321, pp. 245–250, 1998.

73 P. M. Mooney, J. A. Ott, J. O. Chu and J. L. Jordan-Sweet, "X-ray diffraction analysis of SiGe/Si heterostructures on sapphire substrates," *Appl. Phys. Lett.*, vol. 73, pp. 924–926, 1998.

74 S. Christensson, I. Lundstrom and C. Svensson, "Low-frequency noise in MOS transistors - Part I," *Solid State Electron.*, vol. 11, pp. 797–812, 1968.

75 S. J. Mathew, G. Niu, W. B. Dubbelday, J. D. Cressler, J. A. Ott, J. O. Chu, P. M. Mooney, K. L. Kavanagh, B. S. Meyerson and I. Lagnado, "Hole Confinement and Low-Frequency Noise in SiGe pFET's on Silicon-on-Sapphire," *IEEE Electron Dev. Lett.*, vol. 20, pp. 173–175, 1999.

76 W.-C. Lee, Y.-C. King, T.-J. King and C. Hu, "Investigation of Poly-$Si_{1-x}Ge_x$ for Dual-Gate CMOS Technology," *IEEE Electron Dev. Lett.*, vol. 19, pp. 247–249, 1998.

77 T. J. King, J. R. Pfriester, J. D. Scott, J. P. McVittie and K. C. Saraswat, "A polycrystalline SiGe gate CMOS technology," in *IEEE IEDM Tech. Dig.*, pp. 253–256, 1990.

78 N. Kistler and J. Woo, "Symmetric CMOS in fully-depleted silicon-on-insulator using P^+-polycrystalline SiGe gate electrodes," in *IEEE IEDM Tech. Dig.*, pp. 727–730, 1993.

79 T. C. Hsiao, A. W. Wang, K. Saraswat and J. C. S. Woo, "An alternative gate electrode material of fully depleted SOI CMOS for low power applications," in *Proc. 1997 IEEE Int. SOI Conf.*, pp. 20–21, 1997.

80 *IBM Tech. Disc. Bull., March*, 1992.

81 E. Kasper and G. Reitemann, "Can Silicon-Based Heterodevices compete with CMOS for System Solutions ?," in *Future Trends in Microelectronics - The Road Ahead*, pp. 125–132, Wiley Interscience, New York, 1999.

82 M. Mastrapasqua, C. A. King, P. R. Smith and M. R. Pinto, "Functional devices based on real space transfer in Si/SiGe structures," *IEEE Trans. Electron Dev.*, vol. 43, pp. 1671–1677, 1996.

Chapter 7

BICFET, RTD and Other Devices

In addition to the SiGe-HBTs and HFETs discussed in Chapters 5 and 6, considerable work has been done on other SiGe heterostructures. Notable among them are modulation doped field effect transistors, resonant tunnel diodes (RTDs), BICFETs and various optical devices. The MODFET will be covered in Chapter 8 while Si/SiGe optoelectronic devices will be presented in Chapter 10. RTDs and BICFETs are discussed in this chapter. Although many other devices have been reported using SiGe strained layers, the work that has been done is not extensive and their performances have not yet reached a high level.

7.1 Bipolar Inversion Channel FETs

Taylor and Simmons [1] proposed the bipolar inversion channel field effect transistor as a device that would fulfill both the speed requirements and the future needs of scaled devices. It is a three-terminal device, which, although a bipolar transistor in nature, relies upon an inversion channel charge to control current flow. The devices were first fabricated using GaAlAs/GaAs and InGaAs/InAlAs. This concept removed the problems associated with a finite base layer used in a conventional bipolar junction transistor. However, because the influence of the inversion charge on the thermionic barrier and majority carrier flow is strong, the transconductance of the device still maintains the exponential behavior of the BJT. The BICFET has electrical characteristics that are very similar to the BJT. The major difference is that the base hole concentration of the BICFET is very dynamic with respect to applied bias. At low bias the resistance is very high, but it decreases linearly with the applied emitter–base voltage.

340

Extensive theoretical and experimental studies on BICFETs in SiGe materials have been done [2–5]. Taft and Plummer [2] implemented the concept in a SiGe system in order to take advantage of the well established technology base of Si processing. The SiGe BICFET could potentially fulfill two sides of the transistor triangle: high performance (intrinsic speed advantage) and manufacturability (low cost of Si processing). In a BICFET, an inversion channel base is used in place of a charge neutral base. In the absence of the charge neutral base and associated diffusion capacitances and transit times, the speed of the device should be very high with current gain bandwidth products in the range of hundreds of GHz. Furthermore, the doping levels can be high resulting in current densities of the order of 10^6 A/cm^2.

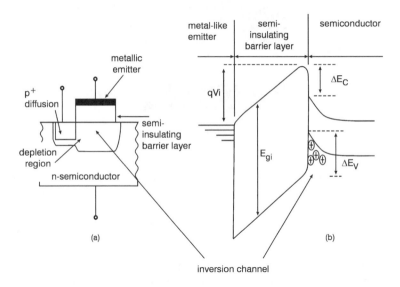

Figure 7.1. (a) Cross-section of a BICFET showing terminal connections and (b) band diagram of a BICFET. After M. E. Mierzwinski, AC Performance of the Ge$_x$Si$_{1-x}$/Silicon Inversion-Base Transistor, *PhD Thesis, Stanford University, 1994.*

The cross-section of a BICFET device is shown in figure 7.1a and the associated band diagram is illustrated in figure 7.1b. The device consists of a wide bandgap semiconductor layer located above a narrow bandgap semiconductor. A metallic emitter is deposited on the wide bandgap layer and provides a source of carriers. The lower layer forms the collector contact.

The carrier flow under active bias is vertically through the device and the wide bandgap layer provides a thermionic barrier to carrier flow between the emitter and the collector. A thin layer of acceptors in the wide bandgap semiconductor increases the zero bias barrier height and controls the turn-on characteristics. Electron flow over the barrier is modulated by hole charge injected into the inversion layer via the adjacent p$^+$ junction as will be discussed below.

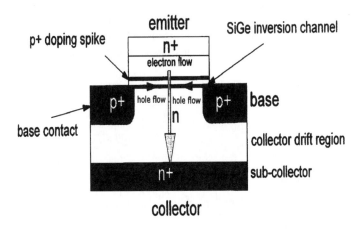

Figure 7.2. Schematic representation of the various components of a BICFET design. The vertical electron flow from the emitter to the collector is controlled by holes injected from the self-aligned p$^+$ base contacts. After M. E. Mierzwinski et al., IEEE IEDM Tech. Dig., 773–776 (1992), copyright © 1992 IEEE.

A heterostructure is required to implement the design of this device because the inversion channel forming the base region can be confined only in a heterojunction. The p-channel device therefore requires large valence band offsets. The cross-section of a typical device showing the direction of current flow is shown in figure 7.2 and its associated zero bias band diagram is shown in figure 7.3. The implanted extrinsic source (or base) regions were self-aligned to the 100 Å undoped Si$_{1-x}$Ge$_x$ base layer. This layer requires a Ge concentration close to 50% to provide the 0.3 eV offset in the valence band. A 10 Å boron delta-doping spike was placed adjacent to the Si$_{1-x}$Ge$_x$ layer. This thin acceptor layer controls the turn-on characteristics of the device. The spike thickness was so small that sufficient band bending does not take place and a neutral base region does not form. The active device is all n-type except for the acceptor layer located above the inversion well. The emitter is formed above the acceptor doping spike and consists of a heavily doped n-type Si layer. The emitter provides the source of electrons and is doped heavily to allow for high current densities without the onset of space-charge-limited flow. The region below the silicon–germanium layer is also doped n-type. However, its doping concentration is small and its thickness is designed to support large voltage drops from the inversion layer to the heavily doped n^+-sub-collector. The voltage drop in this collector drift region provides a region for carrier acceleration and the associated power amplification.

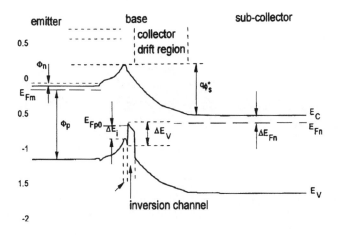

Figure 7.3. Band diagram of a BICFET in vertical bipolar mode of operation taken vertically through the emitter region. The band diagram was obtained by a two-dimensional device simulation of the device with an applied potential of 0.6 V between base and emitter, with the collector grounded to the base. After M. E. Mierzwinski et al., IEEE IEDM Tech. Dig., 773–776 (1992), copyright © 1992 IEEE.

Under normal active bias conditions, the collector is biased positive relative to the grounded emitter (figure 7.3). As a positive bias is applied to the base relative to the emitter, holes are injected from the two self-aligned base junctions into the inversion well at the emitter–base heterojunction interface. These holes compensate for the ionized boron acceptor atoms in the p^+-doping spike, and effectively lower the barrier to electron flow from the emitter as well as to holes from the inversion well into the emitter. Because of the valence band offset at the SiGe to Si interface, the holes see a significantly higher barrier to flow than do the electrons.

The holes move to the $Si_{1-x}Ge_x$ layer and negatively charged boron atoms left behind give rise to a barrier to the flow of electrons from emitter to collector. The barrier height is modulated by the flow of holes from the extrinsic source (or base) into the $Si_{1-x}Ge_x$ layer.

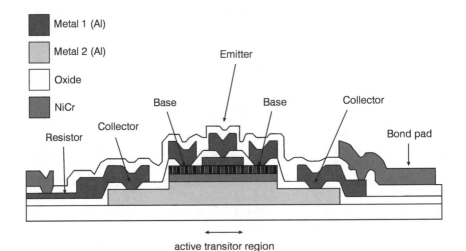

Figure 7.4. Cross-section of a SiGe channel BICFET device after fabrication. After M. E. Mierzwinski et al., IEEE IEDM Tech. Dig., 773–776 (1992), copyright ©1992 IEEE.

The BICFET performance is dependent on the fabrication process. Taft and Plummer [2] used low temperature Si MBE as a method of producing abrupt doping concentrations with high Ge concentrations. Electron beam lithography was used to provide both fine line patterns and close alignment between layers. Working devices with 0.25 μm as-drawn emitters have been fabricated [5]. Figure 7.4 shows a typical device cross-section.

A typical Gummel plot of a 100 μm × 100 μm BICFET with 39% Ge in the strained base layer is shown in figure 7.5. The current gain of this transistor was 1720 at 295 K. The zero-bias intrinsic base resistance was 43 kΩ/sq. The ideality factor of the base current was 2.

Although current gain is mainly dependent on the valence band offset, non-ideal recombination currents in the base can severely affect this gain. Figure 7.6 schematically describes the current paths in the device. Taft *et*

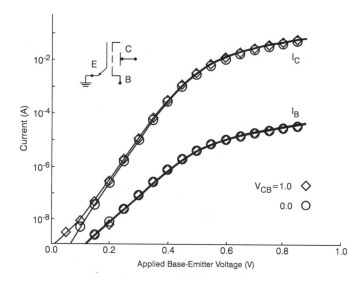

Figure 7.5. Observed Gummel plot with a $Ge_{0.39}Si_{0.61}$ base layer BICFET. Results of simulations are shown by symbols. After M. E. Mierzwinski, AC Performance of the Ge_xSi_{1-x}/Silicon Inversion-Base Transistor, PhD Thesis, Stanford University, 1994.

al. [4] showed that for large area devices, traps at the SiGe interface would not be significant for high current levels, so long as the trap levels were below $10^{12}/cm^2$. In this case, the base current would have an ideality of 1 (58 mV/decade at room temperature). Bulk recombination in the emitter depletion region, on the other hand, would have an ideality of 2. From Gummel plots of large area devices it is apparent that bulk recombination dominates the base current.

Transistors with several other Ge concentrations were fabricated and studied. The turn-on voltage decreased with increase in Ge concentration. There was a small increase in the base current with increasing Ge mole fraction but the increase in the collector current was large. The current gain was found to be very sensitive to Ge concentration. In the devices fabricated by Mierzwinski *et al.* [5], the Ge concentration was 42% and gains as high as 3000 at room temperature and 3500 at 10 K were observed.

By independently contacting base regions of double-base structures as source and drain, the BICFET can also operate as a unipolar device. This

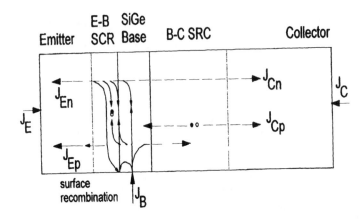

Figure 7.6. *Recombination currents in a SiGe BICFET. After M. E. Mierzwinski,* AC Performance of the Ge_xSi_{1-x}/Silicon Inversion-Base Transistor, *PhD Thesis, Stanford University, 1994.*

scheme allows for an estimation of the base resistance by monitoring the channel conductance. Devices of different base widths (FET channel lengths) were measured as a function of emitter (gate) bias (see figure 7.7). The large variation of the FET source–drain on-resistance is indicative of the strong dependence of bipolar base resistance on emitter–base bias.

A severe "emitter size effect" on current gain has been observed as shown in figure 7.8 [6]. Current gain was quite high for transistors with very large areas, but this performance degraded significantly for smaller devices. Given such a level of peripheral recombination current, the device gain would drop below an arbitrarily set useful level of 100 for devices with areas smaller than $40 \ \mu m^2$. This severely restricted the high frequency response of the transistor, especially if the device were expected to work at minimal power levels.

Figure 7.9 shows the effect of mesa etch depth on current gain. As can be seen, the influence on current gain is not too strong. This is encouraging as it implies the device would tolerate other process variations. The effect of the spacer layer on the device current is also very strong. Figure 7.10 plots the collector current turn-on characteristics for devices grown sequentially with a 50 Å and 100 Å spacer layer between the SiGe heterojunction and the boron doping spike. The doping profile with the highest gain is, as expected, abrupt. An attempt was made to fabricate inverted, or collector-

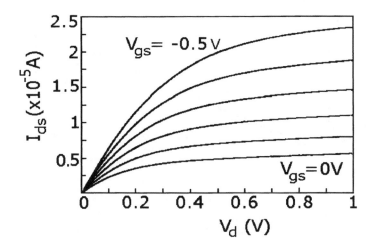

Figure 7.7. Heterojunction FET characteristics. In this mode the emitter terminal is grounded and the collector is biased as the drain. The emitter terminal acts as the gate, modulating the charge in the inversion well. The gate-source voltage V_{GS} was stepped from 0 to -0.5 V in 0.1 V increments. After M. E. Mierzwinski, AC Performance of the Ge_xSi_{1-x}/Silicon Inversion-Base Transistor, PhD Thesis, Stanford University, *1994.*

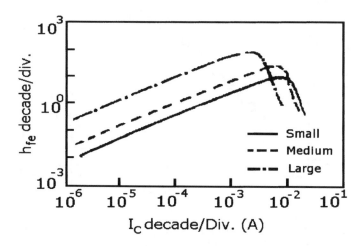

Figure 7.8. Current gain vs. I_c as a function of emitter size.

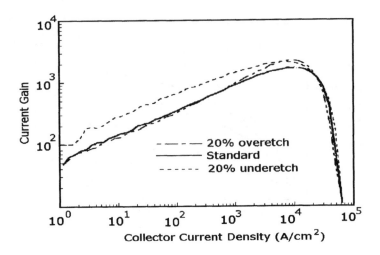

Figure 7.9.   Effect of mesa etch depth on the current gain.   After M. E. Mierzwinski, AC Performance of the Ge_xSi_{1-x}/Silicon Inversion-Base Transistor, *PhD Thesis, Stanford University, 1994.*

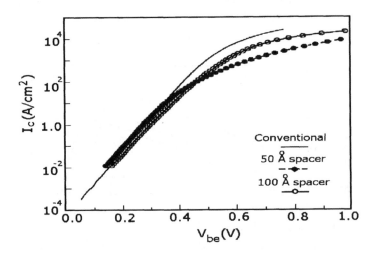

Figure 7.10.   Collector current for devices with a 50 Å and 100 Å spacer layers.   After M. E. Mierzwinski, AC Performance of the Ge_xSi_{1-x}/Silicon Inversion-Base Transistor, *PhD Thesis, Stanford University, 1994.*

up, BICFET structures [5]. In such a structure, the base–collector junction capacitance would be significantly reduced. Since this junction capacitance is a major parasitic, collector-up transistors should demonstrate much better extrinsic performance. In order for such a configuration to function usefully, the extrinsic emitter–base junction must be modified to prevent current flow in the forward-biased p^+–n junction.

7.1.1 AC Response

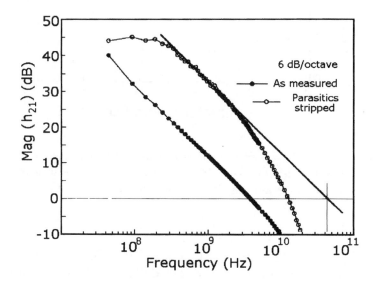

Figure 7.11.   Frequency response of a BICFET. The extrapolated cutoff frequency is 42 GHz. The device was biased with $V_{ce} = 1.5$ V and $I_c = 4.4$ mA with an emitter size of 0.75 μm × 10 μm. After M. E. Mierzwinski, AC Performance of the Ge$_x$Si$_{1-x}$/Silicon Inversion-Base Transistor, *PhD Thesis, Stanford University, 1994.*

The ac response of a BICFET with an emitter area $W \times L = 0.75$ μm × 10 μm, is shown in figure 7.11. The extrapolated cutoff frequency is 42 GHz. Figure 7.12 plots the de-embedded f_T for devices with the same minimum emitter width but varied emitter lengths. It is apparent from these results that the smallest emitter length does not result in the fastest device.

The performance of the BICFET has also been assessed using simulated gate delays [5]. In figure 7.13, published SiGe-HBT delays [7, 8] are

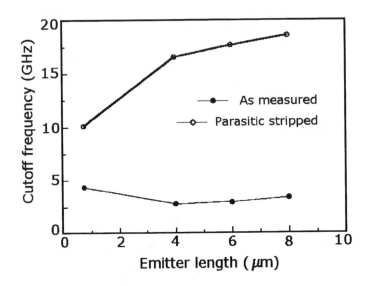

Figure 7.12. Cutoff frequency for devices with minimum emitter width as a function of emitter length. After M. E. Mierzwinski, AC Performance of the Ge_xSi_{1-x}/Silicon Inversion-Base Transistor, *PhD Thesis, Stanford University, 1994.*

compared with BICFET delays. It is noted that the performance of the BICFET lagged behind that of the SiGe-HBT until the lateral geometry was made smaller than 0.3 μm. The BICFET has demonstrated excellent low temperature performance [5], even though the structure was not designed for that environment. When optimized for low temperature, the BICFET may provide a high gain, high speed device capable of integrating with other devices such as optical detectors.

The advantages and disadvantages of the BICFET technology have also been compared with alternative available high performance device technologies. The main classes of devices and materials can be categorized as shown in figure 7.14. In a broad sense, the difficulty in producing each device type can be represented in moving from the upper left down to the lower right quadrant [5]. The relatively simple FET process steps allow for extremely high circuit density and yields, while the multiple levels of non-planar lithography for a bipolar are inherently more difficult to produce.

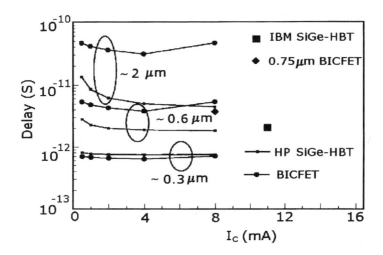

Figure 7.13.   Comparison of simulated gate delays for SiGe BICFETs and SiGe-HBTs.   After M. E. Mierzwinski, AC Performance of the Ge$_x$Si$_{1-x}$/Silicon Inversion-Base Transistor, PhD Thesis, Stanford University, *1994.*

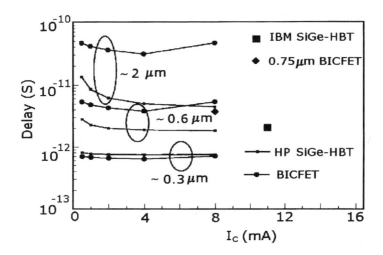

Figure 7.14.   Categories of viable device technologies, categorized by device family and material structure. Unipolar devices and silicon-based transistors are considered easier to fabricate. After M. E. Mierzwinski, AC Performance of the Ge$_x$Si$_{1-x}$/Silicon Inversion-Base Transistor, PhD Thesis, Stanford *University, 1994.*

7.2   Resonant Tunneling Diodes

Heterojunction interfaces grown by MBE can be made atomically abrupt and give rise to potential steps (band discontinuities) in the conduction and valence bands. The conduction and valence band edges as a function of distance inside the materials may be used to confine the motion of electrons and holes for device operation. An electron crossing these potential steps gains a kinetic energy equal to the conduction band discontinuity. Electrons can form standing wave patterns in the direction perpendicular to the layer so that the corresponding kinetic energy is quantized similar to the particle in a box problem in quantum mechanics.

Although the principle of resonant carrier tunneling has been known for a long time, Si-based resonant tunneling devices were reported only in the late 1980s [9, 10]. Gennser *et al.* [11] have made extensive measurements of $I - V$ characteristics of SiGe RTDs over a large temperature range. The potential applications of $Si_{1-x}Ge_x$/Si strain layer heterostructures and superlattices have been studied for the development of electronic [12–14] as well as optical [15, 16] device applications. Applying a bias across a structure consisting of a quantum well and thin barriers surrounding it induces a small current by quantum mechanical tunneling of the carriers through the barriers. The overall current through the device reaches a maximum whenever the energy of the injected carriers is in resonance with the energy of an eigenstate of the quantum well. Thus $I - V$ curves show sharp peaks, and every peak is followed by a region of negative differential resistance (NDR). The RTDs are very useful for high frequency detection and oscillations.

Since valence band offset in SiGe layers under biaxial compression is large, SiGe strained layers grown on a Si(100) substrate are well suited for the study of resonance tunneling of holes. SiGe strained layer RTDs, fabricated in 1988 by Liu *et al.* [9, 10], consisted of a SiGe quantum well separated by Si barriers and were grown on a Si(100) substrate. Tunneling of holes in $Si/Si_{1-x}Ge_x$/Si double barrier diodes on various $Si_{1-x}Ge_x$ buffer layers have also been studied by Rhee *et al.* [10, 17]. The authors fabricated diodes on relaxed buffer layers of different Ge concentrations and were able to distribute the strain between well and barriers in any desired manner. By changing Ge concentration in the buffer layer, the barrier or the quantum well or both can be strained.

Tunneling diodes 25–100 μm in diameter were fabricated using conventional lift-off and mesa-etching techniques and Al was evaporated on mesas to make the contacts. In the experiment, the $Si/Si_{1-x}Ge_x$ heterostructures were grown using Si MBE. Due to the limited critical thickness of the strained $Si_{1-x}Ge_x$ layers, for the realization of device applications, it is essential to grow $Si_{1-x}Ge_x$ heterostructures on a relaxed SiGe buffer layer so that the vertical dimensions of the devices are not restricted to below the critical thickness. The experimental results presented below were obtained by Rhee *et al.* [17] using a 50 μm diameter diode.

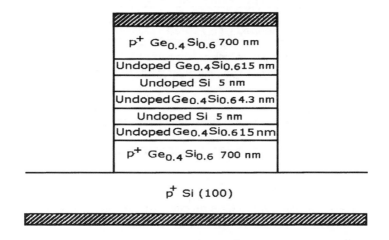

Figure 7.15.   Cross-sectional structure of the double barrier diode with a relaxed $Si_{0.6}Ge_{0.4}$ quantum well. After S. S. Rhee, Studies of SiGe Tunneling Heterostructures Grown by Si Molecular Beam Epitaxy, PhD Thesis, Univ. Calif. at Los Angeles, *1991.*

The cross-sectional structure of the double barrier diode on a relaxed $Si_{0.6}Ge_{0.4}$ buffer is shown in figure 7.15. First, a 7000 Å $Si_{0.6}Ge_{0.4}$ buffer layer was grown on the Si substrate, and then the double barrier structure was grown on the buffer. The buffer layer was doped with B at about 5×10^{18} cm^{-3}. The double barrier structure consists of an undoped 43 Å $Si_{0.6}Ge_{0.4}$ quantum well between two 50 Å Si barriers. On top of the double barrier, a 7000 Å thick unstrained $Si_{0.6}Ge_{0.4}$ contact layer was grown with a B doping concentration of about 5×10^{18} cm^{-3}. 150 Å thick layers in the buffer and the top contact just outside of two Si barriers were intentionally undoped to prevent dopant diffusion into the quantum well structure. In this structure, Si barriers are coherently strained, while the quantum well, the top contact layer, and the buffer are unstrained.

The cross-sectional structure of the double barrier diode on a relaxed Si substrate is shown in figure 7.16. A 5000 Å Si buffer layer was grown on the Si substrate to avoid possible surface contamination and to provide an

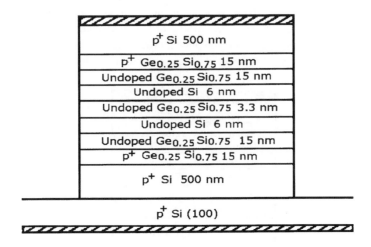

Figure 7.16. Cross-sectional structure of the double barrier diode with a strained $Si_{0.75}Ge_{0.25}$ quantum well on Si substrate. After S. S. Rhee, Studies of SiGe Tunneling Heterostructures Grown by Si Molecular Beam Epitaxy, PhD Thesis, Univ. Calif. at Los Angeles, 1991.

automatically flat surface. The double barrier structure was then grown on the buffer. The buffer layer was doped with B to about 5×10^{18} cm^{-3}. The double barrier structure consists of an undoped 33 Å $Si_{0.75}Ge_{0.25}$ quantum well between two 60 Å Si barriers. On both sides of the double barrier structure, 300 Å coherently strained $Si_{0.75}Ge_{0.25}$ layers were inserted to provide tunneling carriers from the bottom of the barrier. On top of the double barrier, a 5000 Å thick unstrained Si contact layer was grown with a B doping concentration of about 5×10^{18} cm^{-3}. 150 Å thick layers in the spacer layers just outside the Si barriers were intentionally undoped to prevent dopant diffusion into the quantum well structure. In this structure the Si barriers, the top contact layer and the buffer layer are unstrained, while the quantum well and two 300 Å spacer layers are coherently strained.

Figure 7.17 shows the valence band offset and bound state energies of heavy- and light-holes for strained films on a completely relaxed $Si_{0.6}Ge_{0.4}$ buffer layer. In unstrained $Si_{0.6}Ge_{0.4}$ layers, heavy-hole and light-hole bands are degenerate. However, in the strained Si barriers the light-hole band moves upwards due to tensile strain, while the heavy-hole band moves downward. In order to calculate the bound state energies in the quantum well for light- and heavy-holes, the envelope function approximation using different masses of the

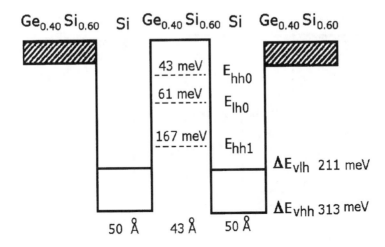

*Figure 7.17. Band diagram of a resonant tunnel diode with a relaxed $Si_{0.6}Ge_{0.4}$ quantum well. Heavy- and light-hole band edges are degenerate in unstrained $Si_{0.6}Ge_{0.4}$ layers. After S. S. Rhee et al., J. Vac. Sci. Technol. B, **7**, 327–331 (1989).*

well and barriers is used. The effective masses of the light- and heavy-holes are estimated using a linear interpolation of effective mass parameters along the $< 001 >$ direction of the warped bulk Si and Ge valence band.

Figure 7.18 shows the valence band offset and bound state energies of heavy- and light-holes for a strained $Si_{0.75}Ge_{0.25}$ quantum well on a Si substrate. The energy separation of the heavy- and light-hole bands is 39 meV. The calculated masses for a strained $Si_{0.75}Ge_{0.25}$ layer are $0.108m_e$ and $0.26m_e$ for light- and heavy-holes, respectively. The estimated values of light- and heavy-hole band offsets are $\Delta E_{vlh} = 146$ meV and $\Delta E_{vhh} = 185$ meV, respectively. For the heavy-hole there are two bound states in the quantum well at energies $E_{hho} = 37$ meV and $E_{hh1} = 137$ meV and for the light-hole only the ground state at $E_{lho} = 104$ meV. All the values are measured downward from the band edge of the unstrained $Si_{0.75}Ge_{0.25}$ quantum well.

The current–voltage and the first and the second derivatives, were measured using a computer controlled tunneling spectroscopy system at 4.2 K, 77 K and room temperature. The $I - V$, dI/dV, and d^2I/dV^2 of the double barrier diode on the relaxed $Si_{0.6}Ge_{0.4}$ are shown in figures 7.19, 7.20, and 7.21, respectively. The inset of figure 7.19 shows the 77 K measurement for higher

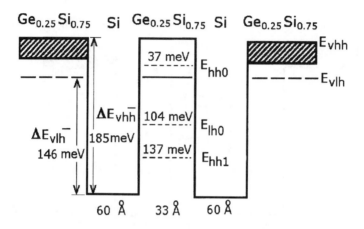

Figure 7.18.   Band diagram of a resonant tunnel diode with a strained $Si_{0.75}Ge_{0.25}$ quantum well on Si. Heavy- and light-hole band edges are split in strained $Si_{0.75}Ge_{0.25}$ layers. After S. S. Rhee, Studies of SiGe Tunneling Heterostructures Grown by Si Molecular Beam Epitaxy, PhD Thesis, Univ. Calif. at Los Angeles, *1991.*

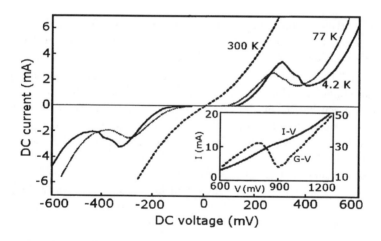

Figure 7.19. $I - V$ characteristics of a 50 μm diameter RTD with a relaxed $Si_{0.6}Ge_{0.4}$ at 4.2 K, 77 K and room temperature. Inset shows the $I - V$ and conductance measurements of an additional peak for high bias at 77 K. After S. S. Rhee et al., J. Vac. Sci. Technol. B, **7**, *327–331 (1989).*

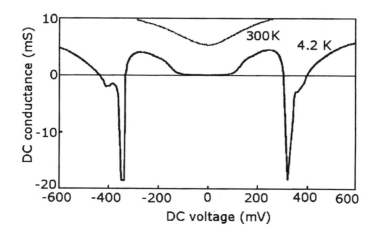

Figure 7.20. Conductance–voltage curves of an RTD with a relaxed $Si_{0.6}Ge_{0.4}$ quantum well at 4.2 K and room temperature. After S. S. Rhee, Studies of SiGe Tunneling Heterostructures Grown by Si Molecular Beam Epitaxy, PhD Thesis, Univ. Calif. at Los Angeles, *1991.*

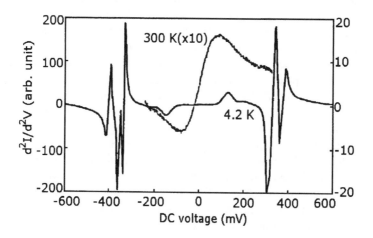

*Figure 7.21. Second derivative of the $I-V$ characteristics of a 50 μm diameter RTD with a relaxed $Si_{0.6}Ge_{0.4}$ at 4.2 K and room temperature showing a peak near the origin due to the resonant tunneling for the heavy-hole ground state. After S. S. Rhee et al., J. Vac. Sci. Technol. B, **7**, 327–331 (1989).*

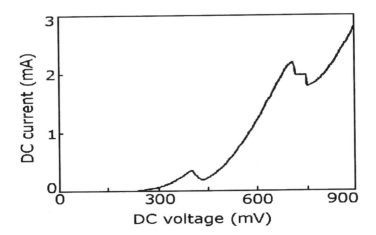

Figure 7.22. *I − V characteristics of a 50 μm diameter RTD with a strained* $Si_{0.75}Ge_{0.25}$ *at 4.2 K. After S. S. Rhee,* Studies of SiGe Tunneling Heterostructures Grown by Si Molecular Beam Epitaxy, PhD Thesis, Univ. Calif. at Los Angeles, *1991.*

bias. No differential negative resistance was observed at room temperature. At 77 K there are two resonant peaks at 270 mV and 900 mV due to the light-hole ground state (E_{lho}) and the heavy-hole first excited state (E_{hh1}), respectively. Another resonance feature is observed from the d^2I/dV^2 data at 170 mV (figure 7.21) and is believed to be due to the heavy-hole ground state (E_{hho}). The current peak-to-valley ratios for NDR for light-hole tunneling were 2.1 and 1.6 at 4.2 K and 77 K, respectively. The oscillations seen in the $I − V$ characteristics are due to the instabilities in the circuit when the resistance becomes negative. The $I − V$ characteristics at 4.2 K of the tunneling diode on the Si substrate are shown in figure 7.22. Two resonant tunneling peaks are shown clearly. The first peak also shows a current peak-to-valley ratio of about 2.1. Noise behavior of a RTD has been studied by Okada *et al.* [18]. The diode produced more noise at 77 K in the resonant tunneling mode than at room temperature as a bulk resistor. The flicker noise varied as $1/f^2$ rather than the $1/f$ commonly observed in other devices.

7.2.1   Resonant Tunneling Hot Carrier Transistors

Double barrier resonant tunneling heterostructures with a negative differential resistance have been studied extensively in III–V materials. Although two-terminal tunneling devices have many potential applications, three-terminal transistors are preferable in many cases [19]. A three-terminal hot-hole transistor has been demonstrated by Rhee *et al.* [20]. The incorporation of NDR with expected operating frequencies in the THz range into existing Si technology could find potential applications in the fields of high speed digital circuits, frequency multipliers, and tunable oscillators/amplifiers. In this section we discuss the realization of a SiGe hot carrier resonant tunneling transistor. The device can also be used for studying hot carriers in SiGe heterostructures.

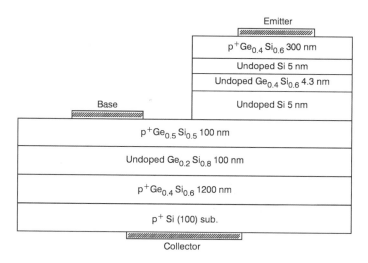

Figure 7.23. Cross-section of a SiGe resonant tunneling hot-hole transistor. After S. S. Rhee et al., IEEE IEDM Tech. Dig., 651–654 (1989), copyright © *1989 IEEE.*

Figure 7.23 shows the cross-sectional schematic of the device. A double barrier structure which consists of two 50 Å Si layers separated by a 43 Å $Si_{0.6}Ge_{0.4}$ quantum well is used as an emitter. A 1.2 μm $Si_{0.6}Ge_{0.4}$ buffer layer acts as a collector, while the collector barrier consists of a 1000 Å $Si_{0.8}Ge_{0.2}$ layer. Between the double barrier quantum well emitter and the collector barrier, a 1000 Å $Si_{0.5}Ge_{0.5}$ base was inserted. The doping concentration was 1×10^{18} cm^{-3} throughout the device except for the collector barrier and the double barrier resonant tunneling structure, which were both undoped. The substrate temperature was held at 530 °C throughout the growth. Two-level mesa structures using selective wet etching and standard photolithography techniques were fabricated for the convenience of making contacts to the emitter and the base. Collector contacts were made on the backside of the wafers and ohmic contacts were obtained by sintering a 1% AlSi alloy.

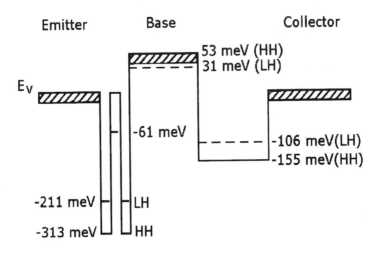

Figure 7.24. Valence band diagram of a SiGe resonant tunneling hot-hole transistor in equilibrium. The valence band edge is degenerate in the relaxed $Si_{0.6}Ge_{0.4}$ After S. S. Rhee et al., IEEE IEDM Tech. Dig., 651–654 (1989), copyright ©1989 IEEE.

The valence band offsets and bound state energy of the light-hole ground state in the quantum well are shown schematically in figure 7.24. All the values are given in reference to the valence band edge of the unstrained $Si_{0.6}Ge_{0.4}$ layers. The collector barrier and the resonant tunneling double barriers in the emitter are subjected to an in-plane tensile strain which causes the heavy-hole band edge to go below the light-hole band edge. In the base, the heavy-hole band edge is above the light-hole band edge due to the compressive strain. In the unstrained $Si_{0.6}Ge_{0.4}$ layers, the light-hole and heavy-hole bands are degenerate. When measured from the collector side, the collector barrier heights are 106 meV and 155 meV for the light- and heavy-holes, respectively. The light- and heavy-holes see the barrier heights of 137 meV and 208 meV, respectively, from the base to the collector. Due to the degenerate light- and heavy-hole bands in the collector, the majority of the current from the collector to the base is due to the light-holes because of the lower light-hole barrier height. On the other hand, a larger portion of the base to the collector current at low temperatures may be due to the heavy-holes. This results from the lifting of the degeneracy in the base region where the heavy-hole band is heavily populated at low temperatures. Thus, the effective barrier height from the collector to the base is 106 meV as seen by the light-holes and 208 meV from the base to the collector as seen by the heavy-holes. An asymmetric current–voltage characteristic between the base and the collector is evidence of the unequal barrier height.

In the double barrier quantum well emitter the barrier heights are 211 meV and 313 meV for the light- and heavy-holes, respectively. There are two bound states in the quantum well for the heavy-hole, but for the light-hole only the ground state at 61 meV is present. The negative differential resistance of the device is due to the light-hole tunneling through the light-hole ground state.

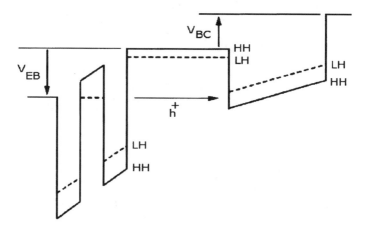

Figure 7.25.   Valence band diagram of a SiGe resonant tunneling hot-hole transistor under bias. After S. S. Rhee et al., IEEE IEDM Tech. Dig., 651–654 (1989), copyright ©1989 IEEE.

Figure 7.25 shows a schematic band diagram under bias. When the emitter is biased positively with respect to the base, holes are injected into the base through the double barrier resonant tunneling emitter with an excess hole energy relative to the valence band maximum of the $Si_{0.5}Ge_{0.5}$ base. The holes injected into the base are then transferred near-ballistically to the collector. The 1000 Å $Si_{0.8}Ge_{0.2}$ collector barrier prevents injection of the holes initiated from the valence band of the base to the collector when V_{bc} is applied while allowing the injected hot holes from the emitter to go through to the collector, if the hot holes have higher energies than the collector barrier height. The injected holes may suffer some elastic and inelastic scattering in the base. In the p-doped unipolar structure, the elastic scattering rate is low because the heavy-holes are effective in screening static impurities. At low temperature, inelastic scattering may be neglected due to the absence of phonon scattering.

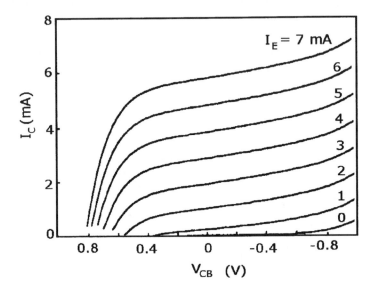

Figure 7.26. Collector current of a SiGe resonant tunneling hot-hole transistor as a function of V_{bc} at 77 K in common–base configuration. After S. S. Rhee et al., IEEE IEDM Tech. Dig., 651–654 (1989), copyright ©1989 IEEE.

The common base mode $I - V$ characteristics of the device at 77 K are shown in figure 7.26 which gives the collector current I_c as a function of collector voltage V_{bc} for several values of I_e beginning from 0 to 7 mA with steps of 1 mA. Since the barrier height of the collector barrier is only 137 meV, as shown in figure 7.24 for light-holes, the thermionic current is dominant at room temperature. At lower temperatures, the thermionic current decreases while the tunneling current becomes more important. Thus, without injection from the emitter, I_c consists of thermionic and tunneling current components going through the triangular barrier formed by V_{bc} and with perhaps some leakage current due to recombination via defects present in the strained films. For a given injection current and a positive V_{bc}, the collector current I_c is related to the current transfer ratio $\alpha = I_c/I_e$ and the leakage current from the base to the collector. When a positive V_{bc} is applied, I_c decreases with the increase of the effective collector barrier height by the bias until V_{bc} reaches a cutoff voltage. Beyond this cutoff voltage the collector current drops sharply to zero. The cutoff voltage increases with I_e.

7.2.2 Electron Tunneling in SiGe-RTDs

Electron resonance tunneling in the SiGe structures has been studied by several workers [21–23]. Ismail *et al.* [21] fabricated the device on a high-quality $Ge_{0.3}Si_{0.7}$ relaxed n-type doped buffer layer grown on a Si(100) substrate. The tunneling structure consisted of a 50 Å Si well sandwiched between two unstrained $Ge_{0.3}Si_{0.7}$ barriers. There were two 150 Å Si layers on both sides of this structure. A phosphorus doped 300 Å SiGe layer and a 40 Å Si cap layer were deposited on the top of the device. At room temperature and 77 K the peak-to-valley ratios (PVR) were 1.2 and 1.5, respectively. There was a finite current at small voltage, which indicates that the Fermi level on both sides of the tunneling junction is within $\sim 1 \, kT$ of the position of the lowest sub-level in the well. The slopes close to zero bias for the two polarities were different, indicating that the RTD was asymmetrical.

RTDs with qualitatively similar structure have been fabricated and studied by Matutinovic-Krstelj *et al.* [23]. Ge concentration in the relaxed buffer layer and in the barrier layers was 0.35, and the thickness of the well was 20 to 50 Å and of the barriers, 40 to 70 Å. The characteristics of the diodes were symmetrical. Two distinct peaks were observed in the current–voltage characteristics at 150 K. As the temperature was lowered, the low-voltage peak disappeared below 50 K. At 4.2 K, the highest available PVR was 2.

7.3 Poly-SiGe Devices

Polycrystalline Si finds wide application in all types of Si integrated circuit technology. In CMOS technology it is used as the gate "metal" in the n- and p-channel MOSFETs, in bipolar and BiCMOS technologies it is used in the polysilicon emitter of the bipolar transistors, and in thin film transistor technology it is used as the substrate in which the transistors are fabricated. In general, other group-IV elements can be combined with Si to produce alloys, such as SiGe or SiGeC, and have been discussed in Chapter 2.

The advantage is that the properties of the material can be tuned by varying the composition of the alloy. For example, the introduction of about 10% Ge into Si produces an alloy with a lower bandgap than bulk Si. This property has been exploited in SiGe heterojunction bipolar transistors, which have been shown in Chapter 5 to operate at higher frequencies than Si-BJTs. While considerable research has been carried out on epitaxial strained SiGe, relatively little research has been done on polycrystalline SiGe and even less on other polycrystalline IV–IV materials. Potential applications of polycrystalline SiGe and SiGeC include:

(i) CMOS: tuning of the work function by 200–300 meV towards midgap and reduced gate depletion due to enhanced dopant activation at low temperatures,

(ii) TFTs: higher mobility and lower thermal budget processing than amorphous or polycrystalline Si,

(iii) BiCMOS: lower thermal budget polysilicon emitters and an increased gain in wide bandgap polycrystalline SiGeC or SiC emitters, and

(iv) Resistors: tuning of temperature coefficient of resistance in polycrystalline SiGe or SiGeC resistors.

King *et al.* [24] have studied the properties of doped poly-SiGe layers, to determine their suitability for use as a gate material. The authors found that SiGe films with Ge mole fractions up to 0.6 were completely compatible with the Si VLSI fabrication process. The work function of the layers decreased with increase in Ge concentration. The decrease was more than 0.3 V for 60% Ge. The use of poly-SiGe as a gate material therefore allows optimization of the channel doping and improved performance of the p-MOS and n-MOS devices in CMOS technology. Another advantage of poly-SiGe as gate material is that the implanted B can be activated by annealing at a lower temperature.

7.3.1   Poly-SiGe TFTs

Polycrystalline thin film transistors (TFTs) are promising for use as high performance pixel and integrated driver transistors for active matrix liquid crystal displays (AMLCDs). Silicon–germanium is a promising candidate for use as the channel material due to its low thermal budget requirements and may enable the integration of circuitry onto the glass substrate, reducing the overall manufacturing cost. King *et al.* [25, 26] have fabricated p-channel MOSFETs in thin poly-SiGe layers deposited using low-pressure CVD. Compared to Si thin film technology, SiGe thin film technology allows use of lower anneal temperatures and shorter anneal times. The MOSFETs with channel length as small as 2 μm showed well-behaved characteristics. This technology has potential application in large-area electronics where high temperature processing is not possible.

The binary nature of the silicon–germanium system complicates the optimization of the channel deposition conditions. SiGe-TFT performance has been improved substantially through the use of a thin Si interlayer to improve the gate–oxide interface [27]. However, little has been done to improve the intrinsic quality of the SiGe channel film itself. Subramanian and Saraswat [28] have reported on optimization studies using response surface characterization done on the low pressure chemical vapor deposition of SiGe and its effect on TFT performance.

7.4 SiGe I²L Logic

High performance bipolar logic circuits are usually realized using ECL which has a relatively low packing density and high power dissipation. The gate delay of integrated injection logic (I²L) circuits is primarily determined by stored charge in parasitic diodes associated with the extrinsic base regions of the I²L gate [29]. A minimum gate delay of 0.8 ns for a layout geometry of 2.5 μm and a fan-out of 3 for self-aligned base–collector structures have been reported [30]. SiGe technology offers the prospect of using bandgap engineering to minimize the stored charge in the parasitic diodes associated with the I²L gate. The lower bandgap of SiGe has a great impact on the propagation delay of I²L. Simulation results suggest that SiGe I²L may be a useful technology in high performance and low power applications such as portable electronic systems [31, 32]. Experimental results on the SiGe I²L (surface-fed and substrate-fed variants) have been reported [33, 34]. The use of heterojunctions can add high speed to the other well known advantages of I²L technology, namely high packing density, low voltage and low power dissipation. Towards the optimization of SiGe I²L circuits a quasi-two-dimensional stored charge model has been developed [34]. The model has been applied to surface-fed and substrate-fed variants of SiGe I²L and the Ge and doping concentrations were varied to determine the important tradeoffs in the gate design.

7.4.1 Fabrication of I²L

SiGe integrated injection logic circuits have been designed and fabricated [34]. The overall architecture used differential epitaxy to produce single-crystal SiGe in an oxide patterned window and poly-SiGe on the oxide layer. Contact to the gate input was made via a p⁺-poly-SiGe layer and contact to the output via low thermal budget polysilicon collector contacts. An important feature of the architecture was the use of a boron implant to create a junction in the Si below the SiGe base in the extrinsic regions of the gate. This had the effect of creating a homojunction in the extrinsic device regions, which reduced charge injection into these regions during gate operation and hence improved switching speeds.

In the substrate-fed architecture, power was injected into the gate from the wafer substrate, whereas in the surface-fed variant power was injected from the surface. The former approach was found to give better performance, but at the cost of more difficult tradeoffs in the technology design. Good agreement between measured and simulated gate delays was obtained, and a minimum gate delay of 20 ps was predicted for a Ge concentration of 16% on 1.2 μm geometry gates. This is nearly ten times faster than the speed of Si-I^{2}L and is approaching the speed of ECL. Even faster speeds may be achievable for sub-micron geometries. This technology has considerable potential for high-speed, low power applications such as mobile communications and portable computers.

7.5   SiGe Sensors

Other interesting applications of SiGe include integrated pressure sensors using Si and SiGe-HBTs. There is considerable interest in fabricating analog sensors which can be easily mass-produced using well established Si processing techniques. One such type of micro sensor is a pressure sensor which typically consists of an etched diaphragm suspended above a Si substrate [35]. However, the basic mechanism underlying Si/SiGe-based pressure sensors is electrical rather than mechanical in origin, i.e., a suspended diaphragm, and is far more compatible with conventional integrated circuit processing.

The bandgap of Si has a negative dependence on increasing pressure, while Ge has a positive dependence on increasing pressure. The values for Si are $dE_g/dP = -2.4 \times 10^{-6}$ eV/kg/cm^2 and for Ge $dE_g/dP = 5.0 \times 10^{-6}$ eV/kg/cm^2 [36]. The introduction of Si and SiGe epitaxial base bipolar technologies allows one to exploit this difference in pressure dependence. For the pressure sensor, it is necessary to have a Si and SiGe base transistor on the same wafer. This can be accomplished with any of the established epi-base bipolar technologies by adding an additional mask level.

The sensor action is obtained by configuring the transistors and resistors as a differential amplifier pair. Note that one transistor has a bulk Si base and the other has a SiGe base. The operation of the sensor is as follows. The emitter–base voltage of both transistors is fixed (and equal) by the circuit configuration. Thus, at zero pressure, the differential output voltage will be positive since the SiGe-HBT has a greater collector current. As the pressure is increased, the relative current of the SiGe device decreases with respect to the Si device.

The collector current of a bipolar transistor depends exponentially on the bandgap of the base region. The collector current of the SiGe or Ge base transistor is an exponentially decreasing function of pressure while that of the Si transistor is an increasing function of pressure. Also the collector current of the SiGe (or Ge) base transistor is higher than that of bulk Si device at zero pressure because its base has a smaller bandgap (1.12 eV vs. 0.66 eV for Si vs. Ge at zero pressure). For a typical sensitivity, a pressure of 1000 kg/cm^2 gives a bandgap change of 5 meV for a Ge device. This translates into an 18% change in collector current over zero pressure. It may be noted that strained SiGe films are not mandatory for operation of the sensor, so that bulk Ge (relaxed) bases could be used in principle. The advantages of the Si/SiGe sensor are:

(i) it is easily compatible with epitaxial bipolar technologies such that any additional electronics could be easily incorporated on the same chip,

(ii) the differential voltage depends only on the differences in collector current between the two transistors, and is thus independent of the current gain. This means that thick (relaxed) SiGe or Ge films can also be used since any increase in base current due to the generated recombination centers associated with the lattice relaxation does not effect the operation,

(iii) the sensitivity of the pressure sensor can be tuned by varying the Ge concentration in the base of SiGe-HBTs,

(iv) the sensor does not require large collector current changes with pressure since the resistor values can be adjusted to be as large as desired (the only requirement is that they be identical), and

(v) because the bandgap is divided by the thermal voltage, it is ideally suited for low temperature applications (e.g., 77 K) where the thermal voltage is reduced by a factor of four.

7.6 Summary

The BICFET provides bipolar-like transconductance by controlling a potential barrier in the emitter with holes injected into an inversion layer located at a heterojunction interface. The dc performance of devices fabricated with electron beam lithography is outstanding. Large area devices demonstrated a current gain at room temperature of over 6500, a record for any BICFET-like device. The collector current threshold voltage depends strongly on doping layer position and is susceptible to variations in growth. However, the lack of a charge neutral base eliminates the diffusion of minority carriers out of the base, providing fast switching time. The implementation of the SiGe BICFET requires a relatively complex fabrication process with relatively small advantages over competing technologies. The conceptual advantages are still valid and, as lateral geometries shrink, BICFETs or a hybrid BJT-BICFET might possibly be a viable technology.

Resonant tunneling diodes grown on various substrates have been discussed. A tunneling diode grown on a (100) relaxed $Si_{0.6}Ge_{0.4}$ buffer layer shows a 2.1 peak-to-valley current ratio at 4.2 K and 1.6 at 77 K. A tunneling diode with a strained $Si_{0.75}Ge_{0.25}$ well on a Si substrate also shows a similar maximum peak-to-valley current ratio at 4.2 K. Double barrier resonant tunneling in a three-terminal device structure in SiGe heterostructures has also been covered. A number of interesting features are evident in the $I - V$ curves. In the current control mode, typical bipolar transistor-like current–voltage characteristics are obtained. The injected carriers are near-ballistically transferred from the emitter to the collector. The device exhibits a controllable negative differential resistance in the current–voltage characteristics. However, the performance of these diodes has not been as impressive as those of the RTDs based on III–V compound semiconductors. Performance of SiGe RTDs has not yet reached a level that can be exploited for commercial applications. However, based on theoretical predictions, a high value of PVR in the range of 10 to 30 is possible at 77 K. The use of SiGe for TFTs, I^2L and pressure sensing devices has been discussed.

7.7   References

1 G. W. Taylor and J. G. Simmons, "The bipolar inversion channel field effect transistor (BICFET) - A new field-effect solid state device:Theory and structure," *IEEE Trans. Electron Dev.*, vol. ED-32, pp. 2345–2367, 1985.

2 R. C. Taft and J. D. Plummer, "Ge_xSi_{1-x}/Silicon Inversion-Base Transistors: Theory of Operation," *IEEE Trans. Electron Dev.*, vol. 39, pp. 2108–2118, 1992.

3 R. C. Taft, J. D. Plummer and S. S. Iyer, "Demonstration of a p-channel BICFET in the Ge_xSi_{1-x}/Si system," *IEEE Electron Dev. Lett.*, vol. 10, pp. 14–16, 1989.

4 R. C. Taft, J. D. Plummer and S. S. Iyer, "Ge_xSi_{1-x}/Silicon Inversion-Base Transistors: Experimental Demonstration," *IEEE Trans. Electron Dev.*, vol. 39, pp. 2119–2126, 1992.

5 M. E. Mierzwinski, J. D. Plummer, E. T. Croke, S. S. Iyer and M. J. Harrell, "AC Characterization and Modeling of the Ge_xSi_{1-x}/Si BICFET," in *IEEE IEDM Tech. Dig.*, pp. 773–776, 1992.

6 C. K. Maiti, "Effect of Emitter Size on the Current Gain in Strained Layer $Si_{1-x}Ge_x$ BICFETs," in *Proc. of 8th Int'l Workshop on Physics of Semiconductor Devices, New Delhi*, pp. 33–35, 1995.

7 G. L. Patton, J. H. Comfort, B. S. Meyerson, E. F. Crabbe, G. J. Scilla, E. De Fresart, J. M. C. Stork, J. Y.-C. Sun, D. L. Harame and J. N. Burghartz, "75-GHz f_T SiGe base heterojunction bipolar transistors," *IEEE Electron Dev. Lett.*, vol. EDL-11, pp. 171–173, 1990.

8 T. I. Kamins, K. Nauka, L. H. Camnitz, J. B. Kruger, J. E. Turner, S. J. Rosner, M. P. Scott, J. L. Hoyt, C. A. King, D. B. Noble and J. F. Gibbons, "High frequency $Si/Si_{1-x}Ge_x$ heterojunction bipolar transistors," in *IEEE IEDM Tech. Dig.*, pp. 647–650, 1989.

9 H. C. Liu, D. Landheer, M. Buchmann and D. C. Houghton, "Resonant Tunneling Diode in the $Si_{1-x}Ge_x$ system," *Appl. Phys. Lett.*, vol. 52, pp. 1809–1811, 1988.

10 S. S. Rhee, J. S. Park, R. P. G. Karunasiri, Q. Ye and K. L. Wang, "Resonant tunneling through a Si/Ge_xSi_{1-x}/Si heterostructure on a GeSi buffer layer," *Appl. Phys. Lett.*, vol. 53, pp. 204–206, 1988.

11 U. Gennser, V. P. Kesan, S. S. Iyer, T. J. Bucelot and E. S. Yang, "Temperature dependent transport measurements on strained $Si/Si_{1-x}Ge_x$ resonant tunneling devices," *J. Vac. Sci. Technol. B*, vol. 9, pp. 2059–2063, 1991.

12 H. Daembkes, H.-J. Herzog, H. Jorke, H. Kibbel and E. Kasper, "The n-channel SiGe/Si modulation-doped Field-Effect transistor," *IEEE Trans. Electron Dev.*, vol. ED-33, pp. 633–638, 1986.

13 H. Temkin, J. C. Bean, A. Antreasyan and R. Leibenguth, "Ge_xSi_{1-x} strained layer heterostructure bipolar transistors," *Appl. Phys. Lett.*, vol. 52, pp. 1089–1091, 1988.

14 T. Tatsumi, H. Hirayama and N. Aizaki, "$Si/Ge_{0.3}Si_{0.7}$ heterojunction bipolar transistor made with Si molecular beam epitaxy," *Appl. Phys. Lett.*, vol. 52, pp. 895–897, 1988.

15 H. Temkin, T. P. Pearsall, J. C. Bean, R. A. Logan and S. Luryi, "Ge_xSi_{1-x} strained-layer superlattice waveguide photodetectors operating near 1.3 μm," *Appl. Phys. Lett.*, vol. 48, pp. 963–965, 1986.

16 T. P. Pearsall, J. Bevk, L. C. Feldman, J. M. Bonar, J. P. Mannaerts and A. Ourmazd, "Structurally induced optical transitions in Ge-Si superlattices," *Phys. Rev. Lett.*, vol. 58, pp. 729–732, 1987.

17 S. S. Rhee, R. P. G. Karunasiri, C. H. Chern, J. S. Park and K. L. Wang, "$Si/Ge_xSi_{1-x}/Si$ resonant tunneling diode doped by thermal boron source," *J. Vac. Sci. Technol. B*, vol. 7, pp. 327–331, 1989.

18 Y. Okada, J. Xu, H. C. Liu, D. Landheer, M. Buchanan and D. C. Houghton, "Noise characteristics of a Si/SiGe resonant tunneling diode," *Solid-State Electron.*, vol. 32, pp. 797–800, 1989.

19 F. Capasso, "Quantum Electron and Optoelectronic Devices," *Microelectronic Engineering*, vol. 19, pp. 909–916, 1992.

20 S. S. Rhee, G. K. Chang, T. K. Carns and K. L. Wang, "SiGe resonant tunneling hot-carrier transistor," *Appl. Phys. Lett.*, vol. 56, pp. 1061–1063, 1990.

21 K. Ismail, B. S. Meyerson and P. J. Wang, "Electron resonant tunneling in Si/SiGe double barrier diodes," *Appl. Phys. Lett.*, vol. 59, pp. 973–975, 1991.

22 J. C. Chiang and Y.-C. Chang, "Resonant tunneling of electrons in Si/Ge strained-layer double-barrier tunneling structures," *Appl. Phys. Lett.*, vol. 61, pp. 1405–1407, 1992.

23 Z. Matutinovic-Krstelj, C. W. Liu, X. Xiao and J. C. Sturm, "Symmetric $Si/Si_{1-x}Ge_x$ electron resonant tunneling diodes with an anomalous temperature behavior," *Appl. Phys. Lett.*, vol. 62, pp. 603–6055, 1993.

24 T.-J. King, J. R. Pfiester and K. C. Saraswat, "A variable work function polycrystalline $Si_{1-x}Ge_x$ gate material for submicrometer CMOS technologies," *IEEE Electron Dev. Lett.*, vol. 12, pp. 533–535, 1991.

25 T.-J. King and K. C. Saraswat, "A Low-Temperature ($\leq$ 550 °C) Silicon–Germanium MOS Thin-Film Transistor Technology for Large-Area Electronics," in *IEEE IEDM Tech. Dig.*, pp. 567–570, 1991.

26 T.-J. King, K. C. Saraswat and J. R. Pfiester, "PMOS transistors in LPCVD polycrystalline silicon germanium films," *IEEE Electron Dev. Lett.*, vol. 12, pp. 584–586, 1991.

27 A. J. Tang, J. A. Tsai, R. Reif and T.-J. King, "A novel poly-silicon-capped poly-silicon-germanium thin film transistor," in *IEEE IEDM Tech. Dig.*, pp. 513–516, 1995.

28 V. Subramanian and K. C. Saraswat, "Optimization of silicon–germanium TFT's through the control of amorphous precursor characteristics," *IEEE Trans. Electron Dev.*, vol. 45, pp. 1690–1695, 1998.

29 H. H. Berger and K. Helwig, "An investigation of the intrinsic delay (speed limit) in MTL/I²L," *IEEE J. Solid-State Circuits*, vol. SC-14, pp. 327–337, 1979.

30 D. D. Tang, T. H. Ning, R. D. Isaac, G. C. Feth, S. K. Wiedmann and H. N. Yu, "Subnanosecond selfaligned I²L/MTL circuits," *IEEE J. Solid-State Circuits*, vol. SC-15, pp. 444–449, 1980.

31 B. Mazhari and H. Morkoc, "Intrinsic gate delay of Si/SiGe integrated injection logic circuits," *Solid-State Electron.*, vol. 38, pp. 189–196, 1995.

32 M. Karlsteen and M. Willander, "Improved switch time of I²L at low power consumption by using a SiGe heterojunction bipolar transistor," *Solid-State Electron.*, vol. 38, pp. 1401–1407, 1995.

33 S. P. Wainwright, S. Hall and P. Ashburn, "Analysis of SiGe Heterojunction Injection Logic Structures Using a Stored Charge Model," in *Proc. ESSDERC'96*, pp. 649–652, 1996.

34 S. P. Wainwright, S. Hall, P. Ashburn and A. C. Lamb, "Analysis of Si:Ge heterojunction integrated injection logic (I²L) structures using a stored charge model," *IEEE Trans. Electron Dev.*, vol. 45, pp. 2437–2447, 1998.

35 "Integrated Pressure Sensor Using Silicon And Germanium Bipolar Transistors." *IBM Tech. Disc. Bull., April 1991.*

36 S. M. Sze, *Physics of Semiconductor Devices.* John Wiley & Sons, New York, 2nd ed., 1981.

Chapter 8

MODFETs

During the last fifteen years, extensive research has been conducted on various aspects of strained SiGe films: growth, transport, devices, and circuits. Major emphasis has been on the heterojunction bipolar transistors where a thin SiGe epitaxial layer has been used as a low bandgap semiconductor in the base. Indeed, SiGe-HBTs have demonstrated excellent high frequency performance with unity current gain cutoff frequency and maximum oscillation frequency exceeding 150 GHz. While SiGe-HBTs have been developed for application in wireless products for personal communication systems and circuits based on SiGe-HBTs are now commercially available, relatively little effort has been given towards enhancement of field effect devices.

Fabrication and performance of silicon-based heterostructure field effect transistors which are compatible with mainstream Si IC have been described in Chapter 6. It has long been established that the best performance of FETs is obtainable in modulation doped devices [1] where the channel and the dopant layers are separated. This chapter discusses the fabrication techniques, dc and high frequency performance of modulation doped field effect transistors. The prospect of combining heterostructure p- and n-type devices for realizing heterostructure complementary MOS (HCMOS) devices which retain the advantages of Si CMOS at a higher level of performance has been a very active area of integration.

In this chapter, an overview of the concept of modulation doping, reported results for SiGe channel p-MODFETs, strained Si n- and p-MOSFETs and MODFETs will be presented. Issues related to material growth and the electrical transport properties of both electrons and holes in strained Si have been discussed in Chapters 2 and 3. This chapter will also cover the technology of fabrication of MODFETs and HMOSFETs with strained Si, strained Ge and strained SiGe channels, the performances attained and possible applications. The state-of-the-art proposed HCMOS is also presented in this chapter.

8.1 Modulation Doping

Modulation doped field effect transistors use buried channels confined in quantum wells defined in undoped regions of the semiconductor and have a separate supply well. It follows that MODFETs are realizable only in heterostructures. Such a configuration reduces the effects of impurity scattering and scattering due to the interface. The mobility attainable is therefore essentially determined by the intrinsic lattice scattering limited mobility and other unavoidable scattering mechanisms such as carrier–carrier scattering. In group-IV heterostructures, strained elemental semiconductors, namely n-type Ge and Si, have the highest lattice scattering limited mobility and do not suffer from alloy scattering. Therefore n-MODFETs realized using such materials should provide very high mobility.

It is of considerable interest to study the performance of MODFETs realized with p- or n-type strained SiGe and strained p-Si. Traditionally the control element in a MODFET is a metal–semiconductor junction. It is possible to use the MOS gate control and yet retain most of the advantages of the MODFET structures. Fortunately such heterostructure MOS devices have proved capable of good performance. The transconductance attainable and the gate to channel capacitance in MODFETs, as in buried channel MOSFETs, depend on the vertical geometry defining the thicknesses and the doping of the heterostructure. The location of the supply layer plays a dominant role.

The development of a Si/SiGe FET technology is very important for the promises it offers to the current Si technology. FETs are easier to fabricate than HBTs because they require fewer processing steps. FETs can be used in a complementary architecture which has significant advantages of ease in circuit design and reduction of the power consumption of digital circuits. Si/SiGe and strained Si can offer greater improvement compared to conventional Si-CMOS. The complementary heterojunction MOS technology may be designed to match the performance of the p- and n-type devices and has the potential to improve the power delay product of CMOS by a factor of four [2].

Recent advances in Si/SiGe heterostructure growth and material properties have resulted in a compatible SiGe technology. Higher mobility for both electrons and holes has been reported for strained Si layers grown on a relaxed SiGe buffer [3, 4]. Figure 8.1 shows the temperature dependence of electron mobility reported for different types of heterostructures [5]. It is expected that the successful realization of both electron and hole quantum

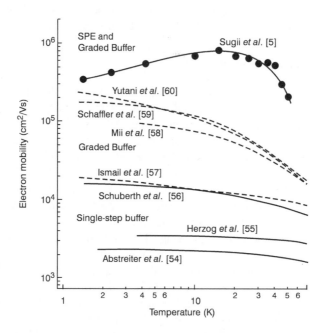

*Figure 8.1. Temperature dependence of electron mobility for a heterostructure with a graded SiGe buffer layer (x = 0.2; 0 ≤ y ≤ x). Data from the literature are shown: - - - - - -, samples with a single-step buffer layer; ———-, samples with a graded buffer layer. After N. Sugii et al., Semicond. Sci. Technol., **13**, A140–A142 (1998).*

wells on the same wafer will enable the integration of both p- and n-type devices. On the basis of computer simulation, several authors [6–10] have predicted excellent high frequency performance for Si/SiGe MODFETs and heterojunction metal oxide semiconductor field effect transistors (HMOSFETs). These achievements are due to the large improvement in the hole transport properties in strained SiGe and both electron and hole transport properties in strained Si.

Yamada *et al.* [10] and Sadek *et al.* [9] have predicted significant velocity overshoot for n-type devices with channel lengths less than 0.2 μm. For self-aligned 0.1 μm n-type MODFETs and HMOSFETs, f_T exceeding 150 GHz was predicted by O'Neill and Antoniadis [7] while f_T of over 100 GHz was predicted for a p-type device. Arafa *et al.* [11] have fabricated a p-type

Si/SiGe MODFET utilizing a 0.1 μm self-aligned gate technology and obtained an f_T of 70 GHz. It is envisaged that low noise operation of MODFETs and HMOSFETs will be possible because they are buried channel devices and therefore not constrained by the severe carrier scattering encountered at the semiconductor–oxide interface of conventional MOSFETs. High performance p- and n-type devices is also significant for the realization of complementary MODFETs and HMOSFETs. Retaining the excellent thermal properties of Si, these emerging device structures give Si/SiGe FET technology the potential to form the basis for low-cost microwave analog and digital circuits.

Modulation doping is widely used in III–V semiconductor devices for realizing high electron mobility transistors (HEMTs) where the growth of a high quality dielectric layer is still a problem. Instead of controlling the channel through an insulated gate, the channel conductivity is controlled by Schottky gates like MESFETs. In contrast to the MOSFET, however, the heterojunction between two semiconductors allows for selective doping of the larger gap material. This modulation doped layer injects the carriers into the undoped channel layer. As a result, a significant reduction in the rate of impurity scattering can be achieved, yielding a higher low field mobility.

This is because the scattering potential due to the ionized impurities which are no longer all around but lie on one side has now to act from a distance. For electrons, this concept is realized by placing the donor atoms in a wider bandgap material called the supply layer. The channel layer has a lower conduction band energy value and is separated from the supply layer by a thin undoped layer called the spacer (figure 8.2). The spacer can be of the same material as the supply layer. It is energetically favorable for the carriers (electrons in this case) to leave their parent atoms and fall into the channel, where they can freely move along the channel with a much higher mobility. For thick channels, the proportion of carriers transferred to the channel depends mainly on the difference of the conduction band edge (ΔE_c) and the spacer thickness. It is less sensitive to the dopant concentration in the supply layer. For narrow channels, the lowest energy level in the well must be calculated quantum mechanically, and the difference between this energy and the edge of the conduction band in the wide bandgap material should replace ΔE_c.

The success of modulation doping in a structure can be judged by measuring the mobility and sheet carrier concentration at different temperatures using conventional Hall measurements. A good confinement of carriers in the channel should result in very little reduction of sheet concentration as the sample is cooled down. Due to the reduction of impurity scattering, a monotonic increase in the mobility should also be observed with lowering of temperature. The supply layer can also be located above the

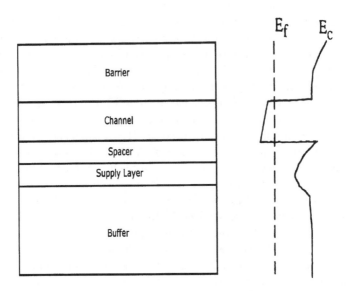

Figure 8.2. Concept of modulation doping demonstrated for electrons in a heterostructure.

channel, which is more advantageous for reducing the parasitic leakage current between the source and drain, as well as for increasing the dc transconductance. For holes, the same concepts apply for the valence band with all potentials changed to negative values.

8.2   SiGe-channel p-MODFETs

Since the invention of the transistor, many types of devices have been proposed [12]. Each innovation attempts to alleviate some of the technical problems associated with the depletion region, reverse breakdown voltage, power consumption and fabrication techniques and also to increase the switching speed, current gain, cutoff frequency, carrier drift velocity, and the mobilities.

The principle of operation of a MODFET is in many ways similar to that of a conventional MOSFET. The main difference between the MODFET and MOSFETs is that in MODFETs the conduction channel is formed in equilibrium by doping the large bandgap material as shown in the schematic cross-sectional diagram of a pseudomorphic SiGe channel MODFET on Si (figure 8.3). The MODFET is, in general, a device structure that has a heavily doped (with either donor or acceptor atoms) layer which is often referred to as the doping spike. The doping spike, rather than the distributed dopants, is useful for providing higher charge carrier density in the channel resulting in a higher transconductance. In addition, the growth of a lightly doped layer on top of the highly doped supply layer results in a Schottky junction with low leakage characteristics and high breakdown voltage.

The placement of the doping spike is very important. The doping spike provides more carriers for charge transport in the channel. The carriers supplied by the doping spike allow a less negative turn-on voltage, and may therefore be referred to as a doping threshold adjustment. If the doping spike is located below the SiGe channel and a negative gate bias is applied to turn the MODFET on, the holes are drawn upward by the electric field, towards the surface of the device, but they are confined, at lower voltages, to the SiGe well. The channel is often physically separated from the doping spike by an i-spacer layer to reduce ionized impurity scattering which would occur if the ionized boron acceptors were placed too close to the channel.

An important feature of the device is the boron p^+ doping spike which makes the device a member of the MODFET family of devices. The source and drain contacts are made directly to the two-dimensional hole gas (2-DHG) while the gate electrode between these two terminals modulates the current. It is also possible to have modulation doping in a MOSFET structure (figure 8.4), which is known as a MOD-MOSFET. In this case an n^+- (or p^+-) polysilicon gate over a thin gate oxide layer and p^+ source and drain control the current. The oxide is also necessary to prevent gate leakage currents and thus allow higher voltage swings. However, the Si/SiO_2 interface introduces additional scattering.

As it is difficult to grow a high quality oxide on a SiGe layer, a Si cap layer is commonly used, separating the SiGe channel from the oxide, leading to a buried channel device. As the cap layer thickness is increased, the oxide/Si interface defect density also decreases [13]. Placing a Si cap between the oxide and the SiGe channel has the advantage of separating the carriers in the channel from the fixed charge at the SiO_2/Si interface, thereby enhancing the mobility of the carriers in the SiGe channel. The Si cap also helps prevent hot carrier degradation of the carriers traveling in the well. However, the Si cap layer itself becomes a channel, as discussed in Chapter 6. In the SiGe MODFET structures, the SiGe channel is the desired main channel of the device because the charge carrier (holes) will have increased mobility in the SiGe and the surface channel is considered parasitic. It is desirable to maximize

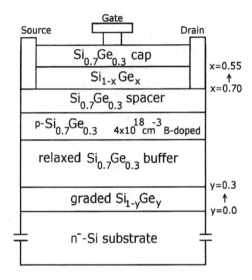

Figure 8.3. A schematic of the layer structure of a SiGe p-MODFET.

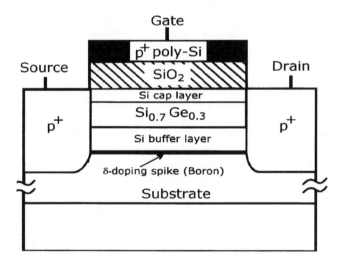

Figure 8.4. A schematic of the layer structure of a SiGe p-MOD-MOSFET grown on Si.

the number of carriers (holes) in the SiGe channel and minimize the number in the Si cap.

8.3   Fabrication of p-MODFETs

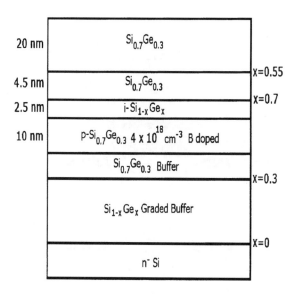

Figure 8.5.   *Layer sequence of a heterostructure strained SiGe channel p-MODFET.*

Figure 8.5 shows a typical layer structure of a p-MODFET on an n-Si substrate [4]. MODFETs are generally fabricated on relaxed $Si_{0.7}Ge_{0.3}$ buffers, which are suitable for the growth of p-type heterostructures. The relaxed $Si_{0.7}Ge_{0.3}$ acts as a "virtual substrate" and may be used for the realization of both p- and n-type MODFETs.

First, a thick (~ 1 μm) step-graded $Si_{1-x}Ge_x$ ($x \rightarrow 0.3$) layer relaxed to the lattice constant of $Si_{0.7}Ge_{0.3}$ is grown. The modulation doped structure consists of a 100 Å B-doped $Si_{0.7}Ge_{0.3}$ supply layer followed by a 25 Å undoped $Si_{0.7}Ge_{0.3}$ spacer layer. The boron doping density is typically 4×10^{18} cm^{-3}. A 450 Å thick graded $Si_{1-x}Ge_x$ channel with Ge mole fraction graded from 0.7 to 0.55 is overlaid by a 200 Å thick $Si_{0.7}Ge_{0.3}$ cap layer. In this structure, a room temperature two-dimensional hole-gas mobility of 700 cm^2/Vs (a factor of two higher than the bulk Si mobility and a factor of 4–5 higher than the mobility of holes in a long-channel MOSFET) and a hole sheet density of 3.5 $\times$ 10^{12} cm^{-2} (2.3×10^{12} cm^{-2}) at room temperature have been reported. At 77 K, the mobility and sheet density were 4000 cm^2/Vs and 2.3×10^{12} cm^{-2}, respectively.

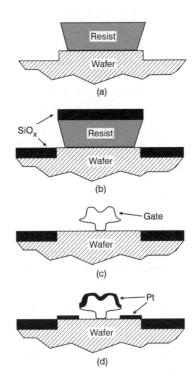

*Figure 8.6.   Fabrication steps for a self-aligned SiGe heterostructure p-MODFET. After I. Adesida et al., Microelectronic Eng., **35**, 257–260 (1997).*

Self-aligned SiGe channel p-MODFETs with various gate lengths ranging from 1 μm down to 0.1 μm have been fabricated. The main process steps are shown in figure 8.6. After wafer cleaning, the mesa that defines the active FET area is patterned with optical lithography (figure 8.6a). The sample is then etched using reactive ion etching in a CF$_4$ plasma and electron beam deposition of SiO$_2$ is performed (figure 8.6b). The subsequent lift-off of this oxide results in a planar sample surface. This oxide gives rise to: (i) better thickness uniformity for the resist used thereafter for the gate lithography and (ii) an insulating floor for pads allowing an accurate measurement of the gate current. After the oxide lift-off, a trilayer resist and electron beam lithography are used to define gate lengths ranging from 0.1 μm to 1 μm. A thinner resist layer is necessary for small geometry devices. T-gate metallization consisting of 200/200/200/2000 Å Ti/Mo/Pt/Au is evaporated and lifted off (figure 8.6c). The metal pads consist of 200/200/2000 Å Ti/Pt/Au. Finally, the devices are sintered at 350 °C for 5 min.

Pearsall and Bean were the first to report on p-type SiGe-MODFETs [14]. At room temperature $g_{m,ext}$ values of 2.5 mS/mm and 3.2 mS/mm were observed for enhancement- and depletion-mode devices, respectively, for a gate length of 3 μm. In 1993 Konig and Schaffler [15, 16] reported on two HFETs with a graded $Si_{1-x}Ge_x$ channel and abulk Ge channel. For the devices with graded-SiGe channel (0.6 μm) Ti/Au gates were used. The maximum $g_{m,ext}$ was found to be 34 mS/mm and 67 mS/mm at 300 K and 77 K, respectively. However, the highest transconductance of 230 mS/mm for a 0.25 μm gate length non-self-aligned MODFET with a $Si_{0.3}Ge_{0.7}$ channel has been reported by Arafa *et al.* [17].

A similar process was used to fabricate a Ge channel device having a better performance. A $g_{m,ext}$ of 125 mS/mm and 290 mS/mm at room temperature and 77 K, respectively, was achieved for recessed 1.2 μm gate length devices. This enhancement is mainly due to the high mobility in the bulk Ge channel reaching 1300 cm^2/Vs at room temperature and 14000 cm^2/Vs at 77 K.

Although a bulk Ge layer has been used as a channel, the mobility in the strained Ge channel is limited by either a high defect density or a high background doping. This was evident from the early saturation of the mobility at cryogenic temperature [18]. The main drawback of a bulk Ge channel is that it requires a high Ge concentration in the relaxed buffer. The large lattice mismatch (4.17%) between Ge and Si restricts the pseudomorphic growth regime of Ge on Si to only a few atomic layers. It is necessary to use a thick step-graded relaxed SiGe buffer with an effective Ge concentration greater than 70% to grow the Ge channel.

Arafa *et al.* [17] have measured a knee voltage lower than 0.2 V at all gate biases, and the turn-on resistance was found to be 1.3 Ω-mm for their devices. Figure 8.7 shows the transfer characteristics of the devices. A maximum extrinsic transconductance of 258 mS/mm and a current density of 60 mA/mm were achieved at a gate bias of 0.2 V and a drain bias of –0.4 V. The device showed a threshold voltage of 470 mV, which is relatively high compared to a conventional MODFET. The authors attributed this high threshold voltage to

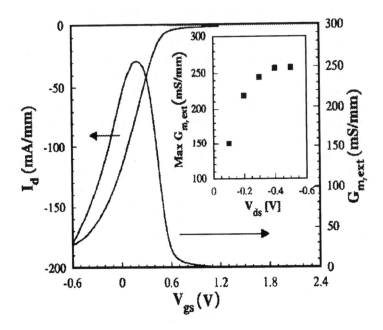

*Figure 8.7. Transfer characteristics of a p-SiGe MODFET (0.1 μm) device. The inset shows the dependence of the maximum extrinsic transconductance on the drain bias. After M. Arafa et al., IEEE Electron Dev. Lett., **17**, 586–588 (1996), copyright ©1996 IEEE.*

the reduction of the source access resistance. The inset shows the dependence of the maximum extrinsic transconductance on the drain bias.

Figure 8.8 shows the transfer characteristics of 0.25 μm and 1 μm gate length devices. A $g_{m,ext}$ of 170 mS/mm was achieved for the 1 μm devices at a drain bias of –0.6 V. This value of $g_{m,ext}$ is 2.5 times higher than for p-MOSFETs with 35 Å oxide operated at a much higher drain bias [19]. A $g_{m,ext}$ of 245 mS/mm was achieved for the 0.25 μm devices operated at only –0.4 V drain bias. The $g_{m,ext}$ value for this gate length is almost the same as that of p-MOSFETs operated at a much higher drain bias. The lower drain bias needed for the case of the self-aligned SiGe p-MODFETs compared to Si p-MOSFETs can be attributed to the fact that the critical field for hole transport is much lower in strained SiGe compared to bulk Si.

Figure 8.9 compares the value of $g_{m,ext}$ of self-aligned devices to state-of-

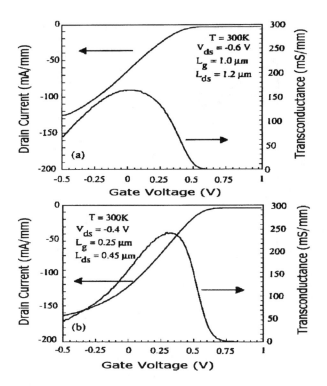

Figure 8.8. Transfer characteristics of typical self-aligned SiGe p-MODFET devices with (a) L_g = *1.0 μm and (b)* L_g = *0.25 μm.   After M. A. Arafa,* Silicon/Silicon–Germanium Modulation Doped Field-Effect Transistors for Complementary Circuit Applications, PhD Thesis, Univ. of Illinois at Urbana-Champaign, *1997.*

the-art p-MOSFETs, with 35 Å gate oxide thickness. It is observed that from the dc $g_{m,ext}$ point of view the deep submicron SiGe self-aligned p-MODFET is not advantageous compared to p-type MOSFETs except that the power consumption in the device is much smaller due to the reduced drain bias.

Figure 8.10 displays the high frequency unity current gain and the maximum available gain for frequencies from 1 GHz to 40 GHz.   By extrapolating these two plots at 20 dB/decade and from the intersection with the 0 dB reference, one determines the unity current gain cutoff frequency and the maximum frequency of oscillation, respectively. At a drain bias of 1 V, an f_T of 70 GHz and an f_{max} of 55 GHz are obtained. Figure 8.11 shows the

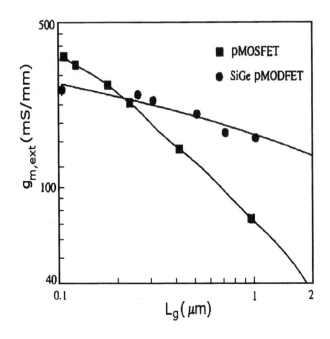

Figure 8.9.   Performance comparison for self-aligned SiGe p-MODFETs and state-of-the-art Si p-MOSFETs with 35 Å gate oxide.   After M. A. Arafa, Silicon/Silicon–Germanium Modulation Doped Field-Effect Transistors for Complementary Circuit Applications, *PhD Thesis, Univ. of Illinois at Urbana-Champaign, 1997.*

gate length (L_g) dependence of the cutoff frequency. It is noted that f_T for all gate lengths, except 0.1 μm, is proportional to $L_g^{-1.1}$ [20]. Some important properties of several experimental p-type MODFETs are shown in table 8.1.

Gluck *et al.* [23] have demonstrated ternary SiGeC channel p-MODFETs with high Ge concentration (45%) directly on Si without using any SiGe buffer layer. The incorporation of small amounts of C (1.2%) into high Ge content layers reduces the compressive strain, resulting in an improved stability and a sufficient valence band discontinuity. The MBE grown modulation doped structure consists of a 300 nm thick Si buffer, followed by an 8 nm thick $Si_{0.54}Ge_{0.45}C_{0.012}$ channel layer, a thin (5 nm) undoped Si spacer layer, a boron-doped (1×10^{19} cm^{-3}) Si supply layer (3 nm) and a 15 nm thick undoped Si cap layer. A multi-level Schottky gate (Ti/Pt/Au: 30/20/120 nm) was formed with varying gate lengths down to 0.75 μm. Room temperature

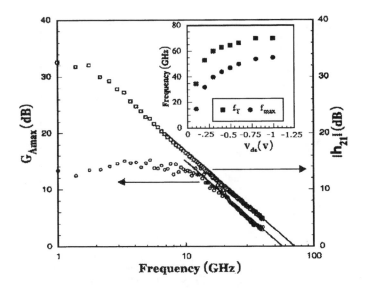

*Figure 8.10. High frequency performance of a typical self-aligned p-MODFET. Measured maximum available gain (G_{max}) and current gain (h_{21}) are shown as a function of frequency. In the inset, the dependence of the unity current gain cutoff frequency and maximum frequency of oscillation on the drain voltage is shown at a fixed gate bias voltage of 250 mV. After M. Arafa et al., IEEE Electron Dev. Lett., **17**, 586–588 (1996), copyright ©1996 IEEE.*

transconductance of $g_m = 57$ mS/mm and a saturated current of 40 mA/mm were reported for the test devices. The higher Hall mobilities (185 cm^2/Vs at 300 K and 2750 cm^2/Vs at 77 K) for the Si$_{0.49}$Ge$_{0.49}$C$_{0.02}$ quantum well in comparison to that of Si$_{0.5}$Ge$_{0.5}$ were attributed to the reduced strain and the increase in bandgap in the ternary alloy.

8.4   Strained Si n-MODFETs

A more than twofold enhancement in electron mobility of the modulation doped strained Si structure promises faster, low power devices compared to conventional NMOS technology. Enhanced electron mobility in strained Si layers compared to bulk Si offers new possibilities for high speed semiconductor devices realized by the n-MODFET [2, 24]. Moreover, the p-MODFETs discussed earlier in this chapter can be integrated with the n-MODFETs

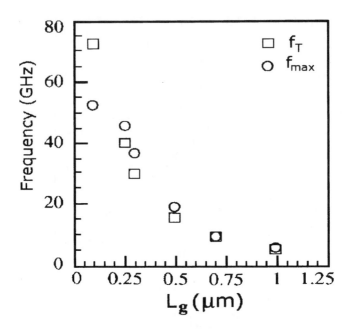

Figure 8.11. The dependence of the unity current gain cutoff frequency and maximum frequency of oscillation on the gate length for SiGe p-MODFETs. After I. Adesida et al., Microelectronic Eng., 35, 257–260 (1997).

leading to a complementary technology that surpasses the widely used CMOS technology in performance.

A variety of n-MODFETs with different ohmic gate configurations has been fabricated [2, 25–33]. A high room temperature $g_{m,ext}$ of 420 mS/mm was reported by Ismail [2] for a 0.4 μm gate length device. An f_T of 40 GHz was reported for the same device, while for a 0.5 μm gate length device an f_T of 33 GHz was reported and was attributed to high source/drain resistances. A self-aligned process should alleviate the series resistance problem. An f_T of over 150 GHz has been predicted for 0.1 μm channel length self-aligned devices [7]. Table 8.2 summarizes the reported results for n-type Si/SiGe heterostructure FETs.

The basic layer structure of the n-MODFET [34] is shown in figure 8.12. The layers are grown by MBE on a p$^-$ substrate (1000 Ω cm) starting with a relaxed SiGe buffer layer whose Ge content is linearly graded to 40%. The core of the layer structure is the 90 Å thick biaxially strained Si channel embedded

Table 8.1. Important properties of some of the reported experimental SiGe p-MODFETs at room temperature.

Ref.	L_{eff} μm	Gate	Doping	$g_{m,ext}$ mS/mm	$g_{m,int}$ mS/mm	f_T/f_{max} GHz
[20]	0.25	Ti/Mo/Au	B	250	285	40/45
[11]	0.1	Ti/Mo/Pt/Au	B	257	295	70/55
[21]	1.0	Ti/Au	B	–	–	5.3/10
[21]	0.7	Ti/Au	B	105	–	9.5/17.8
[17]	0.25	Ti/Pt/Au	B	230	278	24/37
[22]	1.5	Ti/Au	B	95	138	2.1/–
[15]	1.2	Ti/Au	Ga	125	–	–/–
[16]	1.4	Ti/Au	B	34	45	–/–
[16]	0.55	Ti/Au	B	37	60	–/–
[14]	3	Ti	B	2.5	–	–/–

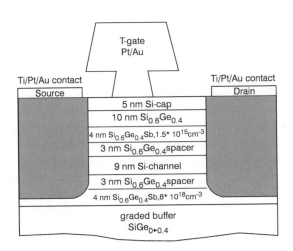

*Figure 8.12. A schematic of the layer sequence of a SiGe/strained-Si n-MODFET. After U. Konig et al., Solid-State Electron., **43**, 1383–1388 (1999).*

Table 8.2. Important properties of some of the reported experimental SiGe n-MODFETs at room temperature.

Ref.	L_{eff} μm	Gate	Doping	Contact	$g_{m,ext}$ mS/mm	$g_{m,int}$ mS/mm	f_T GHz
[33]	0.18	Pt/Au	Sb	P^+ imp.	270	–	46
[32]	0.5	Pt	Sb	Al	130	–	7.5
[31]	0.5	Pt	P	AuSb	390	–	–
[30]	1.5	WSi$_2$		Al	–	–	–
[29]	1.0	Pt/Au	Sb	AuSb	135	–	–
[26]	0.25	Pt	P	AuSb	330	–	–
[28]	1.4	Pt/Au		AuSb	340	380	–
[27]	1.4	Pt/Ti/Au		AuSb	80	88	–
[25]	1.6	Pt/Ti/Au	Sb	AuSb	40	70	2.2

in undoped $Si_{0.6}Ge_{0.4}$ spacers which separate the carrier supply layers from the channel. Due to a Ge content of 40%–45% in the SiGe layers, a high conduction band offset [35] is achieved and therefore the existing quantum well enables sheet carrier densities up to $\sim 6 \times 10^{12}$ cm^{-2} and electron mobilities up to 2700 cm^2/Vs at room temperature.

The device processing of the n-MODFET starts with the formation of the mesa realized by dry etching in a SF_6/O_2 plasma. The ohmic contacts of drain and source are defined by ion implantation. A lift-off process of Ti/Pt/Au forms the metallization layer of the ohmic contacts. The asymmetrically located T-shaped Schottky gate is patterned by electron beam lithography and consists of 500/3000 Å Pt/Au with a gate length of 2500 Å.

As these devices are contemplated for rf wireless applications, the high frequency performance of the n-MODFETs has been measured [34]. An f_T of 43 GHz and f_{max} of 92 GHz for the unrecessed transistor and those produced with a 10 s recess process were reported and are shown in figure 8.13. This performance is comparable to that of GaAs/AlGaAs HEMTs. These frequencies were extrapolated at $V_{DS} = 2.5$ V and operation points of $V_{GS} = -1$ V and $V_{GS} = 0$ V. For the deepest recess the cutoff frequencies of the enhancement MODFET are reduced to an f_T of 30 GHz and f_{max} of 52 GHz due to the lower drain currents and the increased C_{GS}. Figure 8.14 shows the dependence of the cutoff frequencies on the applied V_{DS} for a 10 s recessed device when f_T attains a maximum of 43 GHz at $V_{GS} = 0$ V and $V_{DS} = 1$ V, exactly at the peak of $g_{m,ext}$, and saturates for higher V_{DS}.

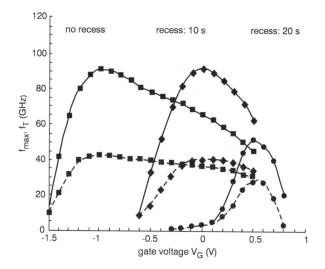

*Figure 8.13. Measured cutoff frequencies f_T (dotted lines) and f_{max} (solid lines) as a function of the gate voltage and recess etching time for the device shown in figure 8.12. After U. König et al., Solid-State Electron., **43**, 1383–1388 (1999).*

8.5 Ge-channel p-MODFETs

The structure of the Ge-channel p-MODFET is shown in figure 8.15. A thick strained Ge hole channel (90 Å) is grown on a strain relaxed SiGe buffer layer with a Ge content of 60%. The channel is followed by a 160 Å $Si_{0.4}Ge_{0.6}$ layer and a 40 Å thick Si cap. Two boron doped supply layers are placed above (8 $\times$ 10^{18} cm^{-3} and 50 Å) and beneath (2×10^{18} cm^{-3} and 50 Å) the channel.

Figure 8.16 shows the maximum drain saturation current of the Ge channel p-MODFET as a function of the gate length. In the case of the hetero-MOSFET, drain current values of up to 650 mA/mm have been achieved at 0.23 μm gate length. For the Ge channel p-MODFET, values of up to 225 mA/mm at $L_g = 0.8$ μm and 300 mA/mm at $L_g = 0.25$ μm were measured. Results achieved by IBM and Daimler-Benz for a $Si_{0.3}Ge_{0.7}$-channel MODFET [11, 17] are shown for comparison. Here the drain current reaches 200 mA/mm at a gate length of 0.1 μm.

Figure 8.17 shows the gate length dependence of the extrinsic

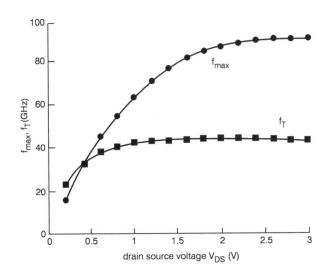

Figure 8.14. f_T and f_{max} values of a 10 s recessed n-MODFET as a function of the drain-source voltage at $V_{GS} = 0$ V. After U. Konig et al., Solid-State Electron., 43, 1383–1388 (1999).

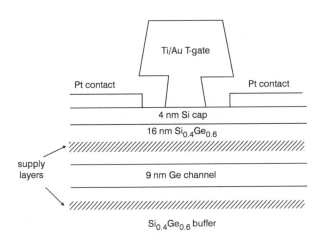

Figure 8.15. A schematic of the layer structure of a Ge-channel p-MODFET. After U. Konig et al., Solid-State Electron., 43, 1383–1388 (1999).

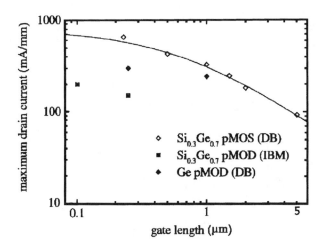

*Figure 8.16. Maximum drain current as a function of gate length for various hetero-FETs. After U. Konig et al., Solid-State Electron., **43**, 1383–1388 (1999).*

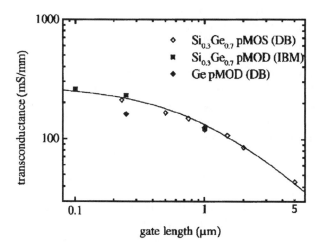

*Figure 8.17. Extrinsic transconductance as a function of gate length for various p-type hetero-FETs. After U. Konig et al., Solid-State Electron., **43**, 1383–1388 (1999).*

transconductance of several p-MODFET devices. The $Si_{0.3}Ge_{0.7}$ channel
MOSFET exhibits values of up to 210 mS/mm at a gate length of 0.23 μm. A
maximum transconductance of 160 mS/mm at $L_g = 0.25$ μm and 125 mS/mm
at $L_g = 0.8$ μm was achieved for the Ge channel p-MODFET.

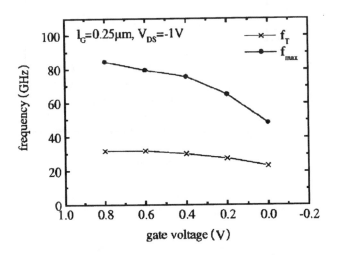

*Figure 8.18. Gate voltage sweep of f_T and f_{max} of the 0.25 μm gate length
Ge channel p-MODFET at $V_{DS} = -1$ V. After U. Konig et al., Solid-State
Electron., **43**, 1383–1388 (1999).*

Results of rf characterization in the frequency range 2–40 GHz for Ge-
channel p-MODFETs are shown in figure 8.18. It is seen that the gate bias
dependence of the f_T at a V_{DS} of –1 V is small and saturates at 32 GHz for
gate voltages higher than 0.6 V. f_{max} increases monotonically for increasing
V_{GS} and shows a maximum value of 85 GHz at $V_{GS} = 0.8$ V.

8.6   Heterojunction Complementary MOSFETs

Heterojunction MOSFETs (HMOSFETs) consist essentially of a modulation
doped structure with a dielectric layer below the gate. The channel in this
case is buried in contrast to conventional   MOSFETs where the channel is
at the surface of the semiconductor. O'Neill and Antoniadis [6] and Sadek *et
al.* [9] have predicted excellent dc and high frequency performance for both
n-type and p-type HMOSFETs. A practical problem of HMOSFETs is oxide
growth which is limited by (i) the restricted maximum temperature (< 680 ^oC)

that SiGe/Si heterostructures can withstand, and (ii) the peculiarities of oxide growth on the compound SiGe material [32, 36]. The latter can be minimized by terminating the   growth with a thin sacrificial Si cap layer as discussed earlier.

Notwithstanding these limitations, a number of investigators [37, 38] have utilized a variety of oxides in fabricating HMOSFETs. These include sputtered oxide, PECVD oxide and wet thermal oxide. Oxidation of SiGe and strained Si using ECR and microwave plasmas has been reported to produce good quality [39–43] and hence may be useful for the fabrication of n-type as well as p-type devices. Towards implementing complementary HMOSFETs, several authors have reported interesting structures using simulation [6–10]. Sadek *et al.* [9] have proposed a planar-type heterostructure for complementary HMOSFETs (CHMOSFETs) as shown in figure 8.19. Figure 8.20 shows the layer structure and energy band diagram for zero, positive and negative gate voltages. A schematic of various process options for a full HCMOS, as suggested by Parker and Whall [44], is shown in figure 8.21.

In figure 8.22, some of the experimental results on f_T and f_{max} achieved for n- and p-HMOSFETs are plotted vs. gate length [11, 33]. The gate length dependence of the cutoff frequencies is clearly visible and simulation (figure 8.23) predicts maximum cutoff frequencies of up to 200 GHz for gate lengths below 0.1 μm or even higher (> 400 GHz) if velocity overshoot is considered [45]. On the other hand, HMOSFET devices with a relaxed gate layout of 0.35 to 0.8 μm have a better performance than Si MOSFETs and benefit from the manufacturability in a standard Si MOS production line.

Si/SiGe heterostructure MOSFETs offer greater improvement compared to conventional Si-CMOS. Besides matching the performance of the p- and n-type devices, stacked p- and n-heterostructure FETs having a strained Si layer as the electron channel and a strained SiGe layer as the hole channel have been demonstrated. The complementary heterojunction MOS (CHMOS) technology has the potential to improve the power delay product of CMOS by a factor of four as shown in figure 8.24.

8.7   Applications of n-MODFETs

n- and p-SiGe HMOSFETs exhibit high performance especially suitable for rf applications.   For the n-MODFETs with tensilely strained Si channels embedded in SiGe layers a maximum $g_{m,ext}$ of 476 mS/mm and cutoff frequencies of f_T of 43 GHz and f_{max} of 92 GHz have been achieved. The best results for p-HFETs were attained for a bulk Ge channel MODFET with an f_T of 32 GHz and an f_{max} of 85 GHz [46].

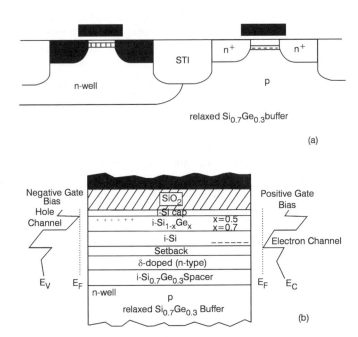

Figure 8.19. (a) Cross-section of a proposed Si/SiGe HCMOS technology and (b) schematic of channel layers and conduction and valence bands for gate bias just above V_T. After M. A. Armstrong et al., IEEE IEDM Tech. Dig., 761–764 (1995), copyright ©1995 IEEE.

Transimpedance amplifiers play an important role in modern communication systems as front-end receivers or low noise amplifiers. A transimpedance amplifier for transforming current signals into amplified voltage signals has been reported [47]. The analog circuit consists of two functional stages (see figure 8.25). The first stage transforms the incoming current into a voltage using a single transistor common-source stage with a drain-to-gate feedback realized by a series connection including a capacitance and a feedback resistor. The second stage amplifies the output voltage of the transimpedance stage with the help of a common-source circuit with two parallel n-MODFETs. The measured transimpedance characteristics of the amplifiers are shown in figure 8.26. It is seen that a transimpedance bandwidth of 1.8 GHz is obtained at a V_{DS} of 5 V and a V_{GS} of –0.2 V.

In order to study the potential of the n-MODFET for digital IC

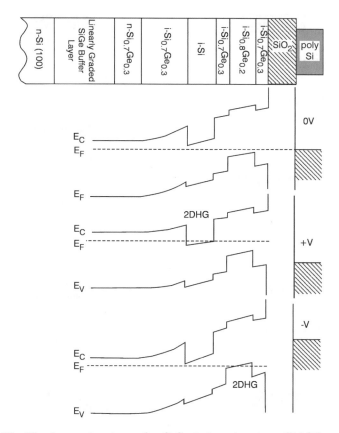

*Figure 8.20. The layer structure of a SiGe heterostructure CMOS and energy band diagram for zero, positive and negative gate voltages. After D. J. Paul, Thin Solid Films, **321**, 172–181 (1998).*

applications, inverter circuits have been realized [34]. The basic inverter consists of two n-MODFETs. The first is the current-driver with source–gate separation $d_{SG} = 0.5$ μm, $W = 100$ μm and a gate length of 1800 or 3000 Å. The second transistor is ungated and therefore works as a resistor in the load path. With an input signal of 600 ps rise time the circuit exhibited a delay of 70 ps for a gate length (L_g) of 3000 Å. Inverters with $L_g = 1800$ Å attained a gate delay of 25 ps and a maximum output voltage swing of 430 mV at an input rise time of 150 ps.

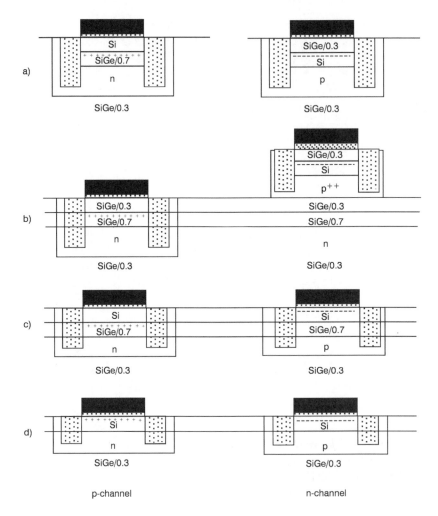

*Figure 8.21. Schematic diagram of different process options for full HCMOS on a blanket coverage virtual substrate of SiGe with terminating composition y = 0.3: (a) involves separate selective epitaxial depositions for the n- and p-channel devices after formation of deep n- and p-wells; (b) involves growth of a stacked p-channel/n-channel structure, followed by selective etching to remove the n-channel layers, hence revealing the p-channel device. The p^{++}-layer can be used for isolation or (c) the n^--channel is formed at the (tensile strained) Si/SiO_2 interface; (d) similar to (c) but both n- and p-channels are formed at the (tensile strained) Si/SiO_2 interface. After E. H. C. Parker and T. E. Whall, Solid-State Electron., **43**, 1497–1506 (1999).*

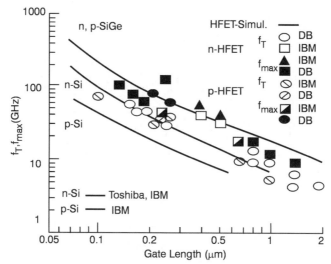

Figure 8.22. Gate length dependence of n- and p-type SiGe-HMOSFETs. After H. Presting and U. Konig, Proc. Current Developments of Microelectronics, 139–150 (1999).

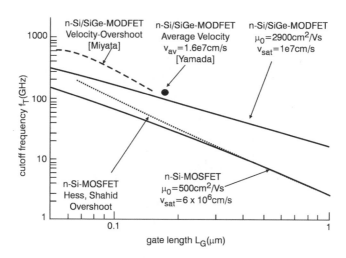

Figure 8.23. Gate length dependence: simulations for n-SiGe HMOSFETs with and without velocity overshoot. After H. Presting and U. Konig, Proc. Current Developments of Microelectronics, 139–150 (1999).

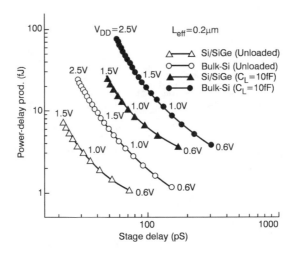

Figure 8.24. Power delay product vs. stage-delay for Si/SiGe HCMOS and bulk Si-CMOS. The corresponding drain bias values are indicated on the curves. After M. A. Armstrong et al., IEEE IEDM Tech. Dig., 761–764 (1995), copyright ©1995 IEEE.

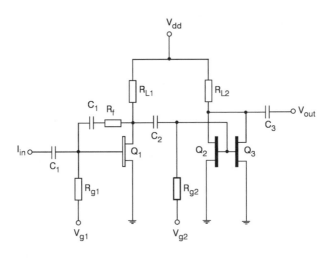

*Figure 8.25. Schematic diagram of a two-stage transimpedance amplifier. After M. Saxarra et al., Electron. Lett., **34**, 499–500 (1998).*

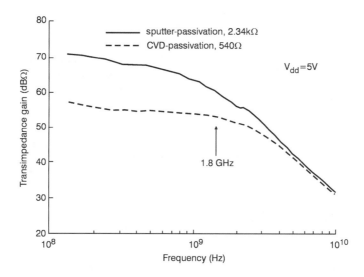

*Figure 8.26. Measured transimpedance characteristics of amplifiers. After M. Saxarra et al., Electron. Lett., **34**, 499–500 (1998).*

8.8 Comparison of Performance

The reported high frequency performance of n-MODFET devices is shown in figure 8.27. Solid and dashed lines are simulation results with the parameters listed in the inset [48]. From the viewpoint of a performance comparison, some of the most exciting $Si_{1-x}Ge_x$ results to date have been achieved with MODFETs grown on virtual substrates [2, 11, 45, 49]. The $Si_{1-x}Ge_x$ n-MODFET has switching times comparable to n-MODFETs in GaAs/AlGaAs devices and larger than GaAs MESFETs. The $Si_{1-x}Ge_x$ p-MODFET is even more impressive as the devices are faster than any other p-channel transistors reported. Room temperature mobility of 2830 cm^2/Vs with a sheet density of 2×10^{12} cm^{-2} for the n-MODFET has been measured. At 77 K, for an n-MODFET, a mobility of 18000 cm^2/Vs with a carrier density of 8×10^{11} cm^{-2} has been reported [3]. For p-MODFETs, mobilities of 1300 cm^2/Vs with a sheet density of 1.5×10^{12} cm^{-2} at room temperature and 14000 cm^2/Vs with a carrier density of 1×10^{12} cm^{-2} at 77 K have been demonstrated [15]. Transconductances in an n-type device with a gate length of 0.25 μm were 330

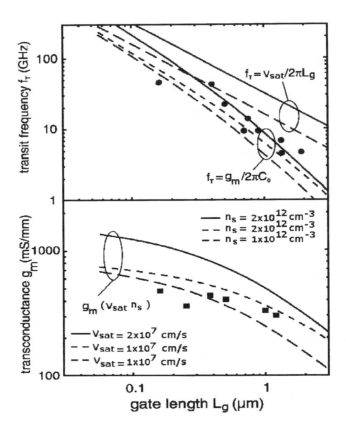

*Figure 8.27. Compilation of the high frequency performance of n-MODFET devices. Both the transconductance g_m and the transit frequency f_T are plotted vs. the gate length L_g. After F. Schäffler, Thin Solid Films, **321**, 1–10 (1998).*

mS/mm at room temperature and 600 mS/mm at 77 K, respectively [2], and that for a 0.1 μm p-MODFET was 237 mS/mm [11].

At low temperature the results are more impressive with two-dimensional electron gas (2DEG) mobilities up to 390000 cm^2/Vs at 400 mK [50] and two-dimensional hole gases (2DHGs) reaching 55000 cm^2/Vs at 4.2 K [51]. The enhancement of mobility in the n-channel is due to reduced intervalley scattering, with the strain splitting the valley degeneracy, and the higher saturation velocity achievable at lower electric fields. The enhancement in the p-channel results from the lower effective mass and Ge-like strain modified valance band structure. Initial modeling of circuit performance is also encouraging. n-MODFETs with loads of 200 fF were shown to exhibit 560 ps delays in NOR gates at 1.1 V compared to 1400 ps delays for the equivalent CMOS [52]. The CMOS had to be run at 3.3 V to achieve the same delay, consuming nine times the power of the $Si_{1-x}Ge_x$ MODFETs. The potential of high-speed complementary logic with such devices looks promising.

8.9 MODHBT

An interesting approach for obtaining a simultaneous enhancement of the vertical and horizontal transport of minority and majority carriers, respectively, in the base region of a bipolar transistor by tailoring of bandgap, doping, and lattice constant has been reported [53]. This may be called a modulation-doped superlattice base heterojunction bipolar transistor (MODHBT).

In a bipolar transistor, the collector and base currents flow perpendicular to each other in the base region. The collector current flows vertically through the base (in a vertical device) and the base current flows laterally along the emitter–base junction. For high frequency applications, the mobility in both of these directions should be high. The transit time of minority carriers in the base is reduced by a high mobility/diffusivity in the vertical direction, while the base series resistance is reduced by high mobility in the lateral direction. To reduce the base series resistance, the use of a modulation doped superlattice structure in the base of a heterojunction bipolar transistor is described below.

The superlattice structure is a multilayered heterojunction formed of alternating layers of wide gap and narrow gap material, each layer having a typical thickness of 50 to 100 Å. In this structure, only the wide gap material is doped. Due to the band discontinuity, mobile carriers move from the wide gap to the narrow gap material and are confined there. Because the carriers are physically separated from the parent atoms, majority carrier mobility is greatly increased, especially at a low temperature where impurity scattering is normally dominant. Such a structure could be fabricated using low temperature epitaxy, i.e., MBE.

The special feature of the MODHBT is that the superlattice structure enhances the vertical transport of minority carriers through the base. In the case of an npn bipolar transistor, the presence of any type of barrier or well in the conduction band of the base material would increase charge storage and reduce the collector current and gain of the device. Even in the case where these barriers are designed to permit resonant tunneling, the available output current is unacceptably low for VLSI applications.

8.10   Summary

The developments in p- and n-MODFETs using Si, Ge and their alloys have established them as frontrunners in high speed devices. MODFETs are intrinsically high mobility and low noise devices. Furthermore complementary structures are possible. For both n- and p-MODFETs, further improvements in performance can be achieved by (i) downscaling and self-aligning the gates of the device, (ii) the utilization of insulating substrates such as SOI, and (iii) the reduction of the high series resistance of the devices via optimization of heterostructure, ohmic contact processing, and device configuration.

In the case of n-MODFETs first circuit applications have been demonstrated: a transimpedance amplifier with a bandwidth of 1.8 GHz and an inverter with a gate delay of 25 ps. This is a milestone towards high speed complementary HFET circuits. However, a significant amount of work needs to be done to bring Si/SiGe MODFETs and related devices to a higher level of performance and commercial production.

8.11   References

1 H. Morkoc, B. Sverdlov and G.-B. Gao, "Strained Layer Heterostructures, and Their Applications to MODFET's, HBT's, and Lasers," *Proc. IEEE*, vol. 81, pp. 493–556, 1993.

2 K. Ismail, "Si/SiGe High-Speed Field-Effect Transistors," in *IEEE IEDM Tech. Dig.*, pp. 509–512, 1995.

3 K. Ismail, S. F. Nelson, J. O. Chu and B. S. Meyerson, "Electron transport properties of Si/SiGe heterostructures: measurements and device implications," *Appl. Phys. Lett.*, vol. 63, pp. 660–662, 1993.

4 K. Ismail, J. O. Chu and B. S. Meyerson, "High hole mobility in SiGe alloys for device applications," *Appl. Phys. Lett.*, vol. 64, pp. 3124–3126, 1994.

5 N. Sugii, K. Nakagawa, Y. Kimura, S. Yamaguchi and M. Miyao, "High electron mobility in strained Si channel of $Si_{1-x}Ge_x/Si/Si_{1-x}Ge_x$ heterostructure with abrupt interface," *Semicond. Sci. Technol.*, vol. 13, pp. A140–A142, 1998.

6 A. G. O'Neill and D. A. Antoniadis, "Deep Submicron CMOS Based on Silicon Germanium Technology," *IEEE Trans. Electron Dev.*, vol. 43, pp. 911–918, 1996.

7 A. G. O'Neill and D. A. Antoniadis, "Investigation of Si/SiGe-Based FET Geometries for High Frequency Performance by Computer Simulation," *IEEE Trans. Electron Dev.*, vol. 44, pp. 80–88, 1997.

8 M. A. Armstrong, D. A. Antoniadis, A. Sadek, K. Ismail and F. Stern, "Design of Si/SiGe Heterojunction Complementary Metal-Oxide Semiconductor Transistors," in *IEEE IEDM Tech. Dig.*, pp. 761–764, 1995.

9 A. Sadek, K. Ismail, M. A. Armstrong, D. A. Antoniadis and F. Stern, "Design of Si/SiGe Heterojunction Complementary Metal-Oxide-Semiconductor Transistors," *IEEE Trans. Electron Dev.*, vol. 43, pp. 1224–1232, 1996.

10 T. Yamada, Z. Jing-Rong, H. Miyata and D. K. Ferry, "In-plane transport properties of $Si/Si_{1-x}Ge_x$ structure and its FET performance by computer simulation," *IEEE Trans. Electron Dev.*, vol. 41, pp. 1513–1522, 1994.

11 M. Arafa, K. Ismail, J. O. Chu, B. S. Meyerson and I. Adesida, "A 70-GHz f_T low operating bias self-aligned p-type SiGe MODFET," *IEEE Electron Dev. Lett.*, vol. 17, pp. 586–588, 1996.

12 E. B. Stoneham, "The search for the fastest three-terminal device," *Microwaves*, pp. 55-60, 1982.

13 S. S. Iyer, P. M. Solomon, V. P. Kesan, A. A. Bright, J. L. Freeouf, T. N. Nguyen and A. C. Warren, "A Gate-quality dielectric system for SiGe Metal-Oxide-Semiconductor Devices," *IEEE Electron Dev. Lett.*, vol. EDL-12, pp. 246–248, 1991.

14 T. P. Pearsall and J. C. Bean, "Enhancement and Depletion-Mode p-Channel Ge_xSi_{1-x} Modulation-Doped FETs," *IEEE Electron Dev. Lett.*, pp. 308–310, 1986.

15 U. Konig and F. Schaffler, "p-Type Ge-Channel MODFET's with High Transconductance Grown on Si Substrates," *IEEE Electron Dev. Lett.*, vol. 14, pp. 205–207, 1993.

16 U. Konig and F. Schaffler, "p-type SiGe channel modulation doped field-effect transistors with post-evaporation patterned submicrometre Schottky gates," *Electronics Lett.*, vol. 29, pp. 486–488, 1993.

17 M. Arafa, P. Fay, K. Ismail, J. O. Chu, B. S. Meyerson and I. Adesida, "DC and RF Performance of 0.25 μm p-Type SiGe MODFET," *IEEE Electron Dev. Lett.*, vol. 17, pp. 449–451, 1996.

18 C. M. Engelhard, D. Tobben, M. Aschauer, F. Schaffler, G. Abstreiter and E. Gornik, "High mobility 2-D hole gases in strained Ge channels on Si substrates studied by magnetotransport and cyclotron resonance," *Solid-State Electron.*, vol. 37, pp. 949–952, 1994.

19 Y. Taur, S. Cohen, S. Wind, T. Lii, C. Hsu, D. Quinlan, C. Chang, D. Buchanan, P. Agneilo, Y. Mii, C. Reeves, A. Acovic and V. Kesan, "High transconductance 0.1 μm pMOSFET," in *IEEE IEDM Tech. Dig.*, pp. 901–904, 1992.

20 I. Adesida, M. Arafa, K. Ismail, J. O. Chu and B. S. Meyerson, "Submicrometer P-type SiGe Modulation-doped Field-Effect Transistors for High-speed Applications," *Microelectronic Engineering*, vol. 35, pp. 257–260, 1997.

21 M. Arafa, P. Fay, K. Ismail, J. O. Chu, B. S. Meyerson and I. Adesida, "High-Speed P-Type SiGe Modulation-Doped Field-Effect Transistors," *IEEE Electron Dev. Lett.*, vol. 17, pp. 124–126, 1996.

22 M. Arafa, K. Ismail, P. Fay, J. O. Chu, B. S. Meyerson and I. Adesida, "High transconductance p-type SiGe modulation-doped field-effect transistor," *Electron. Lett.*, vol. 31, pp. 680–681, 1995.

23 M. Gluck, U. Konig, W. Winter, K. Brunner and K. Eberl, "Modulation-doped $Si_{1-x-y}Ge_xC_y$ p-type hetero-FETs," *Physica E*, vol. 2, pp. 768–771, 1998.

24 K. Ismail, J. O. Chu and M. Arafa, "Integrated enhancement- and depletion-mode FETs in modulation-doped Si/SiGe heterostructures," *IEEE Electron Dev. Lett.*, vol. 18, pp. 435–437, 1997.

25 H. Daembkes, H.-J. Herzog, H. Jorke, H. Kibbel and E. Kasper, "The n-channel SiGe/Si modulation-doped Field-Effect transistor," *IEEE Trans. Electron Dev.*, vol. ED-33, pp. 633–638, 1986.

26 K. Ismail, B. S. Meyerson, S. Rishton, J. Chu, S. Nelson and J. Nocera, "High-Transconductance n-Type Si/SiGe Modulation-Doped Field-Effect Transistors," *IEEE Electron Dev. Lett.*, vol. 13, pp. 229–231, 1992.

27 U. Konig and F. Schaffler, "Si/SiGe modulation doped field-effect transistor with two electron channels," *Electronics Lett.*, vol. 27, pp. 1405–1407, 1991.

28 U. Konig, A. J. Boers, F. Schaffler and E. Kasper, "Enhancement mode n-channel Si/SiGe MODFET with high intrinsic transconductance," *Electronics Lett.*, vol. 28, pp. 160–162, 1992.

29 U. Konig, A. J. Boers and F. Schaffler, "N-channel Si/SiGe MODFETs: effects of rapid thermal activation on the DC performance," *IEEE Electron Dev. Lett.*, vol. 14, pp. 97–99, 1993.

30 T. N. Jackson, S. F. Nelson, J. O. Chu and B. S. Meyerson, "Undoped SiGe heterostructure field effect transistors," *IEEE Trans. Electron Dev.*, vol. 40, pp. 2104–2105, 1993.

31 K. Ismail, S. Rishton, J. O. Chu, K. Chan and B. S. Meyerson, "High-performance Si/SiGe n-type modulation-doped transistors," *IEEE Electron Dev. Lett.*, vol. 14, pp. 348–350, 1993.

32 V. I. Kuznetsov, R. V. Veen, E. van der Drift, K. Werner, A. H. Verbruggen and S. Radelaar, "Technology for high-performance n-channel SiGe modulation-doped field-effect transistors," *J. Vac. Sci. Technol. B*, vol. 13, pp. 1353–1354, 1995.

33 M. Gluck, T. Hackbart, U. Konig, A Hass, G. Hock and E. Kohn, "High f_{max} n-type Si/SiGe MODFETs," *Electronics Lett.*, vol. 33, pp. 335–337, 1997.

34 U. Konig, M. Zeuner, G. Hock, T. Hackbarth, M. Gluck, T. Ostermann and M. Saxarra, "n- and p-type SiGe HFETs and circuits," *Solid-State Electron.*, vol. 43, pp. 1383–1388, 1999.

35 R. People, "Physics and Applications of Ge_xSi_{1-x}/Si Strained Layer Heterostructures," *IEEE J. Quantum Elec.*, vol. QE-22, pp. 1696–1710, 1986.

36 F. K. LeGoues, R. Rosenberg, T. Nguyen, F. Himpsel and B. S. Meyerson, "Oxidation studies of SiGe," *J. Appl. Phys.*, vol. 65, pp. 1724–1728, 1989.

37 V. P. Kesan, S. Subbanna, P. J. Restle, M. J. Tejwani, J. M. Aitken, S. S. Iyer and J. A. Ott, "High Performance 0.25 μm p-MOSFETs with Silicon Germanium Channels for 300K and 77K Operation," in *IEEE IEDM Tech. Dig.*, pp. 25–28, 1991.

38 P. W. Li, E. S. Yang, Y. F. yang, J. O. Chu and B. S. Meyerson, "SiGe pMOSFET's with Gate Oxide Fabricated by Microwave Electron Cyclotron Resonance Plasma Processing," *IEEE Electron Dev. Lett.*, vol. 15, pp. 402–405, 1994.

39 M. Mukhopadhyay, S. K. Ray, C. K. Maiti, D. K. Nayak and Y. Shiraki, "Electrical properties of oxides grown on strained SiGe layer at low temperatures in a microwave oxygen plasma," *Appl. Phys. Lett.*, vol. 65, pp. 895–897 (see also Vol. 66(12), p.1566, March 20, 1995 issue), 1994.

40 M. Mukhopadhyay, S. K. Ray, D. K. Nayak and C. K. Maiti, "Ultrathin Oxides Using N_2O on Strained $Si_{1-x}Ge_x$," *Appl. Phys. Lett.*, vol. 68, pp. 1262–1264, 1996.

41 M. Mukhopadhyay, S. K. Ray, C. K. Maiti, D. K. Nayak and Y. Shiraki, "Properties of SiGe oxides grown in a microwave oxygen plasma," *J. Appl. Phys.*, vol. 75, pp. 6135–6140, 1995.

42 L. K. Bera, S. K. Ray, M. Mukhopadhyay, D. K. Nayak, N. Usami, Y. Shiraki and C. K. Maiti, "Electrical Properties of N_2O/NH_3 Plasma Grown Oxynitride on strained Si," *IEEE Electron Dev. Lett.*, vol. 19, pp. 273–275, 1998.

43 L. K. Bera, M. Mukhopadhyay, S. K. Ray, D. K. Nayak, N. Usami, Y. Shiraki and C. K. Maiti, "Oxidation of strained Si in a microwave electron cyclotron resonance plasma," *Appl. Phys. Lett.*, vol. 70, pp. 217–219, 1997.

44 E. H. C. Parker and T. E. Whall, "SiGe heterostructure CMOS circuits and applications," *Solid-State Electron.*, vol. 43, pp. 1497–1506, 1999.

45 R. Hagelauer, T. Ostermann, U. Konig, M. Gluck and G. Hock, "Performance estimation of Si/SiGe hetero-CMOS circuits," *Electronics Lett.*, vol. 33, pp. 208–210, 1997.

46 H. Presting and U. Konig, "State and Applications of Si/SiGe High Frequency and Optoelectronic Devices," in *Proc. Current Developments of Microelectronics*, pp. 139–150, 1999.

47 M. Saxarra, M. Gluck, J. N. Albers, D. Behammer, U. Langmann and U. Konig, "Transimpedance amplifiers based on Si/SiGe MODFETs," *Electronics Lett.*, vol. 34, pp. 499–500, 1998.

48 F. Schäffler, "Si/Si$_{1-x}$Ge$_x$ and Si/Si$_{1-y}$C$_y$ heterostructures: materials for high-speed field-effect transistors," *Thin Solid Films*, vol. 321, pp. 1–10, 1998.

49 D. J. Paul, "Silicon–Germanium Strained Layer Materials in Microelectronics," *Advance Materials*, vol. 11, pp. 191–204, 1999.

50 K. Ismail, M. Arafa, K. L. Saenger, J. O. Chu and B. S. Meyerson, "Extremely high electron mobility in Si/SiGe modulation-doped heterostructures," *Appl. Phys. Lett.*, vol. 66, pp. 1077–1079, 1995.

51 Y.-H. Xie, D. Monroe, E. A. Fitzgerald, P. J. Silverman, F. A. Thiel and G. P. Watson, "Very high mobility two-dimensional hole gas in Si/GeSi/Ge structures grown by molecular beam epitaxy," *Appl. Phys. Lett.*, vol. 63, pp. 2263–2265, 1993.

52 U. Konig, M. Gluck, A. Gruhle, G. Hoch, E. Kohn. B. Bozon, D. Nuernbergk, T. Ostermann and R. Hagelauer, "Design Rules for n-type SiGe Hetero FETs," *Solid-State Electron.*, vol. 41, pp. 1541–1547, 1997.

53 *IBM Tech. Disc. Bull., August,* 1989.

54 G. Abstreiter, H. Brugger, T. Wolf, H. Jorke and H. J. Herzog, "Strain-Induced Two-Dimensional Electron Gas in Selectively Doped Si/Si$_x$Ge$_{1-x}$ Superlattices," *Phys. Rev. Lett.*, vol. 54, pp. 2441–2444, 1985.

55 H.-J. Herzog, H. Jorge and F. Schaffler, "Two-dimensional electron gas properties of symmetrically strained Si/Si$_{1-x}$Ge$_x$ quantum well structures," *Thin Solid Films*, vol. 184, pp. 237–245, 1990.

56 G. Schuberth, F. Schaffler, M. Besson, G. Absteriter and E. Gornik, "High electron mobility in modulation-doped Si/SiGe quantum well structures," *Appl. Phys. Lett.*, vol. 59, pp. 3318–3320, 1991.

57 K. Ismail, B. S. Meyerson and P. J. Wang, "High electron mobility in modulation-doped Si/SiGe," *Appl. Phys. Lett.*, vol. 58, pp. 2117–2119, 1991.

58 Y. J. Mii, Y. H. Xie, E. A. Fitzgerald, D. Monroe, F. A. Thiel, B. Weir and L. C. Feldman, "Extremely high electron mobility in Si/Ge$_x$Si$_{1-x}$ structures grown by molecular beam epitaxy," *Appl. Phys. Lett.*, vol. 59, pp. 1611–1613, 1991.

59 F. Schaffler, D. Tobben, H.-J. Herzog, G. Abstreiter and B. Hollander, "High-electron-mobility Si/SiGe heterostructures: influence of the relaxed SiGe buffer layer," *Semicond. Sci. and Technol.*, vol. 7, pp. 260–266, 1992.

60 A. Yutani and Y. Shiraki, "Hybrid MBE growth and mobility limiting factors of n-channel Si/SiGe modulation-doped systems," *J. Crystal Growth*, vol. 175-176, pp. 504–508, 1997.

Chapter 9

Contact Metallization on Strained Layers

Semiconductor–metal interfaces play a crucial role in modern electronic and optoelectronic devices. SiGe and related group-IV alloys have shown a great potential for next generation devices and circuits for VLSI/ULSI applications. Though Al and Al–Si have been successfully used in Si devices, they are not the best choices for contacts to group-IV alloy films. For applications of poly-SiGe as gate material, the interaction of the SiGe alloys with noble/refractory metals should also be investigated for low resistance contacts and as metal–semiconductor diodes.

The choice of metals for ohmic contacts for strained layers should satisfy several requirements. First, the composition of the unreacted alloys must remain unchanged after contact reactions. Second, a single compound, not a mixture of compounds (e.g., silicides and germanides), should be in contact with the alloy film. Third, the consumption of the alloy film during reaction must be small since the thicknesses of the strained layers are limited by the critical thickness. In table 9.1 important material properties of commonly used metals for microelectronic device fabrication using Si and Ge are presented.

In this chapter, the formation and characterization of silicides (using Pt, Pd, Ir and Ti on SiGe and SiGeC) using various analytical tools such as x-ray diffraction, Rutherford backscattering and Auger electron spectroscopy will be discussed. We describe Schottky barrier diodes using Ti, Pt and Pd on p-type SiGe, SiGeC, strained Si and GeC films. Experimental results on barrier heights, ideality factor and energy distribution of the interface state density for various diodes and simulation results on forward current–voltage characteristics of Schottky diodes on strained Si are presented.

Table 9.1. Material properties for metals.

Property	Al	Au	Pt	Ni	Cr
Molecular weight (amu)	26.98	196.96	195.09	58.69	52.02
Density (g/cm^3)	2.699	19.288	21.452	8.903	7.19
Melting point (^{o}C)	659.4	1062.2	1768	1454	1875
Oxidation potential (V)	1.66	does not oxidize	does not oxidize	0.25	does not oxidize
Work function at vacuum (eV)	4.25	5.1	5.7	5.1	4.5
Schottky barrier to n-Si (eV)	0.69	0.79	0.9	0.61	0.61
Schottky barrier to p-Si (eV)	0.38	0.25		0.51	0.50
Schottky barrier to n-Ge (eV)	0.48	0.59		0.49	
Schottky barrier to p-Ge (eV)		0.3			

9.1 Contacts on SiGe

Silicides are widely used materials for different process types in the production of electronic devices. For contacts in microelectronic devices, titanium silicide is commonly used. However, its use is complicated by the existence of two crystalline phases. The high resistivity C49-TiSi$_2$ forms first on heating Ti on Si above 500 ^{o}C, and an additional heating above 700 ^{o}C is needed to transform C49 into the low resistivity C54-TiSi$_2$. The growth temperature of strained Si$_{1-x}$Ge$_x$ and related films is typically around 600 ^{o}C. In order to avoid strain relaxation, the silicidation temperature should not exceed the film growth temperature.

The disilicide of cobalt is also a promising contact material for submicron technology with the advantage of a low specific resistivity as well as a low preparation temperature. Whereas TiSi$_2$ at present is commonly used in applications, CoSi$_2$ has clear advantages in the future downscaling of microelectronic devices.

Metal-germanosilicides have been used for making metallic contacts with Si$_{1-x}$Ge$_x$ and other alloys, and knowledge about the formation and stability of thin metal-germanosilicide films is essential for such application. Thermally induced metal/$Si_{1-x}Ge_x$ reactions have been investigated, such as Pt [1–6], Pd [7, 8], W [9], Ni [10], Ti [11–16], Zr [17], and Co [18, 19].

The above studies revealed that during the metal–$Si_{1-x}Ge_x$ reaction, Pd and Pt preferentially react with Si, resulting in Ge segregation [7, 20]. This creates defects that pin the Fermi level near midgap leading to a high Schottky barrier height [7]. To avoid Ge segregation, a Si sacrificial layer was used on top of the SiGe film [21]. Various degrees of Ge segregation and/or the formation of a segregated layered structure were observed in the reactions of these metal/$Si_{1-x}Ge_x$ systems. Pt-silicide is widely used as a low resistance contact to Si. Furthermore, Pt-silicide-germanide contacts to SiGe have application to IR detectors and other devices [21, 22].

Schottky contacts are needed for rectification of electrical signals, mixing of microwave signals and optical detection. For the fabrication of Schottky diodes, surface preparation for metal deposition is very important. In most cases, the departure of the ideality factors of the diodes from unity is due to the presence of an interfacial layer between the metal and the semiconductor [11]. The nonidealities are mostly due to the states associated with the defects near the surface of the semiconductor. These defects act as recombination centers giving rise to excess current which causes deviation from the ideal thermionic emission behavior at low voltage and low temperature.

For Schottky contacts, the general requirement is to adjust the junction parameters, such as barrier height, and ideality factor and to control their reproducibility and stability. Thermal annealing influences the interface and pinning position of the Fermi level which in turn affects the barrier height of the Schottky junction [23].

Several methods may be used to characterize the Schottky contacts for barrier height, reverse bias saturation current density, and the diode nonideality factor. Graphical analysis of the $I - V$ characteristic is used to find J_s, from which the contact barrier height is derived. Another method for analysis of the $I - V$ data uses mathematical curve-fitting to the Schottky diode thermionic emission model to extract the three parameters of interest.

9.1.1 Al–SiGe

In Si technology, aluminum is commonly used to form ohmic contacts to p-Si and highly doped n-Si and Schottky contacts to lightly doped n-Si due to its low cost and ease of fabrication. However, Al and Si form an alloy and not a stable silicide like Ti or W. Al/p-strained-$Si_{1-x}Ge_x$ Schottky contacts have been reported and their electrical properties were measured by Jiang *et al.* [24]. The effect of the Ge fraction and the strain relaxation on

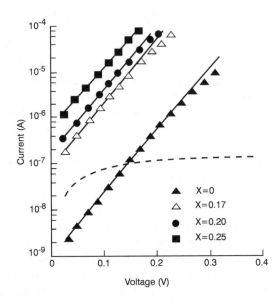

*Figure 9.1. I − V characteristics for Al/p-Si$_{1-x}$Ge$_x$ Schottky contacts (x = 0, 0.17, 0.20, 0.25). After R. L. Jiang et al., Appl. Phys. Lett., **68**, 1123–1125 (1996).*

Schottky barrier heights and the influence of Si sacrificial cap layers on the properties of Schottky contacts were investigated. The electrical properties of Al/p-Si$_{1-x}$Ge$_x$ Schottky contacts were characterized by current–voltage measurements. The $I − V$ characteristics were analyzed with a thermionic emission model [25]. Schottky barrier height and ideality factor were extracted from forward $I − V$ curves. The effective Richardson constant was estimated by using a linear dependence on the Ge composition.

Figure 9.1 shows the forward $I − V$ characteristics of four Si capped Al/Si$_{1-x}$Ge$_x$ samples with x = 0, 0.17, 0.20, and 0.25, respectively, and the reverse $I − V$ characteristics of the sample with x = 0.20 at room temperature. The reverse current density is as low as 7.0×10^{-5} A/cm^2 at a 0.4 V reverse bias. Figure 9.2 shows the variation of Schottky barrier height and the bandgap for strained Si$_{1-x}$Ge$_x$ as a function of the Ge fraction. The Schottky barrier height is seen to decrease with increasing Ge fraction x.

In order to investigate the influence of strain relaxation on the properties of Schottky barriers, films with Ge fraction 0.2 and thickness 500, 1800 and 3500 Å were used. The reverse $I − V$ curve for the 500 Å thick sample is almost

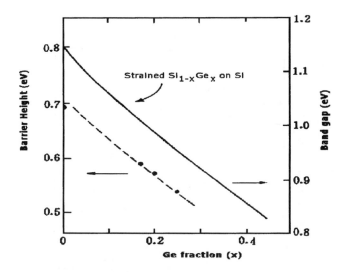

Figure 9.2. Schottky barrier height and bandgap for strained $Si_{1-x}Ge_x$ as a function of Ge fraction. After R. L. Jiang et al., Appl. Phys. Lett., **68,** *1123–1125 (1996).*

linear and similar to that for the Al/p-Si substrate. The reason is that the depletion region has crossed the 500 Å SiGe layer and reached the Si substrate. For 1800 Å (fully strained) and 3500 Å thick (almost fully relaxed) samples, the $q\Phi_b$ values estimated from the forward $I-V$ curves are 0.575 and 0.63 eV, respectively. The difference is 0.055 eV, close to the 0.07 eV difference between the bandgaps of the above-mentioned two SiGe samples [9, 26]. These results suggest that the barrier height increases with an increasing relaxation rate of SiGe strained layers. The Fermi level at the Al/p-relaxed $Si_{1-x}Ge_x$ interface is still pinned at about 0.43 eV below the conduction band.

Figure 9.3a shows the Auger depth profile for a $Si_{0.8}Ge_{0.2}$ epitaxial layer with Si cap layer. Due to the Ge diffusion in the $Si_{0.8}Ge_{0.2}$ layer, the Si cap layer became a $Si_{1-y}Ge_y$ layer in which y decreases from $x = 0.2$ at about 100 Å to $x = 0.1$ at the surface. Figure 9.3b shows the Auger depth profile of Al-$Si_{1-y}Ge_y$-$Si_{1-x}Ge_x$ after alloying. It shows that Al alloys preferentially with Si. The Si in the cap layer is sacrificed to form the Al–Si alloy and the actual interface nearly reaches the $Si_{0.8}Ge_{0.2}$ layer. This is similar to the observation reported by Liou et al. [7] who investigated the interfacial reactions and Schottky barriers of Pt and Pd on epitaxial $Si_{1-x}Ge_x$ alloys. It

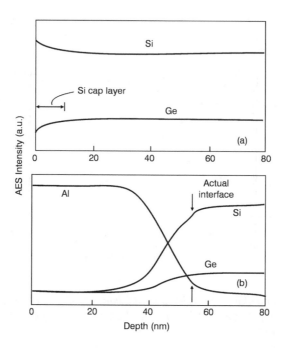

*Figure 9.3. Auger depth profile of (a) Si cap (100 Å) and $Si_{0.8}Ge_{0.2}$ structure and (b) Al–Si sacrificial cap and $Si_{0.8}Ge_{0.2}$ structure after alloying. After R. L. Jiang et al., Appl. Phys. Lett., **68**, 1123–1125 (1996).*

has been found that during the Pt, $Pd–Si_{1-x}Ge_x$ reaction, Pt and Pd react preferentially with Si, resulting in Ge segregation and Fermi level pinning.

9.1.2   Ti–SiGe

Refractory metal silicides such as $TiSi_2$, WSi_2, $TaSi_2$ and $MoSi_2$ have attracted much attention in microelectronic devices due to their low resistivity and high temperature stability required for VLSI/ULSI interconnects. Among various silicides, $TiSi_2$ possesses the lowest resistivity (~ 12.4 $\mu\Omega$-cm) [27], good high temperature stability and excellent compatibility with Si processing technology and is widely used for submicron CMOS contacts.

It has been reported from kinetic studies of Ti/SiGe systems that Si and Ge are the dominant moving species during thermal reaction. Resistance of the resulting silicide formed at about 650 °C is stable and low due to the formation of C54 of $TiSi_2$, along with $Ti(SiGe)_2$ and $TiGe_2$ phases [12].

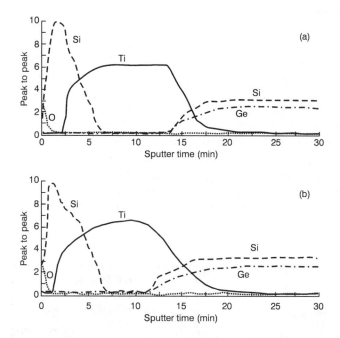

*Figure 9.4. AES depth profiles of (a) as-deposited and (b) 430 °C annealed Ti/Si$_{0.30}$Ge$_{0.70}$ samples. After J. B. Lai and L. J. Chen, J. Appl. Phys., **86**, 1340–1345 (1999).*

For Ti/Si$_{0.3}$Ge$_{0.7}$ samples, AES profiles of as-deposited samples showed that the oxygen level is below the detection limit of the AES, i.e., about 1 at.% as shown in figure 9.4a. AES profiles of the Ti/Si$_{0.3}$Ge$_{0.7}$ sample annealed at 430 °C are shown in figure 9.4b. The profiles show significant intermixing between the metal thin film and substrate. An amorphous interlayer, less than 20 Å in thickness, is observed to form in all as-deposited samples and also in samples annealed at 430 °C. The intermixing is more pronounced between Si and Ti atoms than between Ge and Ti. It is thought that, although Ge atoms diffuse faster than Si at the same temperature, the Ti–amorphous interlayer and amorphous–Si$_{0.6}$Ge$_{0.4}$ interfaces hindered the diffusion of Ge since the atomic size of Ge is larger than that of Si. Ge segregation was also observed.

The structures of the titanium germanosilicide films that formed in Ti/Si$_{1-x}$Ge$_x$ systems were determined from high resolution x-ray diffraction analysis. The XRD scan of the Ti/Si$_{0.6}$Ge$_{0.4}$ samples annealed at 600 °C shows peaks corresponding to diffraction planes of C49-Ti$(Si_{1-x}Ge_x)_2$ and the planes of C54-Ti$(Si_{1-x}Ge_x)_2$ as shown in figure 9.5. In the Ti–Si reaction, C49-TiSi$_2$ is the precursor to the low resistivity C54-TiSi$_2$ phase. Apparently the addition of 40% Ge to the Si did not alter this reaction path. The XRD scans of the samples annealed above 650 °C contained peaks corresponding to diffraction from the planes of C54-Ti$(Si_{1-y}Ge_y)_2$. There was no indication of the presence of C49-Ti$(Si_{1-y}Ge_y)_2$ for $y = 0.4$ samples. The peak deviates slightly because of the larger lattice parameter of the Ti–SiGe phase than of Ti–Si phases.

On the other hand, the XRD scan of the Ti/Si$_{0.3}$Ge$_{0.7}$ sample annealed at 600 °C contained peaks corresponding to the Ti$_6$(Si$_{1-z}$Ge$_z$)$_5$ and the C54-Ti$(Si_{1-y}Ge_y)_2$ planes. For annealing above 650 °C, the XRD spectra contained peaks corresponding to diffraction from the planes of C54-Ti$(Si_{1-y}Ge_y)_2$, as seen in figure 9.6. The C49-Ti$(Si_{1-y}Ge_y)_2$ was no longer present in the reaction path of Ti–Si$_{1-x}$Ge$_x$ for $x = 0.7$.

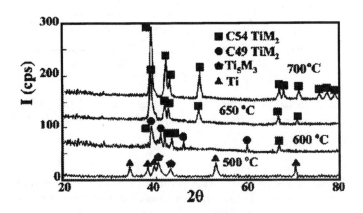

Figure 9.5. Glancing-angle XRD data of Ti/Si$_{0.6}$Ge$_{0.4}$ samples annealed at 500, 600, 650 and 700 °C. M represents Si$_{1-x}$Ge$_x$. After J. B. Lai and L. J. Chen, J. Appl. Phys., 86, 1340–1345 (1999).

Schottky barrier heights (SBHs), contact resistivities and solid-phase reactions at the interface of Ti/Si$_{0.8}$Ge$_{0.2}$/Si(100) systems have been investigated by several workers [2, 28]. The Schottky barrier heights of Ti/Si$_{0.8}$Ge$_{0.2}$/Si(100) and Ti/Si(100) diodes are shown in figure 9.7, as a function of annealing temperature. In the case of p-type diodes, very low SBHs are obtained for Ti/Si$_{0.8}$Ge$_{0.2}$/Si below 300 °C. However, the lowering of SBHs compared with Ti/Si is larger than the value of 0.1 eV expected from

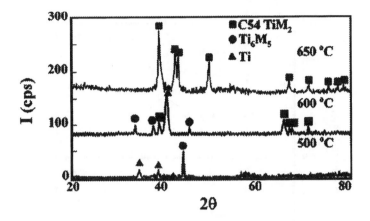

Figure 9.6. Glancing-angle XRD data of Ti/$Si_{0.3}Ge_{0.7}$ samples annealed at 500, 600, 650 and 700 °C. M represents $Si_{1-x}Ge_x$. After J. B. Lai and L. J. Chen, J. Appl. Phys., 86, 1340–1345 (1999).

the E_g of $x = 0.2$. Moreover, it should be noted that SBHs lower than those of Ti/Si are also obtained for n-type diodes above 460 °C. The sum of the SBHs of n- and p-Ti/$Si_{0.8}Ge_{0.2}$/Si diodes in this temperature range has a constant value of about 0.9 eV, which is smaller than the bandgap of $Si_{0.8}Ge_{0.2}$, 1.0 eV. These facts suggest that the Ge fraction at the interfaces is increased by solid-phase reactions.

In order to study the solid-phase reaction between Ti and $Si_{1-x}Ge_x$, XPS measurements were performed using Ti(35 Å)/$Si_{0.5}Ge_{0.5}$/Si(100) samples. Figure 9.8 shows (a) Si 2p and (b) Ge 2p XPS spectra on Ti(35 Å)-$Si_{0.5}Ge_{0.5}$/Si(100) surfaces annealed at various temperatures. Although the Si 2p peak can be observed even for as-deposited samples, the Ge 2p peak appeared above 400 °C. In as-deposited and 300 °C-annealed samples, there are chemical shifts of the Si 2p peak by ~ 0.2 eV compared with that of the Si substrate. In samples annealed at 400 and 500 °C, one observes chemical shifts of the Si 2p and Ge 2p peaks, whose values are ~ 0.4 eV and ~ 0.7 eV from the peaks of the Si and Ge substrates, respectively. The chemical shift of the Si 2p peak due to silicidation by annealing at 700 °C was reported to be ~ 0.4 eV [28]. These results indicate that Ti atoms preferentially react with Si atoms below 300 °C and the reaction with Ge atoms starts above 400 °C, which suggests that a Ge-rich layer is formed at the Ti/$Si_{1-x}Ge_x$ interface.

Figure 9.9 shows an XRD spectrum taken from Ti/$Si_{0.8}Ge_{0.2}$/Si(100)

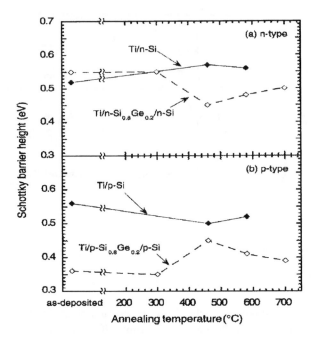

*Figure 9.7. Changes in Schottky barrier heights for (a) n- and (b) p-type Schottky diodes of Ti/Si(100) and Ti/Si$_{0.8}$Ge$_{0.2}$/Si(100), as a function of annealing temperature. After J. Kojima et al., Appl. Surf. Sci., **117/118**, 317–320 (1997).*

samples annealed at 650 °C. It should be noted that all peaks are observed between the diffraction angles expected from C54-TiSi$_2$ and C54-TiGe$_2$, as indicated by arrows. As C54-TiSi$_2$ and C54-TiGe$_2$ are totally miscible and the lattice constant has a linear relationship with the Ge concentration, by representing the composition of this solid solution as Ti(Si$_{1-y}$Ge$_y$)$_2$, the Ge fraction y is evaluated to be 0.12 from the lattice spacings of C54-TiSi$_2$ and C54-TiGe$_2$. The lower values of SBHs for Si$_{0.8}$Ge$_{0.2}$ than those for Si observed above 500 °C are thought to originate from the low work function of C54-Ti(Si$_{1-y}$Ge$_y$)$_2$. From the viewpoint of applications, it can be concluded that the Si$_{1-x}$Ge$_x$/Si structure is a favorable candidate in achieving low resistivity contacts for both n$^-$- and p$^+$-Si.

The measured and calculated contact resistivities for n$^-$- and p$^+$-Si$_{0.8}$Ge$_{0.2}$ are shown in figure 9.10 as a function of doping concentration. The solid and

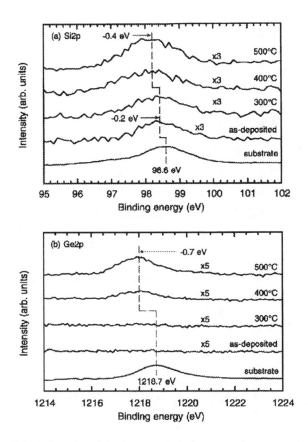

Figure 9.8. (a) Si 2p and (b) Ge 2p XPS spectra of as-deposited, 300, 400 and 500 °C-annealed Ti/Si$_{0.5}$Ge$_{0.5}$/Si(100) systems. The spectra for Si and Ge substrates are also shown for comparison. After J. Kojima et al., Appl. Surf. Sci., 117/118, 317–320 (1997).

dashed lines are calculated curves for n$^-$- and p$^+$-type contacts, respectively. The measured contact resistivities of Ti/Si$_{0.8}$Ge$_{0.2}$ annealed at 460 °C are very close to the calculated values for n$^-$- and p$^+$-SiGe. In the 580 °C annealed case, however, the contact resistivities for n$^-$- and p$^+$-Si$_{0.8}$Ge$_{0.2}$ are low as expected. Since the interfacial reaction proceeds below 600 °C, the dopant atoms would accumulate at the interface if the diffusion of dopant atoms into Ti/Si$_{1-x}$Ge$_x$ layers is sufficiently slow. The lowering in contact resistivity at 580 °C is due to the segregation of dopant atoms at the interface.

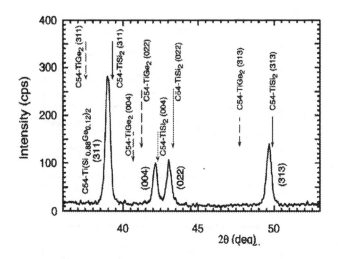

*Figure 9.9. An XRD pattern of the $Ti/Si_{0.8}Ge_{0.2}/Si(100)$ system annealed at 650 °C. The solid and dashed arrows indicate the peaks of bulk $C54$-$TiSi_2$ and $C54$-$TiGe_2$, respectively. After J. Kojima et al., Appl. Surf. Sci., **117/118**, 317–320 (1997).*

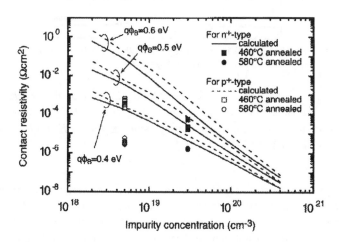

*Figure 9.10. The dependence of contact resistivities on doping concentration for $Ti/Si_{0.8}Ge_{0.2}/Si(100)$. After J. Kojima et al., Appl. Surf. Sci., **117/118**, 317–320 (1997).*

9.1.3 Ir–SiGe

Since the demonstration of Schottky barrier infrared detectors in Si, there has been considerable interest in Schottky diodes with low barrier heights on p-type Si. PtSi/p-Si diodes have been successfully used as infrared detectors operating in the 3–5 μm atmospheric window. Many investigations have been carried out to extend the application of Si based infrared detectors into the atmospheric window 8–14 μm [21, 22, 29]. IrSi, which has the lowest barrier height to p-type Si, is one of the most extensively studied for this application. However, the barrier height with p-Si is not low enough to cover the entire window. It has been shown that the barrier height of the Schottky diode can be further reduced by using a strained $Si_{1-x}Ge_x$ layer instead of Si [21, 22]. Moreover, the barrier height can be tuned by adjusting the amount of Ge in the strained layer.

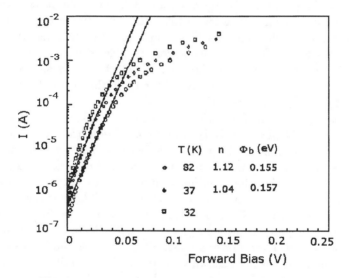

Figure 9.11. $I - V$ *characteristics for Ir/strained* p-$Si_{0.86}Ge_{0.14}$ *Schottky contacts. Extracted Schottky barrier heights and ideality factors are also shown. After O. Nur et al., Semicond. Sci. Technol., 10, 551–555 (1995).*

In order to utilize Schottky barrier diodes of $Si_{1-x}Ge_x$ as infrared detectors, Ir/strained p-$Si_{1-x}Ge_x$ have been reported [30]. $I - V$ curves of Ir/strained p-$Si_{0.86}Ge_{0.14}$ diodes at several temperatures are shown in figure 9.11. The corresponding SBH and ideality factors are shown in the inset. The barrier height of the Ir/strained p-Si$_{0.86}$Ge$_{0.14}$ Schottky diode is considerably less than that of the Ir/p-Si junction. In the Ir/strained p-Si$_{0.86}$Ge$_{0.14}$ samples, the current densities are much higher at any given forward bias than those in Ir/p-Si samples, directly indicating a much smaller Schottky barrier [30].

9.2   Contacts on SiGeC

Studies on several metal/SiGeC contact systems such as Al [31], Co [32], Ti [33], Pt [34], and W [35, 36] have been reported. It has been observed that Schottky contacts to p-type SiGeC show compositional dependence in the barrier height, while Schottky contacts to n-type SiGeC show no compositional dependence, an indication of Fermi level pinning [37]. Incorporation of C into the SiGe epitaxial layer has been shown in References 31 and 33 to reduce epitaxial layer stress and relaxation during contact formation. Eyal *et al.* [33] have shown that the introduction of C into SiGe layers causes a marked decrease in strain relaxation in SiGe occurring during the Ti silicide–germanide reaction, indicating that SiGeC may prove to be a material which has advantages over SiGe for contacts using Ti. The study of metal/SiGeC contacts and barrier heights is one area in which understanding is still incomplete. It is thus important to understand the phase formation of metal/silicide-germanides in SiGeC.

Carbon is found to inhibit the strain relaxation process as well as delay the formation of the C54 phase of TiSi$_2$. Upon complete silicidation, a decrease of Ge concentration in the silicide-germanide/epi-layer and accumulation of C atoms at the interface have been found. To avoid such complexities associated with the thermal reactions between metal and group-IV alloys, use of a thin Si sacrificial layer on top of the strained SiGe or SiGeC layer is common [21].

The phase formation of Co and Ti silicide-germanides on SiGeC epilayers has been shown to proceed through phase sequences identical to those of Si and SiGe, but at lower reaction rates [32, 33]. Similar to silicide-germanide formation in SiGe, Ge has been shown to segregate from the silicide-germanide during formation, accumulating on the epilayer side of the interface. Both Donaton and Eyal report that C also segregates from the silicide-germanide to the silicide-germanide/epilayer interface during formation at temperatures in the 600 °C range, perhaps causing reduced reaction rates in the silicide-germanide by impeding diffusion of reactants.

Characterization of the phase transformation sequence of Pt silicide-germanides Pt$_x$(SiGeC)$_y$ shows it to be similar to that of Pt on Si, although the formation temperature of Pt$_x$(SiGeC)$_y$ differs significantly from that of Pt$_x$Si. Completely reacted Pt$_x$(SiGeC)$_y$ is uniform in composition and is similar in

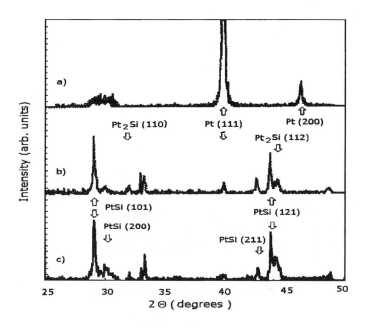

*Figure 9.12. XRD spectra of $Pt_x(SiGeC)_y$ phase with respect to RTA time: (a) XRD spectra after 20 s, 600 °C RTA showing elemental Pt, (b) XRD spectra after 80 s, 600 °C RTA showing a mixture of Pt, $Pt_2(SiGeC)$ and $Pt(SiGeC)$, and (c) XRD spectra after 160 s, 600 °C RTA showing completely reacted $Pt(SiGeC)$. After J. J. Peterson et al., Solid-State Electron., **43**, 1725–1734 (1999).*

smoothness to the underlying SiGeC layers, having an average roughness between 10 and 20 Å. Figure 9.12 shows XRD spectra of the $Pt_x(SiGeC)_y$ phase transformation for 600 °C annealed samples. Figure 9.12a shows that after 20 s of RTA the Pt film shows some evidence of $Pt_2(SiGeC)$ formation, but still shows a close match to the X-ray spectrum for Pt. Figure 9.12b shows the spectrum after 80 s of 600 °C anneal by RTA showing that the reacted layer contains a mixture of $Pt_2(SiGeC)$ and $Pt(SiGeC)$ with a slight Pt signal still present. Figure 9.12c shows the spectrum after annealing for 160 s; in this curve the silicide appears to be completely converted to the $Pt(SiGeC)$ final phase. Additional annealing shows little change in the silicide composition, indicating the $Pt(SiGeC)$ phase is the final phase. Rutherford backscattering spectrometry measurements on these samples show that Ge segregation from the Pt silicide-germanides does occur, with excess Ge accumulating at the silicide-germanide/SiGeC interface.

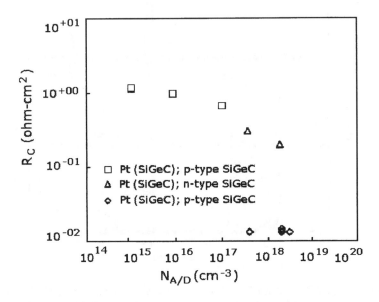

Figure 9.13.   Contact resistance vs.   contact layer doping for Pt(SiGeC) contacts to n-type and p-type SiGeC. Open square: contacts to p-type SiGeC. Open triangle: contacts to n-type SiGeC. Open diamond: contacts to p-type SiGeC. After J. J. Peterson et al., Solid-State Electron., 43, 1725–1734 (1999).

Figure 9.13 displays a plot of contact resistance vs. doping for Pt(SiGeC) to p-type SiGeC contacts and for Pt(SiGeC) to n-type SiGeC contacts, respectively. As expected, increasing the contact doping decreases the contact resistance. A 10× increase in contact doping level results in a 3–4× reduction in the contact resistance. Figure 9.13 also shows that highly n-type doped Pt(SiGeC) ohmic contacts have a smaller contact resistance than comparably doped Pt(SiGeC) to p-type SiGeC contacts. As a comparison to Si, these data show that Pt(SiGeC) to SiGeC contacts will require substantially higher contact doping levels to achieve contact resistance values equivalent to those for PtSi contacts to Si.

The influence of carbon on W/SiGeC-p/Si(100)-p Schottky diodes has been investigated by Serpentini *et al.* [36]. Figure 9.14 shows barrier heights extracted from the reverse $I-V$ characteristics assuming a thermionic emission mechanism. Barrier height increases with the increase in carbon content, at least up to 1%. This increase cannot be explained only by the band gap increase of 24 meV per 1 at.% C measured by the photoluminescence technique [38]. An interfacial layer formed during W deposition on the epilayer is likely to be responsible for the drastic increase of barrier height by charge accumulation. Moreover, the related charge density could also increase with carbon incorporation.

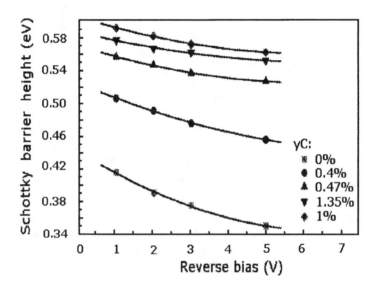

Figure 9.14. Reverse bias dependence of effective barrier height on carbon content (y) varying from 0 to 1.35% and x = 10%. After M. Serpentini et al., J. Vac. Sci. Technol. B, 16, 1686–1688 (1998).

9.3 Contacts on $Ge_{1-y}C_y$

Ohmic contact measurements of Al, Au and W metallization to p-type epitaxial $Ge_{0.9983}C_{0.0017}$ grown on a (100) Si substrate by MBE have been reported by Shao *et al.* [39]. The contacts were annealed at various temperatures and values of specific contact resistance were measured.

Figure 9.15 shows a plot of the total resistance measured between metal contacts as a function of the contact spacing between them. Three parameters, namely sheet resistance, contact resistance and transfer length, were extracted

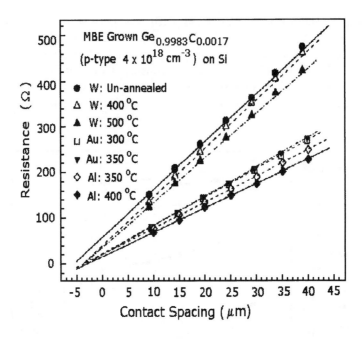

*Figure 9.15. The total resistance R_T measured between various metal contact pads in the TLM structure as a function of the contact separation for p-$Ge_{0.9983}C_{0.0017}$ on Si. After X. Shao et al., IEEE Electron Dev. Lett., **18**, 7–9 (1997), copyright ©1997 IEEE.*

from a least-squares interpolation line of the data. These results demonstrate low resistance ohmic contacts for Al and Au on $Ge_{0.9983}C_{0.0017}$, with values of specific contact resistance of the order of 10^{-5}–10^{-6} Ω cm, after suitable annealing. For an annealing temperature of 300 °C, the Al contacts were found to be under-annealed. When the annealing temperature was raised to 450 °C or above, strong reactions of both Al and Au with the $Ge_{0.9983}C_{0.0017}$ epilayer occurred, which resulted in a highly irregular morphology with numerous metallic islands. Optimal anneal temperatures based on resistances for Al and Au occurred within the 350–400 °C window.

Due to the high melting point of W, the heat of reaction of W–GeC is relatively large compared to Au or Al. This was evident as no morphological reaction was observed even after an anneal at 650 °C. It was expected that W could form a Schottky barrier on GeC due to the difference between the W work function and the electron affinity of GeC using small carbon percentages.

9.4 Contacts on $Si_{1-y}C_y$

Cobalt silicide ($CoSi_2$) has become one of the most promising silicides for deep submicron technology [40, 41], because of its linewidth independent resistivity and minimum lateral growth. Epitaxial CoSi layers are more desirable because of better thermal stability and smooth interfaces than polycrystalline ones. The key factor to form epitaxial rather than polycrystalline CoSi is the suppression of native oxide. This is because Co is thermodynamically stable on oxide [42, 43], so that a thin oxide between Si and Co not only blocks the silicidation but also limits the reaction to local weak spots; this roughens the CoSi surface and creates a polycrystalline structure. The suppression of native oxide is also the most important factor to achieve an atomically smooth interface and good performance [44].

Teichert *et al.* [45] have reported on cobalt silicide formation on $Si_{0.999}C_{0.001}$ (001) layers. Cobalt films were deposited by electron beam evaporation under UHV conditions and silicide formation was performed using rapid thermal annealing. RBS measurements were performed in order to study the phase formation sequence for SiC samples annealed isochronally in the temperature range between 500 and 700 °C. In figure 9.16, the RBS spectra are plotted for films on $Si_{0.999}C_{0.001}$ and on Si, in each case for three samples: after deposition of Co, and after RTP treatment at 500 °C and 700 °C. The comparison of the spectra taken from samples on different substrate layers after the same annealing treatment demonstrates that for the selected reaction stages there is almost no influence of the incorporated carbon. However, this means only that for these particular temperatures the silicide reaction has progressed nearly to the same degree for both substrate types.

The transformation from CoSi to $CoSi_2$ is accompanied by a spatial broadening of the Co distribution and by a decrease of the mean Co concentration in the silicide film, causing broadening and decreasing of the Co signal in the RBS spectrum. For comparison the expected values for pure CoSi and $CoSi_2$ films located at the sample surface and simulated by RUMP have been inserted into the plot shown in figure 9.16.

9.5 Metal/strained-Si Interface

Schottky contacts play an important role in the operation of semiconductor devices for various electronic and optoelectronic applications. Strained Si has shown its potential in applications for novel band-engineered heterostructure devices. Heterostructure Schottky diodes on strained Si using Ti [11] and Pt [6] have been reported. In general, the electrical characteristics of a

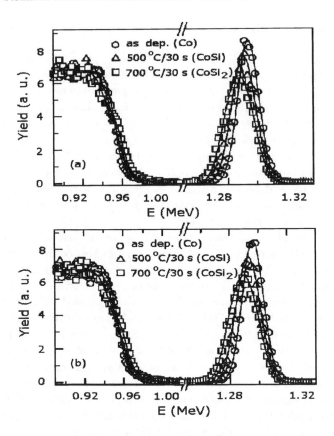

*Figure 9.16. RBS spectra for films on Si (a) and on $Si_{0.999}C_{0.001}$ (b) substrate layers; symbols represent measured data and lines RUMP simulations. After S. Teichert et al., Solid-State Electron., **43**, 1051–1054 (1999).*

Schottky diode are controlled mainly by its interface quality. Interface state distributions of Pt/p-strained-Si [3] and TiW/SiGeC [46] Schottky contacts have been reported. The energy distribution of the interface states has been computed using the measured current–voltage and capacitance–voltage data.

The computed energy band diagram of a Pt/p-strained-Si Schottky diode with an interfacial oxide layer thickness of 5 Å and a series resistance of 10 Ω is shown in figure 9.17. In a practical metal–semiconductor contact, there is always a native insulating layer of oxide, typically 10–20 Å, on the surface of the semiconductor [47]. The current density across such a Schottky diode can

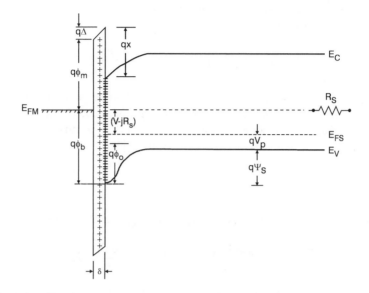

Figure 9.17. Energy band diagram of a Pt/p-strained-Si Schottky contact with an interfacial oxide layer and a series resistance.

be expressed as [3]

$$j = A^*T^2 \exp\left(\frac{-q\phi_b}{kT}\right)\exp\left(\frac{qV - jR_s}{kT}\right)$$ (9.1)

where A^* is the effective Richardson constant, T is temperature in K, ϕ_b is barrier height, k is the Boltzmann constant, R_s is the series resistance present in the device, V is the applied bias and q is the electronic charge. From the band diagram (figure 9.17) the surface potential of the device is given by

$$\psi_s = \phi_b - v_p - (V - jR_s)$$ (9.2)

where v_p is the difference between the Fermi level and the valence band maximum and is given by

$$v_p = \frac{kT}{q}\ln\left(\frac{N_v}{N_a}\right)$$ (9.3)

where N_v is the effective valence band density of states and N_a is acceptor concentration.

Using equations 9.1, 9.2 and 9.3 one can express the current equation for the Schottky diode as

$$j = A^*T^2 \exp\left(\frac{-q\psi_\text{s}}{kT}\right)\exp\left(\frac{-qv_\text{p}}{kT}\right) \tag{9.4}$$

When an incremental (ac) voltage δV is applied across a device, there are corresponding changes in surface potential $\delta\psi_\text{s}$, in space charge density δQ_sc and in interface state charge density δQ_it. The change in surface potential $\delta\psi_\text{s}$ due to change in the current density is given by [48]

$$\delta j = -\frac{jq}{kT}\delta\psi_\text{s} \tag{9.5}$$

The ac changes in charge densities δQ_sc and δQ_it can be expressed as [48]

$$\delta Q_\text{sc} = \left(\frac{\partial Q_\text{sc}}{\partial\psi_\text{s}}\right)\delta\psi_\text{s} \qquad \delta Q_\text{it} = \left(\frac{\partial Q_\text{it}}{\partial\psi_\text{s}}\right)\delta\psi_\text{s} \tag{9.6}$$

At low frequencies, interface states respond to the ac signal and the low frequency capacitance is given by

$$C_\text{LF} = \frac{\left(q\epsilon_\text{s}N_\text{a}/2\psi_\text{s}\right)^{1/2} + qD_\text{it}}{1 + qjR_\text{s}/kT + (\delta/\epsilon_\text{i})\left[\left(q\epsilon_\text{s}N_\text{a}/2\psi_\text{s}\right)^{1/2} + qD_\text{it}\right]} \tag{9.7}$$

where δ is the thickness of the interfacial layer, D_it is the interface state number density and ϵ_i and ϵ_s represent the permittivity of the interfacial oxide layer and that of the semiconductor, respectively.

At high frequency, the interface states cannot follow the ac signal and the capacitance can be expressed as

$$C_\text{HF} = \frac{\left(q\epsilon_\text{s}N_\text{a}/2\psi_\text{s}\right)^{1/2}}{1 + qjR_\text{s}/kT + (\delta/\epsilon_\text{i})\left(q\epsilon_\text{s}N_\text{a}/2\psi_\text{s}\right)^{1/2}} \tag{9.8}$$

From equations 9.7 and 9.8 one gets

$$D_\text{it} = \frac{1}{q}\frac{C_1}{C_\text{HF}}\frac{C_\text{HF} - C_\text{LF}}{C_\text{LF} - C_1}\left(q\epsilon_\text{s}N_\text{a}/2\psi_\text{s}\right)^{1/2} \tag{9.9}$$

where $C_1 = \epsilon_i/\delta$. However, for a Schottky diode the native oxide at the interface is about 5–10 Å and $C_1 \gg C_{\text{LF}}$ and one obtains

$$D_{\text{it}} \approx \left(q\epsilon_s N_a/2\psi_s \right)^{1/2} \frac{C_{\text{LF}} - C_{\text{HF}}}{qC_{\text{HF}}} \tag{9.10}$$

Thus knowing the experimental values of C_{LF}, C_{HF} and ψ_s one obtains the value of the interface state density, D_{it}.

The forward logarithmic $I - V$ characteristics of a typical Pt/p-strained-Si Schottky contact at room temperature is shown in figure 9.18. Using equation 9.4 the surface potential ψ_s is calculated from the measured $I - V$ data. Figure 9.19 shows the plot of the calculated surface potential (ψ_s) with applied voltage. It is seen from figure 9.19 that the surface potential decreases linearly until a critical voltage (V_c) after which the voltage drop across the series resistance becomes comparable with the applied voltage and $\psi_s(J_c,V_c)$ presents the corresponding surface potential. The barrier height obtained is 0.48 eV [6].

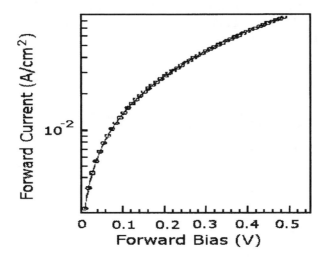

Figure 9.18. Forward $I-V$ characteristics of a Pt/p-strained-Si Schottky diode at room temperature.

Figure 9.20 shows the plots of measured $C - V$ data at 10 kHz and 1 MHz. It is seen from the plot that the low frequency capacitance increases with the applied voltage while the high frequency capacitance remains almost constant. The interface state density, D_{it}, has been calculated from the C_{LF} (measured

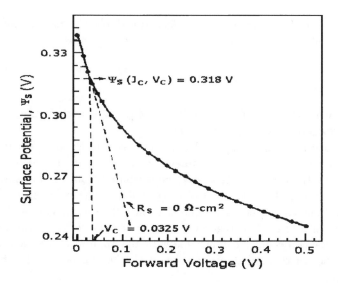

Figure 9.19. Variation of surface potential (ψ_s) with forward voltage (V) for a Pt/p-strained-Si Schottky diode. The dashed line shows the variation in the absence of the series resistance.

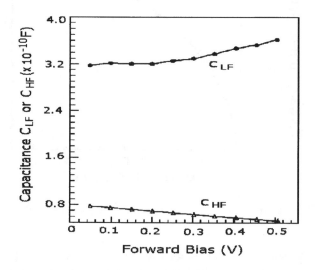

Figure 9.20. Experimental low and high frequency capacitance as a function of bias voltage for a Pt/p-strained-Si Schottky diode.

at 10 kHz) and C_{HF} (measured at 1 MHz) values. The energy distribution of the interface states is shown in figure 9.21. It is seen from the distribution that the interface state density decreases with increase in energy from the edge of the valence band. A minimum value (7.7×10^{11} cm^{-2}/eV) of the interface state density near the mid-gap is obtained and is comparable to the values obtained using MOS capacitors on strained Si [49, 50].

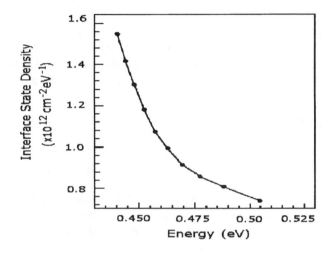

Figure 9.21. Energy distribution of interface state density of a Pt/p-strained-Si Schottky diode.

9.6 Summary

Formation and characterization of noble/refractory metal silicides (Pt, Pd and Ti on SiGe, SiGeC, SiC, GeC and strained Si) using x-ray diffraction, Rutherford backscattering and Auger electron spectroscopy have been discussed. Different phase transformations are observed during silicide formation on SiGe and other alloys. The metallization of SiGe and other group-IV alloys is complicated compared to that of bulk Si.

In the metal–SiGe reaction, metals react preferentially with one component of the alloy, leading to a compositional change in the unreacted alloy. The compositional change results from the formation of a ternary compound within a narrow range of homogeneity. Ternary metal–SiGe phase diagrams are also required to predict the final phases of metal/SiGeC ternary reactions. Cobalt silicide formation on $Si_{0.999}C_{0.001}$ (001) substrate layers has been discussed. X-ray diffraction studies of rapid thermal anneal silicidation of Pt on SiGeC indicate that the reaction proceeds from elemental Pt to $Pt_2(SiGeC)$ and ends in the Pt(SiGeC) phase, analogous to Pt/Si silicides. Suitable low resistance ohmic contacts to p-GeC epitaxial layers grown on Si substrates using pure metallic contacts have been achieved using Al, Au and W metals.

Electrical characterization over a wide range of temperature to determine Schottky diode parameters showed that the barrier heights decrease with decrease in temperature and increase in Ge mole fraction. Extracted ideality factors have values greater than unity and are found to increase with decrease in temperature for all metal–material systems discussed. A method for the estimation of metal/strained-Si interface state density has been discussed.

9.7   References

1 Q. Z. Hong and J. W. Mayer, "Thermal reaction between Pt thin films and Si_xGe_{1-x} alloys," *J. Appl. Phys.*, vol. 66, pp. 611–615, 1989.

2 D. Dentel, L. Kubler, J. L. Bischoff, S. Chattopadhyay, L. K. Bera, S. K. Ray and C. K. Maiti, "Molecular beam epitaxial growth of strained $Si_{1-x}Ge_x$ layers on graded $Si_{1-y}Ge_y$ for Pt-Silicide Schottky Diodes," *Semicond. Sci. Technol.*, vol. 13, pp. 214–219, 1998.

3 S. Chattopadhyay, L. K. Bera, S. K. Ray, P. K. Bose and C. K. Maiti, "Extraction of Interface State Density of Pt/p-strained-Si Schottky Diode," *Thin Solid Films*, vol. 335, pp. 142–145, 1998.

4 S. Chattopadhyay, L. K. Bera, S. K. Ray, P. K. Bose, D. Dentel, L. Kubler, J. L. Bischoff and C. K. Maiti, "Determination of Interface State Density of PtSi/strained-$Si_{1-x}Ge_x$/Si Schottky Diodes," *J. Mat. Sci.: Electronic Mater.*, vol. 9, pp. 403–407, 1998.

5 S. Chattopadhyay, L. K. Bera, S. K. Ray, P. K. Bose and C. K. Maiti, "Low Temperature Characteristics of Pt/p-Strained-Si Schottky Diodes," in *Proc. of 9th Int'l Workshop on Physics of Semiconductor Devices, New Delhi*, pp. 621–623, 1997.

6 S. Chattopadhyay, L. K. Bera, S. K. Ray and C. K. Maiti, "Pt/p-strained-Si Schottky diode characteristics at low temperature," *Appl. Phys. Lett.*, vol. 71, pp. 942–945, 1997.

7 H. K. Liou, X. Wu, U. Gennser, V. P. Kesan, S. S. Iyer, K. N. Tu and E. S. Yang, "Interfacial reactions and Schottky barriers of Pt and Pd on epitaxial $Si_{1-x}Ge_x$ alloys," *Appl. Phys. Lett.*, vol. 60, pp. 577–579, 1992.

8 A. Buxbaum, M. Eizenberg, A. Raizman and F. Schaffler, "Compound formation at the interaction of Pd with strained layers $Si_{1-x}Ge_x$ epitaxially grown on Si(100)," *Appl. Phys. Lett.*, vol. 59, pp. 665–667, 1991.

9 V. Aubry, F. Meyer, P. Warren and D. Dutartre, "Schottky Barrier Heights of W on $Si_{1-x}Ge_x$ Alloys," *Appl. Phys. Lett.*, vol. 63, pp. 2520–2522, 1993.

10 R. D. Thompson, K. N. Tu, J. Angillelo, S. Delage and S.S. Iyer, "Interfacial reaction between Ni and MBE grown SiGe alloys," *J. Electrochem Soc.*, vol. 135, pp. 3161–3163, 1988.

11 S. Chattopadhyay, L. K. Bera, K. Maharatna, S. Chakrabarti, S. K. Ray and C. K. Maiti, "Schottky diode characteristics of Ti on strained-Si," *Solid-State Electron.*, vol. 41, pp. 1891–1893, 1997.

12 O. Thomas, S. Delage, F. M. d'Heurle and G. Scilla, "Reaction of titanium with germanium and silicon–germanium alloys," *Appl. Phys. Lett.*, vol. 54, pp. 228–230, 1989.

13 O. Thomas, F. M. d'Heurle and S. Delage, "Some titanium germanium and silicon compounds: Reaction and properties," *J. Mater. Res.*, vol. 5, pp. 1453–1462, 1990.

14 D. B. Aldrich, Y. L. Chen, D. E. Sayers, R. J. Nemanich, S. P. Ashburn and M. C. Ozturk, "Effect of composition on phase formation and morphology in $Ti-Si_{1-x}Ge_x$ solid phase reactions," *J. Mater. Res.*, vol. 10, pp. 2849–2863, 1995.

15 D. B. Aldrich, Y. L. Chen, D. E. Sayers, R. J. Nemanich, S. P. Ashbun and M. C. Ozturk, "Stability of C54 titanium germanosilicide on a silicon–germanium alloy substrate," *J. Appl. Phys.*, vol. 77, pp. 5107–5114, 1995.

16 J. B. Lai, C. S. Liu, L. J. Chen and J. Y. Cheng, "Formation of amorphous interlayers by solid-state diffusion in Ti thin films on epitaxial SiGe layers on silicon and germanium," *J. Appl. Phys.*, vol. 78, pp. 6539–6542, 1995.

17 Z. Wang, D. B. Aldrich, R. J. Nemanich and D. E. Sayers, "Electrical and structural properties of zirconium germanosilicide formed by a bilayer solid state reaction of Zr with strained $Si_{1-x}Ge_x$ alloys," *J. Appl. Phys.*, vol. 82, pp. 2342–2348, 1997.

18 W.-J. Qi, B. Z. Li, W. N. Huang, Z. Q. Gu, H.-Q. Lu, X.-J. Zhang, M. Zhang, G.-S. Dong, D. C. Miller and R. G. Aitken, "Solid state reaction of Co, Ti with epitaxially-grown $Si_{1-x}Ge_x$ film on Si(100) substrate," *J. Appl. Phys.*, vol. 77, pp. 1086–1092, 1995.

19 Z. Wang, D. B. Aldrich, Y. Y. L. Chen, D. E. Sayers and R. J. Nemanich, "Silicide formation and stability of Ti/SiGe and Co/SiGe," *Thin Solid Films*, vol. 270, pp. 550–560, 1995.

20 H. Kanaya, Y. Cho, F. Hasegawa and E. Yamaka, "Preferential PtSi formation in thermal reaction between Pt and $Si_{0.80}Ge_{0.20}$ MBE layers," *Jap. J. Appl. Phys.*, vol. 29, pp. L850–L852, 1990.

21 X. Xiao, J. C. Sturm, S. R. Parihar, S. A. Lyon, D. Meyerhafer, S. Palfrey and F. V. Shallcross, "Silicide/Strained $Si_{1-x}Ge_x$ Schottky-Barrier Infrared Detectors," *IEEE Electron Dev. Lett.*, vol. 14, pp. 199–201, 1993.

22 H. Kanaya, F. Hasegawa, E. Yamaka, T. Moriyama and M. Nakajima, "Reduction of the barrier height of silicide/p-$Si_{1-x}Ge_x$," *Jap. J. Appl. Phys.*, vol. 28, pp. L544–L546, 1989.

23 M. O. Aboelfotoh, "Temperature dependence of the Schottky-barrier height of tungsten on n-type and p-type silicon," *J. Appl. Phys.*, vol. 67, pp. 51–55, 1990.

24 R. L. Jiang, J. L. Liu, J. Li, Y. Shi and Y. D. Zheng, "Properties of Schottky contact of Al on SiGe alloys," *Appl. Phys. Lett.*, vol. 68, pp. 1123–1125, 1996.

25 S. M. Sze, *Physics of Semiconductor Devices*. John Wiley & Sons, New York, 2nd ed., 1981.

26 D. Dutartre, P. Warren, I. Berbezier and P. Perret, "Low temperature silicon and $Si_{1-x}Ge_x$ epitaxy by rapid thermal chemical vapor deposition using hydrides," *Thin Solid Films*, vol. 222, pp. 52–56, 1992.

27 J. Engqvist, U. Jansson, J. Lu and Carlsson, "C49/C54 phase transformation during chemical vapor deposition TiSi$_2$," *J. Vac. Sci. Technol. A*, vol. 12, pp. 161–168, 1994.

28 J. Kojima, S. Zaima, H. Shinoda, H. Iwano, H. Ikeda and Y. Yasuda, "Interfacial reactions and electrical characteristics in Ti/SiGe/Si(100) contact systems," *Appl. Surf. Sci.*, vol. 117/118, pp. 317–320, 1997.

29 T. L. Lin, A. Ksendzov, S. M. Dejewski, E. W. Jones, R. W. Fathauer, T. N. Krabach and J. Maserjian, "SiGe/Si heterojunction internal photoemission long-wavelength infrared detectors fabricated by molecular beam epitaxy," *IEEE Trans. Electron Dev.*, vol. 38, pp. 1141–1144, 1991.

30 O. Nur, M. R. Sardela, M. Willander and R. Turan, "Schottky barrier heights of Ir/p-Si and Ir/strained p-Si$_{1-x}$Ge$_x$ junctions," *Semicond. Sci. Technol.*, vol. 10, pp. 551–555, 1995.

31 J. Mi, A. Gupta, C. Y. Yang, J. Zhu, P. K. L. Yu, P. Warren and M. Dutoit, "Properties of Schottky contacts of aluminum on strained Si$_{1-x-y}$Ge$_x$C$_y$ layers," *Appl. Phys. Lett.*, vol. 69, pp. 3743–3745, 1996.

32 R. A. Donaton, K. Maex, A. Vantomme, G. Langouche, Y. Morciaux, A. St. Amour, and J. C. Sturm, "Co silicide formation on SiGeC/Si and SiGe/Si layers," *Appl. Phys. Lett.*, vol. 70, pp. 1266–1268, 1997.

33 R. Eyal, R. Brener, R. Beserman, M. Eizenberg, Z. Atzmon, D. J. Smith and J. W. Mayer, "The effect of carbon on strain relaxation and phase formation in the Ti/Si$_{1-x-y}$Ge$_x$C$_y$/Si contact system," *Appl. Phys. Lett.*, vol. 69, pp. 64–66, 1996.

34 J. J. Peterson, C. E. Hunt and M. Robinson, "Schottky and ohmic contacts to doped Si$_{1-x-y}$Ge$_x$C$_y$ layers," *Solid-State Electron.*, vol. 43, pp. 1725–1734, 1999.

35 M. Mamor, C. Guedj, P. Boucaud, F. Meyer and D. Bouchier, "Schottky Diodes on Si$_{1-x-y}$Ge$_x$C$_y$ Alloys," in *Mat. Res. Soc. Symp. Proc.*, vol. 379, pp. 137–141, 1995.

36 M. Serpentini, G. Bremond, V. Aubry-Fortuna, F. Meyer and M. Mamor, "Influence of carbon on the electrical properties of W/SiGeC-p/Si(100)-p Schottky diodes," *J. Vac. Sci. Technol. B*, vol. 16, pp. 1684–1686, 1998.

37 F. Meyer, M. Mamor, V. Aubry-Fortuna, P. Warren, S. Bodnar, D. Dutartre and J. L. Regolini, "Schottky barrier heights on IV–IV compound semiconductors," *J. Electron. Mater.*, vol. 25, pp. 1748–1753, 1996.

38 C. L. Chang, A. St. Amour and J. C. Sturm, "Effect of carbon on the Valance Band Offset of $Si_{1-x-y}Ge_xC_y$/Si Heterojunctions," in *IEEE IEDM Tech. Dig.*, pp. 257–260, 1996.

39 X. Shao, S. L. Rommel, B. A. Orner, P. R. Berger, J. Kolodzey and K. M. Unruh, "Low Resistance Ohmic Contacts to p-$Ge_{1-x}C_x$ on Si," *IEEE Electron Dev. Lett.*, vol. 18, pp. 7–9, 1997.

40 K. K. Ng and W. T. Lynch, "The impact of intrinsic series resistance on MOSFET scaling," *IEEE Trans. Electron Dev.*, vol. ED-34, pp. 503–511, 1987.

41 C. P. Chao, K. E. Violette, S. Unnikrishnan, M. Nandakumar, R. L. Wise, J. A. Kittl, Q. Z. Hong and I. C. Chen, "Low resistance Ti or Co salicided raised source/drain transistors for sub-0.13-μm CMOS technologies," in *IEEE IEDM Tech. Dig.*, pp. 103–106, 1997.

42 W. D. Chen, Y. D. Cui and C. C. Hsu, "Interaction of Co with Si and SiO_2 during rapid thermal annealing," *J. Appl. Phys.*, vol. 69, pp. 7612–7619, 1991.

43 A. E. Morgan, K. N. Ritz, E. K. Broadbent and A. S. Bhansali, "Formation and high-temperature stability of $CoSi_x$ films on an SiO_2 substrate," *J. Appl. Phys.*, vol. 67, pp. 6265–6268, 1990.

44 A. Chin, W. J. Chen, T. Chang, R. H. Kao, B. C. Lin, C. Tsai and J. C.-M. Huang, "Thin oxides with in-situ native oxide removal," *IEEE Electron Dev. Lett.*, vol. 18, pp. 417–419, 1997.

45 S. Teichert, M. Falke, H. Giesler, G. Beddies, H.-J. Hinneberg, G. Lippert, J. Griesche and H. J. Osten, "Silicide reaction of Co with $Si_{0.999}C_{0.001}$," *Solid-State Electron.*, vol. 43, pp. 1051–1054, 1999.

46 M. Zamora, G. K. Reeves, G. Gazecki, J. Mi and C. Y. Yang, "Measurement of interface states parameters of $Si_{1-x-y}Ge_xC_y$/TiW Schottky contacts using schottky capacitance spectroscopy," *Solid-State Electron.*, vol. 43, pp. 801–808, 1999.

47 E. H. Rhoderick and R. H. Williams, *Metal–Semiconductor Contacts*. Clarendon Press, Oxford, 1988.

48 P. Chattopadhyay, "Capacitance Technique for the Determination of Interface State Density of Metal-Semiconductor Contact," *Solid-State Electron.*, vol. 39, pp. 1491–1493, 1996.

49 L. K. Bera, S. K. Ray, D. K. Nayak, N. Usami, Y. Shiraki and C. K. Maiti, "Electrical properties of oxides grown on strained Si using microwave N_2O plasma," *Appl. Phys. Lett.*, vol. 70, pp. 66–68, 1997.

50 L. K. Bera, M. Mukhopadhyay, S. K. Ray, D. K. Nayak, N. Usami, Y. Shiraki and C. K. Maiti, "Oxidation of strained Si in a microwave electron cyclotron resonance plasma," *Appl. Phys. Lett.*, vol. 70, pp. 217–219, 1997.

Chapter 10

Si/SiGe Optoelectronics

Elemental silicon has been widely used for realization of photodetectors and photovoltaics. Both Si and Ge have found wide applications in APDs. Silicon based optoelectronic materials are crystalline Si, SiGe, SiGeC and other IV–IV alloys, poly- and amorphous Si, SiGe and porous Si. A basic limitation of Si and Ge arises from the fact that they have an indirect bandgap and therefore quantum efficiency is low. Some of the important optical properties of Si and Ge are presented in table 10.1.

Table 10.1. Optical properties for Si and Ge.

Property	Si	Ge
Transparent regions (μm)	1.1–6.5	1.8–15
Dielectric constant	11.9	16
Refractive index (optical)	3.455	4.001
Optical-phonon energy (eV)	0.063	0.037
Phonon mean free path (Å)	76 (electron) 55 (hole)	105

10.1 Optoelectronic Devices: Principle

Optoelectronic devices can be divided into three groups: (i) devices that convert electrical energy into optical radiation (LED and laser), (ii) devices that detect optical signals (photodetectors), and (iii) devices that convert optical radiation into electrical energy (photovoltaic cell). An optoelectronic

440

system contains in addition optical waveguiding structures and modulation methods, electrical waveguides where necessary and electronic systems.

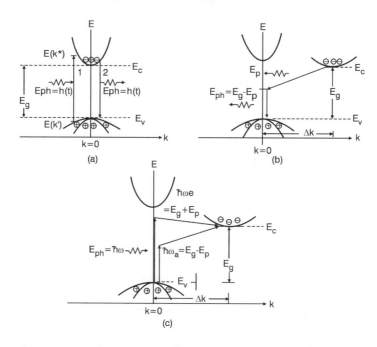

Figure 10.1.   Illustration of band-to-band absorption and recombination process in (a) direct bandgap semiconductor, (b) and (c) indirect bandgap semiconductor.

Energy states between which transitions may take place are those in the conduction and valence bands and the impurity states within the bandgap. Basic transitions in a semiconductor may be classified as: (i) interband (between conduction and valence band), (ii) conduction or valence band to donor or acceptor and donor to acceptor, and (iii) intraband involving states in the conduction or valence band. Figure 10.1 shows band-to-band absorption and recombination processes in direct and indirect bandgap semiconductors. Figure 10.2 indicates the nature of the near bandgap absorption spectra of some semiconductors.

The variation of Ge mole fraction permits control of optical bandgap, absorption coefficient and refractive index [1]. It has been demonstrated that infrared light (wavelength $> 1.2 \ \mu$m) can be waveguided, detected, emitted and switched in the SiGe system [2]. The effects of the introduction of Ge into

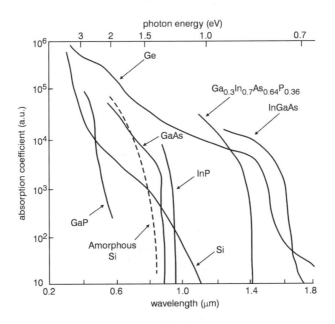

Figure 10.2. Near bandgap absorption spectra.

Si have been discussed in Chapter 3 in respect of band structure, band offset, interface types and splitting.

Deficiencies in the Si/SiGe system for optical applications are the indirect bandgap, limited tunability, and relatively small absorption coefficient. The critical thickness and absorption coefficient product is also small in SiGe alloys for 1.33 μm and 1.55 μm, necessitating the use of MQW and microcavities. Although Si/SiGe has a low luminescence efficiency, enhancement is possible with annealing and incorporation of sulfur, carbon, erbium and indium. Absence of electro-optic and acousto-optic effects in Si/SiGe is another hurdle and leads to lossy carrier induced change in refractive index and field induced absorption modulation. A strategy for dealing with the indirectness in the bandgap of Si is to use a quantum-confined heterostructure, where bandgap and strain engineering are used [3].

Silicon based photonic components for the 1.3–1.5 μm fiber optic communications wavelengths have recently claimed wide attention. The challenges one has to meet in Si-based electro-optical devices are efficient light-emitting diodes, lasers, optical amplifiers, and electro-optical modulators operating within the 0.8–1.6 μm wavelength range, a range that includes the popular 1.3 and 1.55 μm fiber-optic wavelengths. Several heterostructures of Si with SiGe, a:SiGe, SiGeC, SiGeSn, SiSnC and SiC have been proposed. SiGeSn has potential as a direct-gap semiconductor which provides motivation for continued GeSn research. Gurdal *et al.* [4] grew a metastable $Ge_{1-x}Sn_x/Ge$ strained layer superlattice ($< 24\%$ Sn) on a Ge wafer using temperature-modulated MBE at 150 ^{o}C.

The advantages of SiGeC/Si for optoelectronic devices are adjustable strain, bandgap, and band offsets, all of which come from tailoring the $Si_{1-x-y}Ge_xC_y$ composition. It is felt that SiGeC will offer the same band offsets as SiGe/Si with less strain. Substitution of C atoms for Ge atoms in SiGe/Si compensates for Ge-induced strain because of carbon's smaller atomic size. Perfect compensation (zero strain) occurs when the choice of mole fractions x and y gives a lattice match to Si.

In this chapter, we present an overview of trends and progress made in group-IV heterostructures for optoelectronics. Topics will include devices demonstrated using SiGe, SiGeC (an alloy that can be lattice matched to Si), and direct bandgap, strained heterostructures of GeSn upon SiGe/Si. We shall focus on monolithic optoelectronic integration, a technique that promises low-cost, reliable, high performance silicon-based optoelectronic integrated circuits (OEICs). In general, OEICs contain both passive guided-wave components, such as bends, splitters, couplers, and detectors, and active or controlled elements such as laser diodes, routers, switches and modulators.

10.2   Optical Detectors Using SiGe Alloys

Due to a low dark current, favorable ionization coefficients, and compatibility with Si integrated circuit technology, Si photodiodes are widely used as photodetectors. The use of Si and $Si_{1-x}Ge_x$ alloys permits the realization of novel optoelectronic devices operating in the 0.8–1.3 μm wavelength regime. In addition, 1.3 μm Si waveguide modulators using the thermo-optic effect and free-carrier absorption have been realized. These results show promise for an integrated preamplifier-detector for an all-Si-based receiver at short and long wavelengths.

The earliest practical application of SiGe alloys has been in optical detectors. Optical detectors may be classified as photoconductors, photodiodes and phototransistors. Performance measures of a photodetector are gain, frequency response and speed. Responsivity of photodiodes and phototransistors may be enhanced by incorporating an avalanche multiplication region and also by confining the light.

10.2.1   Quantum Efficiency, Responsivity and Noise

Quantum efficiency is a measure of the number of carriers generated per incident photon and is defined as

$$\eta = I_{\text{opt}} h\nu / q P_{\text{opt}} \tag{10.1}$$

where P_{opt} is the incident power at frequency ν and I_{opt} is the photogenerated current. The internal gain by transistor action or carrier multiplication may be considered separately. The responsivity R is defined as

$$R = I_{\text{opt}} / P_{\text{opt}} \tag{10.2}$$

Contributions to the noise arise from low frequency flicker noise, thermal noise in resistors and devices, shot noise in junctions, generation-recombination noise and also the shot noise due to photoemission. The mean square current per Hz that determines the shot noise may be expressed as

$$I^2 = 2q(I_{\text{opt}} + I_{\text{b}} + I_{\text{d}})G^2 \tag{10.3}$$

where G is the internal gain and I_{b} and I_{d} are the currents due to background radiation and thermal generation of electron–hole pairs. The thermal noise is contributed by the device resistance, external load resistance and the input resistance of the following amplifier. When the electric field is high enough to cause carrier multiplication, an excess noise appears which depends on the ratio of the ionization coefficients of electrons and holes.

10.2.2   Gain, Frequency and Time Response of APDs

A semiconductor photodiode is typically a p-n junction diode geometrically arranged so that the junction or portions of the diode close to the junction can be illuminated. The shortest response times are obtained if the light is absorbed mainly within the junction field region. The incoming photons create electron–hole pairs which move in the field in opposite directions and give rise to a signal current. The electrons and holes traversing the high field region have impact ionization probabilities α and β respectively which are functions of the electric field. In the avalanche multiplying region, the equations applying for space–time growth of electrons and holes are given by

$$\frac{1}{v_{\text{n}}} \frac{dn}{dt} = \alpha l_{\text{n}} + \beta l_{\text{p}} - \frac{dn}{dx} + g(x,t) \tag{10.4}$$

$$\frac{1}{v_{\text{p}}} \frac{dp}{dt} = \alpha l_{\text{n}} + \beta l_{\text{p}} + \frac{dp}{dx} + g(x,t) \tag{10.5}$$

where n and p are the input electron and hole density entering the depletion region. v_n and v_p are the saturation drift velocities of electrons and holes, respectively. l_n and l_p are the electron and hole diffusion lengths, respectively, and $g(x,t)$ is the rate at which the electron–hole pairs are generated in the depletion region, and is given by

$$g(x, t) = \alpha_0 \phi_0(t) \exp(-\alpha_0 x) \tag{10.6}$$

where $\phi_0(t)$ is the incident flux and α_0 is the absorption coefficient. Considering sinusoidal excitation one can express the electron density and hole density at a distance x as a sum of exponentials. The boundary conditions are then utilized to find the spatial variation of the carriers.

The refractive index of fully strained $Si_{1-x}Ge_x$ layers, with compositions $x = 0.20$, 0.25, 0.30, and 0.33, has been measured using spectroscopic ellipsometry as a function of wavelength in the 0.9–1.7 μm range [5]. The dependence of the refractive index on wavelength, for these compositions, is similar to that of crystalline Si. Its value for any wavelength is lower than that of relaxed $Si_{1-x}Ge_x$ of the same composition. The experimental values of the refractive index of the fully strained $Si_{1-x}Ge_x$ were fitted to the expression

$$n_{SiGe}(x, \lambda) = n_{Si}(\lambda) + (1.16 - 0.26\lambda)x^2 \tag{10.7}$$

This expression is applicable for wavelengths from 0.9 to 1.7 μm and compositions $x < 0.33$.

The variation of bandgap with Ge concentration (figure 10.3) and band offsets have been measured using XPS. The refractive index and absorption coefficient of $Si_{1-x}Ge_x$ for the energy range from 1.5 eV to 6.0 eV are shown in figure 10.4. Early absorption experiments in Si/SiGe superlattices verified the decreased bandgap obtained from these alloys. Use of superlattices permitted an extension of the absorption cutoff wavelength to over 1.3 μm. Janz *et al.* [6] have measured, using waveguide mode profile technique, the index of refraction of pseudomorphic $Si_{1-x}Ge_x$ films (with $x = 0.01$ to 0.1) grown on Si at wavelengths $\lambda = 1310$ nm and $\lambda = 1550$ nm. The index of refraction was found to be significantly larger for light polarized parallel to the growth direction than for light polarized in the plane of the epilayer. Various types of optical detectors studied in Si/SiGe systems are photoconductors, junction diodes, photo-transistors and photo-FET, avalanche photodiodes, multiple quantum well APD, micro-cavity photodiodes and photo-IMPATT.

Detection of infrared light of wavelength 1.2 to 1.6 μm is important as the loss in fiber optical communication systems in this wavelength region is minimum. Silicon has a cutoff wavelength of about 1.10 μm and is transparent

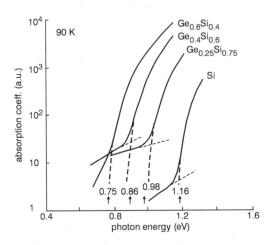

Figure 10.3. Normalized photocurrent spectra for different SiGe samples. The energies indicate the lowest quadratic threshold for each sample.

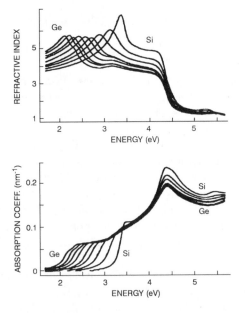

Figure 10.4. Refractive index and absorption coefficient of $Si_{1-x}Ge_x$ for the energy range from 1.5 eV to 6 eV.

in the wavelength region 1.30–1.60 μm. The SiGe absorption edge shifts to the red with increasing Ge mole fraction. This shift offers a means for absorbing 1.3–1.6 μm light by choosing $x > 0.3$. As the absorption coefficient of SiGe alloy is small, critical thickness limits the absorbing layer thickness. The major aims of SiGe optoelectronics are to achieve high responsivity, low noise currents, fast response and integration with the existing Si electronics. The SiGe/Si heterodiode is a good choice for the detection of 1.30–1.60 μm light at 300 K [7]. Most of the reported studies include:

(i) a waveguided MSM photodiode [8]. A responsivity of 0.2 A/W was measured at 1.3 μm with a 500 pA/μm^2 dark current at 5 V bias,

(ii) waveguided p-i-n photodetectors [9, 10] with 50% internal quantum efficiency at 1.3 μm and 200 nA dark current at –15 V in a 10 μm × 750 μm device, and

(iii) a p-i-n diode for 1.3 μm wavelength with 50% internal quantum efficiency, 200 ps impulse response and 10 pA/μm^2 dark current at 15 V bias [2, 11, 12].

$Si_{1-x}Ge_x$ rib waveguide avalanche photodetectors for 1.3 μm application and strained layer superlattice waveguide photodetectors have also been reported [13, 14]. Camera arrays involving Si Schottky barrier photoemissive detectors [15] and SiGe/Si p-n heterojunction photodiodes have been demonstrated [16]. PtSi/SiGe Schottky photodetectors have been proposed for detection of infrared radiation of wavelength up to 10 μm [17, 18]. A summary of reported SiGe/Si photodetectors is presented in table 10.2.

If a graded layer of SiGe is grown on a Si substrate such that the top layer is Ge and a p-i-n structure realized, one could have achieved a photodetector for this range. The principal difficulty experienced relates to the critical thickness which is small for Ge fractions more than 0.25 as required. This can be overcome by using an Si/SiGe superlattice which is stable because the average value of the Ge concentration is not large. For ensuring the necessary absorption one needs to increase the superlattice (SL) period. SiGe p-i-n detectors were reported as early as in 1986 [13]. The waveguide type structure that has achieved the best overall performance is the $Si_{1-x}Ge_x$ rib waveguide avalanche photodiode [14, 27].

A schematic diagram of the detector (SiGe 60 Å, Si 290 Å, thickness of SL = 0.65 μm) is shown in figure 10.5. The detectors are illuminated from the edge to increase the optical path length and light propagates parallel to the heterojunction interfaces. The photogenerated carriers, however, travel perpendicularly to the optical signal across the thin depletion region. Separated absorption multiplication avalanche photodiodes (SAMAPD) have found wide application. The excess multiplication noise in Si is small because

Table 10.2. Summary of SiGe/Si photodetectors. After L. Colace et al., IEEE J. Quantum Electron., 35, 1843–1852 (1999).

Structure	Efficiency	Dark current	Ref.
Ge p-i-n	$\eta_{ext} = 40\%$ at $\lambda = 1.45$ μm	50 ma/cm^2	[19]
SiGe MQW	$\eta_{ext} = 59\%$ at $\lambda = 1.3$ μm	–	[20]
photoconductor	$\eta_{int} = 300\%$ at $\lambda = 1.3$ μm		
SiGe MQW	$\eta_{int} = 40\%$ at $\lambda = 1.3$ μm	7 mA/cm^2	[13]
p-i-n waveguide			
SiGe MQW APD	$\eta_{ext} = 400\%$ at $\lambda = 1.3$ μm	–	[14]
waveguide			
SiGe MQW	$\eta_{ext} = 1\%$ at $\lambda = 1.3$ μm	60 mA/cm^2	[12]
SiGeC	$\eta_{ext} = 1\%$ at $\lambda = 1.3$ μm	7 mA/cm^2	[21]
SiGe MQW	$\eta_{ext} = 11\%$ at $\lambda = 1.3$ μm	1 mA/cm^2	[22]
p-i-n waveguide	$\eta_{int} = 40\%$ at $\lambda = 1.3$ μm		
SiGe SL p-i-n	$\eta_{ext} = 25\%$ at $\lambda = 0.98$ μm	50 μA/cm^2	[23]
Ge p-n diode	$\eta_{ext} = 12.6\%$ at $\lambda = 1.3$ μm	0.15 mA/cm^2	[24]
SiGeC p-i-n	$\eta_{ext} = 2\%$ at $\lambda = 1.3$ μm	0.1 mA/cm^2	[25]
MSM Ge	$\eta_{ext} = 23\%$ at $\lambda = 1.3$ μm		[26]
	$\eta_{int} = 89\%$ at $\lambda = 1.3$ μm		

the ionization of holes is small compared to that of electrons. A SiGe absorption region and Si multiplication region would be ideally suited. A schematic diagram of a separate absorption and multiplication SAMAPD is shown in figure 10.6.

10.2.3 High Frequency Photodetectors

A heterojunction phototransistor (HPT) is an optical detector that provides an internal gain. With the base open at dc, the ac impedance may be chosen to control the frequency response. The current due to photogenerated carriers is amplified by the transistor action. In an HPT, the emitter has a wider bandgap than the base. This results in a high emitter injection efficiency. The computation of gain, frequency response and noise of an HPT can be carried out in much the same way as for an HBT by including the carrier generation term in the continuity equation [28] for the minority carriers in the base.

Photodetectors for high frequency applications include heterojunction phototransistors which can be integrated with an HBT amplifier [28–30],

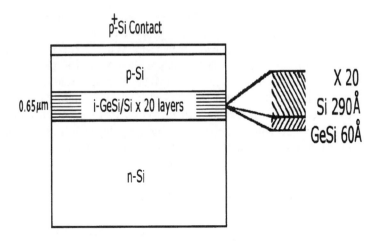

*Figure 10.5. Schematic diagram of the structure of a SiGe p-i-n photodetector. The intrinsic region consists of an undoped $Si_{1-x}Ge_x/Si$ 20 period superlattice. After H. Temkin et al., Appl. Phys. Lett., **48**, 963–965 (1986).*

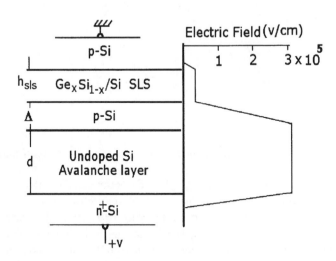

*Figure 10.6. Structure of a separated absorption and multiplication avalanche photodetector. After S. Luryi et al., IEEE Electron Dev. Lett., **7**, 104–106 (1986), copyright ©1986 IEEE.*

photo-IMPATTs, where the optical signal is used to modulate the reverse saturation current, and photoparametric amplifiers where detection and parametric amplification actions are integrated. When light is incident on an APD mounted in a microwave circuit like an IMPATT, the photodetected current acts as an additional source. If the IMPATT is operated in a constant current mode, as it usually is, and the optical input is unmodulated, the dc voltage across the diode will decrease when the optical excitation amplitude is increased. This will result in a change in resonant frequency. If the optical signal is amplitude modulated, the output power is expected to vary as the system works as an APD with a resonant load.

10.2.4   MSM Photodetectors

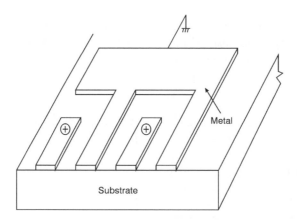

Figure 10.7. Schematic diagram of a metal-semiconductor-metal photodiode.

A planar metal-semiconductor-metal photodetector consists of interdigitated metal fingers forming Schottky diodes on a semiconductor surface. A schematic view of an MSM photodetector is shown in figure 10.7. Photogenerated carriers within the semiconductor are collected by means of an electric field applied between the electrodes. The planar structure makes it IC compatible and at the same time helps achieve a low capacitance and high speed. However, fundamental limiting factors in the performance of MSMs are dark current and quantum efficiency. It is difficult to obtain low dark current in an MSM because the lateral current flow can be significantly influenced by the semiconductor surface. An increase of the depletion layer width with bias is used to enhance the collection efficiency of the detector. The degradation in the quantum efficiency of MSMs is caused by electrode shadowing. The effect of contact shadowing may be partially avoided by using submicron electrodes. The speed of the MSM-PD is limited by a combination of the transit time of carriers and the RC charging time.

One approach to increase the speed of Si MSM photodiodes is to decrease the carrier lifetime through ion implantation. Sharma *et al.* [31] reported that both the quantum efficiency and the bandwidth of Ni-Si-Ni MSMs could be significantly enhanced by using ion implantation of $^{19}F^+$ to create a damaged, highly absorbing region $\sim 1~\mu m$ below the Si surface. The quantum efficiency at 860 nm and 1060 nm was 64% and 23%, respectively. It was not clear whether photoconductive gain played a role in the responsivity. Dutta *et al.* [32] have used a similar approach and obtained a bandwidth as high as 6 GHz, compared to 300 MHz for unimplanted devices, but at 3 to 5 times lower quantum efficiencies ($\sim 10\%$).

The indirect bandgap of Si poses an additional challenge in that it causes low absorption coefficients and, consequently, long absorption lengths. A solution to this problem is to operate at short wavelengths (~ 400 nm) where the absorption length is only 100 nm. In this wavelength region, the transit-time component of the bandwidth is determined by the spacing between the fingers instead of the absorption depth. Liu *et al.* [33] have reported bandwidths as high as 110 GHz with "short-wavelength" Si MSMs that had finger separation and width of 200 nm. The quantum efficiency of these MSM photodiodes was approximately 12%. Insulating substrates (SOI/SOS) have been used to extend the wavelength range in the near infrared region [34, 35]. The motivation is to prevent carriers generated out of the shallow depletion region from slowly diffusing to the junction where they contribute to the optical signal but they also greatly reduce the bandwidth.

Liu *et al.* [36] fabricated MSM photodiodes on SIMOX substrates; the top Si layer was thinned to 100 nm. These photodiodes achieved a very high bandwidth of 140 GHz at 780 nm, although the efficiency was only about 2%. Wang *et al.* [35] have reported Si MSMs with finger width and spacing down to 200 nm fabricated on silicon-on-sapphire substrates. The authors observed that the pulse response was relatively independent of wavelength but the quantum efficiency decreased from approximately 20% at $\lambda = 400$ nm to $< 1~\%$ for $\lambda \approx 900$ nm.

To increase the optical path length for absorption while maintaining short carrier transit distances, use of textured surfaces has been proposed [37, 38]. The basic idea is to use scattering to "trap" light inside the semiconductor, an idea first used to enhance the conversion efficiency of solar cells. Internal and external quantum efficiencies of 61% and 25%, respectively, have been reported for a textured photodiode [38].

Ultrahigh speed MSM photodetectors in Si and SiGe have been reported [39]. They are very important as optical interconnects and for optical communication, and can provide knowledge about carrier transport in these nanoscale devices. MSM photodetectors on Si with 250 Å finger spacing and width have been fabricated on p-type Si wafers (10^{17} cm^{-3}). Nanoscale interdigitated mesa fingers were defined using electron beam lithography and a lift-off. Coplanar transmission line contacts were made to the detectors for high-speed measurements. An electro-optic sampling system consisting of a wavelength tunable femtosecond Ti:sapphire laser and a LiTaO$_3$ sampling crystal was used to measure the impulse response of the detectors [39].

Simulation of the photoresponse of Si$_{1-x}$Ge$_x$ MSM-PDs has been reported [40]. It has been shown that the photoresponse increases with increasing Ge mole fraction. However, the dark current of SiGe detectors is found to be higher than in Si photodetectors. Responsivity increases with increase of the absorption layer thickness underneath the metal fingers.

A MSM-PD with 2000 Å finger spacing and width was found to have an impulse response full width at half maximum of 3.7 ps and a 3 dB bandwidth of 110 GHz which is believed to be the fastest Si photodetector reported so far (figure 10.8). Investigation of similar MSM-PDs in SiGe and quasi-ballistic transport in these nanoscale detectors is being pursued. MSM-PDs with 2000 Å and 3000 Å finger spacing and width were tested. When the laser wavelength was 725 nm, both detectors had a full width at half maximum response time of 11 ps, independent of the finger structure. This is because the light absorption depth in Si at the wavelength is 5 μm and a large number of carriers are generated deep inside the semiconductor bulk. The diffusion time of the carriers from the bulk to the metal electrodes determines the device response time. However, when the laser wavelength was reduced to 400 nm and the light absorption depth became 0.1 μm, the detectors of 3000 Å and 2000 Å finger spacing and width showed, respectively, response times of 5.5 ps and 3.7 ps and bandwidths of 75 GHz and 110 GHz.

The response time measured using a 400 nm light pulse agrees with a one-dimensional Monte Carlo study of Si MSM-PDs indicating that the detectors are transit-time limited. Furthermore, the Monte Carlo study showed that with proper scaling Si MSM-PDs with 250 Å finger spacing and width can achieve 1 ps response time and 440 GHz bandwidth.

Colace *et al.* [41] have demonstrated that Ge on Si photodetectors for the near infrared region (NIR) exhibit promising characteristics as components for

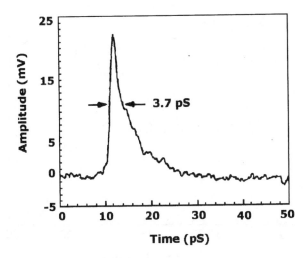

Figure 10.8. Impulse response of a Si MSM-PD with 2000 Å finger spacing and width. The full width at half maximum is 3.7 ps and the 3 dB bandwidth is 110 GHz.

optical communications. Responsivity and speed of the fabricated detectors are compatible with distributed communication systems. The authors have reviewed the recent results on Ge-based near infrared photodetectors grown on Si. Figure 10.9 shows the spectral response of an experimental Ge MSM device biased at 0.2 V in comparison with a Si p-i-n detector, a SiGe superlattice (SL), and an InGaAs p-i-n photodiode. It is apparent that bulk Ge detectors exhibit a photoresponse extending beyond the third window of optical fiber communications (1.55 μm), a clear advantage over SiGe SL diodes.

10.2.5   Schottky Photodetectors

Metal-semiconductor contacts (Schottky barrier diodes) can be used as a very efficient photodetector. The main advantage of a Schottky barrier device is that it is a majority carrier device. It does not suffer from minority carrier storage and removal problems and one can expect high speed and bandwidth. The temporal response, speed and frequency bandwidth of detectors are controlled by the transit time of the carriers through the absorption region and external circuit parameters. In high speed diodes the absorption region is between 0.2 and 0.5 μm which ensures full depletion of the region even at

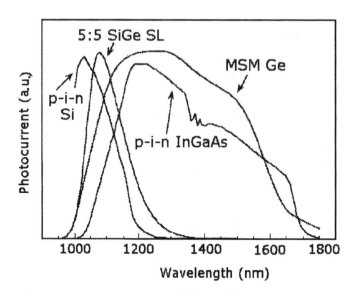

Figure 10.9. Spectral response of MSM Ge photodetectors, compared with Si, SiGe:SL and InGaAs p-i-n photodiodes. After L. Colace et al., IEEE J. Quantum Electron., 35, 1843–1852 (1999), copyright ⓒ1999 IEEE.

low values of reverse bias and both electrons and holes can travel at their respective saturation velocities. Schottky barrier photodetectors can operate in two modes:

(i) when $q\phi_b < h\nu < E_g$, i.e., the energy of the incident photon flux is higher than the corresponding Schottky barrier height but smaller than the bandgap energy of the semiconductor, electrons will be photoexcited in the metal and surmount the barrier by thermionic emission. Emitted electrons will transit through the semiconductor and will be collected at the contact electrodes. The process extends the spectral range towards the red (as it absorbs an energy lower than the bandgap energy) but decreases device speed as the thermionic process is a slow one.

(ii) when $h\nu > E_g$, photon flux penetrates through the semi-transparent metal layer and gets absorbed in the semiconductor. The photo–generated electron–hole (e–h) pairs move in opposite directions due to the existing electric field with their respective saturation velocities and are collected at the electrodes. This is the most efficient mode of operation of Schottky diodes and is similar to that of high-speed p-i-n diodes.

Absorption of infrared radiation of 8–12 μm in the atmosphere is small and this wavelength range is important for night vision applications. Group II–VI compound semiconductor IR sensors are very sensitive in this wavelength range. An IR sensor for a large-scale CCD needs to be integrated monolithically on Si substrates. HgCdTe can be used for IR CCDs only in a hybrid form and thus cannot be used for a large-scale CCD sensor. PtSi/p-Si Schottky diodes are used in a large-scale CCD but it works in the 3–5 μm range. IrSi/p-Si Schottky diodes have a cutoff wavelength of 7.3 μm [17].

As discussed in Chapter 9, PtSi/Si$_{1-x}$Ge$_x$ Schottky diodes are very promising for sensing far infrared radiation as they have a smaller barrier height compared to PtSi/Si or IrSi/Si Schottky diodes and can be monolithically integrated on Si substrates. PtSi/Si$_{1-x}$Ge$_x$ and PtSi/Si Schottky photodetectors were simulated in the wavelength range 2–8 μm [42]. It was shown that the PtSi/Si$_{1-x}$Ge$_x$ photodetectors offer superior responsivity and higher cutoff wavelength compared to conventional PtSi/Si Schottky photodetectors. Responsivity, dark current and cutoff wavelength of a Si$_{1-x}$Ge$_x$ p-i-n photodetector increase with increasing Ge mole fraction in the absorbing i-layer and cover a wavelength range of 1.10–1.50 μm as the Ge mole fraction increases from 0.0 to 0.75. The high responsivity, low dark current (in the range of nA) and low capacitance observed indicate that the PtSi/SiGe Schottky detectors are good candidates for infrared light detection in the wavelength range 1.30–1.50 μm.

Infrared detectors for long wavelength have applications in night vision. Intersubband absorption in SiGe multiple quantum well structures offers a method of fabricating such detectors [43]. Figure 10.10 shows the structure of an infrared detector using silicide on SiGe. The change in spectral response with incorporation of Ge is shown in figure 10.11.

10.2.6 HIP Photodetectors

Recently interest has been raised on mid-IR SiGe detectors fabricated as large area focal plane arrays for thermal imaging applications in the 3–5 μm and 8–12 μm regime to replace the commercially available silicide Schottky barrier detectors such as Pt:Si and Ir:Si. The concept of using a hetero-internal photoemission (HIP) structure in place of the PtSi/Si and Si$_{1-x}$Ge$_x$ HIP was reported by Lin and Maserjian [44]. The structure consists of p$^+$-Si$_{1-x}$Ge$_x$ layers grown by MBE on p-type Si substrates. The operating principle involves absorption in the degenerately doped Si$_{1-x}$Ge$_x$ layer, primarily by free carrier absorption, and subsequent internal photoemission of photoexcited holes over the Si$_{1-x}$Ge$_x$/Si heterojunction barrier into the Si substrate. The authors

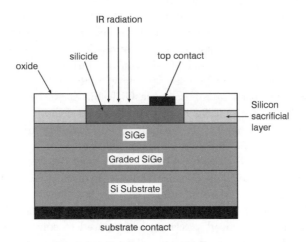

Figure 10.10. Schematic structure of a silicide/SiGe photodetector. After X. Xiao et al., IEEE IEDM Tech. Dig., 125–128 (1992), copyright ©1992 IEEE.

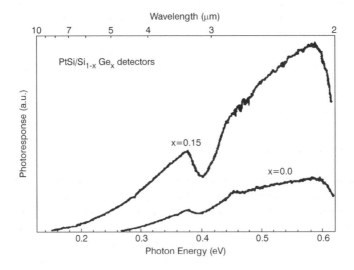

Figure 10.11. Spectral response of a Schottky barrier PtSi/Si$_{0.85}$Ge$_{0.15}$ detector compared with the response of a similar PtSi/Si detector at 77 K. After X. Xiao et al., IEEE IEDM Tech. Dig., 125–128 (1992), copyright ©1992 IEEE.

achieved cutoff wavelengths between 2 and 4 μm for Ge concentrations in the range 0.2–0.4, respectively, and for a p-doping level of $\sim 10^{20}$ cm^{-3}.

Detectors operating on the principle of hetero-internal photoemission (HIP) from photoexcited holes of a highly doped $Si_{1-x}Ge_x$ quantum well, have lower quantum efficiencies than comparable III–V detectors such as HgCdTe and InSb. The advantages of SiGe-HIP detectors are that they can be fabricated on large area Si substrates with good homogeneity, fill factor, and a perfect thermal match to a mounted Si readout circuit. Moreover, the use of SiGe has the advantage of a wavelength tunable multi-color detector where the cutoff wavelength can be easily adjusted by the choice of Ge content and/or doping level. Si/SiGe focal plane arrays have already been demonstrated with 256×256 and 400×400 pixels with excellent homogeneity, low dark current and external efficiencies around 0.75% with a cutoff wavelength of 9.3 μm and monolithic CCD readout circuitry. These arrays exhibited a high degree of pixel-to-pixel uniformity ($< 1\%$ rms variation) and the overall performance at 50 K was reported to be superior to that of IrSi arrays [16].

The external quantum efficiency of HIP photodetectors is determined by the amount of absorption in the Si layers and the efficiency with which the photogenerated holes surmount the heterojunction barrier. There is a tradeoff involving the thickness of the $Si_{1-x}Ge_x$ layers. Thin layers yield improved barrier transport because scattering is less detrimental but there is also less absorption in thin layers. Park *et al.* [45] have proposed using stacks of thin (< 50 Å) $Si_{0.7}Ge_{0.3}$ layers to enhance the absorption without degrading the transport efficiency. Strong response in the wavelength range 2–20 μm was achieved; the quantum efficiencies were 4% and 1.5% at 10 μm and 15 μm, respectively.

Using a PtSi/SiGe/Si HIP Jimenez *et al.* [46] have demonstrated a tunable response. The relative heights of the two barriers, the PtSi/SiGe Schottky barrier and the SiGe/Si heterojunction barrier, are very sensitive to the applied voltage. It was found that the effective barrier height could be varied from 0.3 eV at zero bias to 0.12 eV at 2.4 V bias for a 450 Å thick $Si_{0.8}Ge_{0.2}$ layer. However, the tunability is achieved at the price of reduced quantum efficiency at the shorter wavelengths because the carriers must traverse the $Si_{0.8}Ge_{0.2}$ layer prior to being collected.

10.2.7 Ge Photodiodes

Growth of high quality Ge layers on Si substrates is a practical and cost-effective route to monolithic Si-based optoelectronics. Good quality Ge/Si can act as a virtual substrate for GaAs growth for fabricating LEDs, lasers

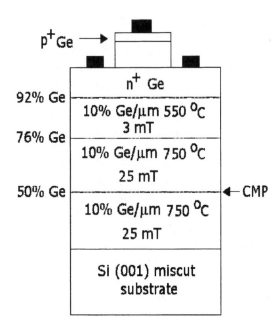

Figure 10.12. A schematic showing the optimized relaxed graded buffer growth sequence with the Ge mesa photodiode on top. After S. B. Samavedam et al., Appl. Phys. Lett., **73**, *2125–2127 (1998).*

and solar cells on Si substrates. Ge-based diodes can be used as high speed and high quantum yield photodetectors in the 1–1.6 μm range for optical communications. The integration of Ge photodetectors on Si substrates is advantageous for various Si-based optoelectronics applications. Samavedam *et al.* [24] have fabricated integrated Ge photodiodes on a graded optimized relaxed SiGe buffer on Si.

Figure 10.12 shows a schematic of the growth structure of the optimized relaxed buffer graded to bulk Ge with a p-n mesa photodiode on top. The contacts to the n^+ Ge layer were made by etching and patterning different-sized square mesas with sides ranging from 95 μm to 250 μm. The p^+ contact was patterned on top of the mesas. The contacts to both the p^+- and n^+ Ge were Ti/Pt.

Figure 10.13 shows the room temperature $I - V$ characteristics of Ge photodiodes with different mesa areas. At low forward voltages (< 0.3 V) the

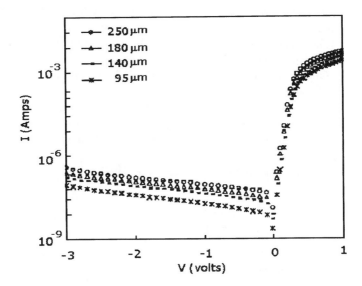

Figure 10.13. I − V characteristics of Ge square mesa diodes of different sizes. After S. B. Samavedam et al., Appl. Phys. Lett., **73**, *2125–2127 (1998).*

diodes exhibit an ideality factor of $n = 1.1$, with the slope of the $I − V$ curve decreasing at higher biases due to high series resistance from the top-contact geometry. A series resistance R_s of 26 Ω was observed for the larger diodes (250 μm). The 95 μm diodes had a larger R_s of 55 Ω probably due to the smaller p-contact area on top of the mesa. The reverse current approximately scales with the active area of the device, indicating that the leakage current is primarily from the bulk of the device rather than from the surface or edge effects. This behavior is not unexpected since the reverse current from diffusion processes is expected to be large due to the small bandgap of Ge. Under a reverse bias of –1 V the reverse current densities range from $J_s = 0.15$–0.22 mA/cm^2 for different sized devices.

10.3   QW Photodiodes

Quantum wells using SiGe alloy layers are of two types, type I confining mainly holes and type II confining electrons. It is possible to place type I and type II in adjacent regions by a choice of the compositions. Each quantum well has a series of subbands, e.g., HH and LH in the valence band and Δ_2 and

Δ_4 subbands in the conduction band. Compared to a bulk material where transitions occur between the bands or between impurity defined states and the bands, in heterostructures transitions may occur between the series of subbands in the same well or adjacent wells and may be utilized for optical applications. One may also convert HBT and HFET to photoactive devices. Some of these devices are discussed below.

Quantum well infrared photodetectors (QWIP) have been studied by several workers [47–51]. Spurred by the promise of uniform high-detectivity arrays, research has begun on SiGe QWIP diodes that sense mobile carriers created during intersubband absorption of light. Figure 10.14 shows the subband structure and the detection process for QWIPs.

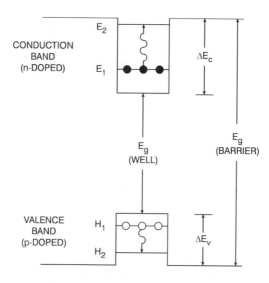

*Figure 10.14. Quantum well band structure illustrated for both electron (for n-type) and hole (for p-type) interband absorption. After B. F. Levine, J. Appl. Phys., **74**, R1–R81 (1993).*

Recently some reports [52, 53] on $Si_{1-x}Ge_x$/Si quantum wells have appeared. The higher density of states in SiGe subbands suggests that SiGe QWIPs are inherently superior to AlGaAs QWIPs. For p-type $Si_{1-x}Ge_x$/Si QWIPs normal incidence absorption is strong because strong mixing of heavy- and light-hole states enables hole intersubband transitions. Valence band technology is preferred for 8–14 μm SiGe/Si QWIPs because it allows normal incidence of light on the detectors. The polarization of normal light is always perpendicular to the growth direction of the quantum well (QW) layers. Kruck *et al.* [51] have reported a normal incidence QWIP that operates in the wavelength range 3–8 μm with a peak responsivity of 76 mA/W and a detectivity as high as 2×10^{10} cm (Hz W)$^{1/2}$.

Although low noise and good responsivity have been realized, a long path in the waveguided diode is needed due to the low absorption coefficient. This long path tends to raise the parasitic capacitance of the distributed diodes. It becomes difficult to get better responsivity at higher wavelengths as the stability of strained SiGe QWs decreases rapidly as the Ge fraction increases. As an alternative to SiGe multiple quantum wells, bandgap narrowing in heavily doped Si allows detection at 1.3 μm [54].

10.3.1 p-i-n Photodiode

A p-i-n photodiode consists of a single p-n junction which responds to photon absorption. It is the most popular photodetector due to ease of fabrication and excellent optical characteristics. With regard to responsivity, ideally each incident photon should result in the charge of one electron flowing in the external circuit. However, in practical devices, there are several physical effects such as incomplete absorption, recombination, reflection from the semiconductor surface and contact shadowing which tend to decrease the responsivity.

Vonsovici *et al.* [55] have performed room temperature photocurrent measurements on SiGe/Si double heterostructure p-i-n diodes grown by selective epitaxy in order to determine the bandgap energy variation as a function of Ge content. Figure 10.15 shows a schematic cross-section of the device that was used to measure the photocurrent spectra. The structures were realized by selective growth on (100) oriented n-Si substrates patterned by a SiO$_2$ mask. The layer sequence is shown in figure 10.15. Figure 10.16 shows the square-root of the measured photoresponse for the three samples investigated.

High dc responsivity and a wide frequency response have been achieved in a p-i-n Si/SiGe superlattice detector at long wavelength by growing the structure on a SOI substrate. SOI structures are useful for Si-based integrated optoelectronics since the buried oxide layer forms a low index confinement region that permits effective waveguiding in the Si overlayer. Si/SiGe superlattice p-i-n integrated rib waveguides and detectors have been fabricated [11, 12]. They exhibited low reverse leakage current (10–30 pA/μm at a 15 V reverse bias) and 50% internal quantum efficiency at 1.1 μm with an impulse

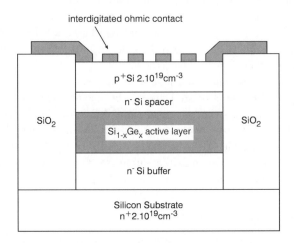

*Figure 10.15. Schematic cross-section of the p-i-n Si/SiGe/Si device structure. After A. Vonsovici et al., IEEE Trans. Electron Dev., **45**, 538–542 (1998), copyright ©1998 IEEE.*

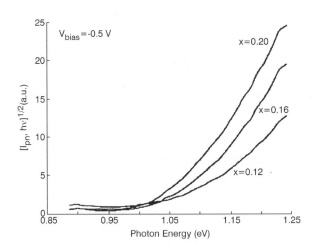

*Figure 10.16. Photocurrent spectrum of the Si/SiGe/Si p-i-n device. After A. Vonsovici et al., IEEE Trans. Electron Dev., **45**, 538–542 (1998), copyright ©1998 IEEE.*

response time of 200 ps and frequency response bandwidth of 1–2 GHz. The detector device geometry consisted of a nominally undoped Si/SiGe multiple quantum well absorbing region grown by MBE.

10.3.2 Microcavity Photodiodes

There is a fundamental tradeoff between quantum efficiency and bandwidth of a photodiode. A large value of quantum efficiency requires the use of a thick ($> 1\ \mu$m) absorption region while a large bandwidth dictates the use of a thin absorbing layer. In SiGe strained layers, the thickness must remain within the critical thickness. A microcavity photodiode provides an elegant solution to the problem. A resonant cavity structure increases the absorption through multiple reflections between two parallel mirrors in a Fabry–Perot cavity whose length is less than one wavelength [56].

The cavity consists of a dielectric stack anti-reflection window, a thin absorbing layer of SiGe MQM placed inside a resonant cavity formed by the wide bandgap window and reflecting mirrors. The bottom reflecting mirror is formed by a superlattice distributed Bragg reflector. It is common for these Bragg reflectors to have a reflectivity $> 99\%$. The quarter wave superlattice reflector is made up of alternate layers of high and low refractive index. The incident light is made to bounce back and forth between the reflectors and thus the absorption is considerably enhanced. Figure 10.17 shows the simulated and experimental Si/SiGe mirror designed to operate at 800 nm. When these mirrors are used for visible and near infrared, the reflectivities are reduced due to absorption. Incorporating similar mirrors (~ 600 nm) in a resonant cavity photodiode, Murtaza *et al.* [57] have reported a microcavity photodiode with a peak quantum efficiency of 67% at 608 nm. In order to obtain the same quantum efficiency with a conventional Si p-i-n structure about 6 μm absorption region thickness would be needed. The frequency response of a Si/SiGe photodiode having a resonance at 780 nm and an n-doped mirror is shown in figure 10.18. The bandwidth is approximately 15 GHz, a significant improvement ($\sim 10\times$) compared to conventional Si p-i-n photodiodes.

Ishikawa *et al.* [58] have developed a technique for fabricating Si/SiO$_2$ Bragg reflectors by multiple separation-by-implanted oxygen (SIMOX). This approach has the advantage that the Si layers, including the top layer of the mirror, are single crystal. The resonant cavity photodiodes with Si/SiGe mirrors are more compatible with standard Si IC processing. Conventional poly-Si deposition techniques yield photodiodes with high current ($\sim 1\ \mu$A at 1 V) and low quantum efficiencies [59]. Bean *et al.* [60] have developed a MBE technique for growth directly on SiO$_2$ that yields superior device quality poly-

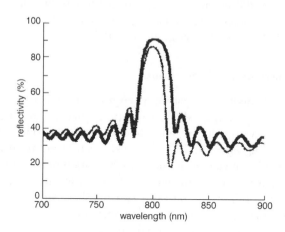

Figure 10.17. Simulated and measured reflectivities of a forty-five period $Si_{0.67}Ge_{0.33}(250$ Å$)/Si(800$ Å$)$ Bragg mirror. After S. S. Murtaza et al., Electronics Lett., 30, 315–316 (1994).

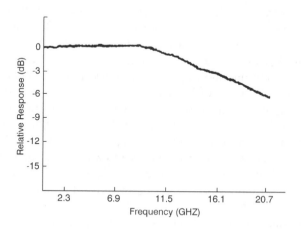

Figure 10.18. Frequency response of a $Si_{1-x}Ge_x/Si$ resonant cavity photodiode. After S. S. Murtaza et al., IEEE Photonics Technol. Lett., 8, 927–929 (1996), copyright ©1996 IEEE.

Si films. p-i-n photodiodes fabricated using this technique show 100× decrease in dark current and an improvement in peak external quantum efficiency to 50%. Figure 10.19 shows the dark current vs. bias voltage. At bias voltages of –10 V and –25 V, the dark current is 60 nA and 500 nA, respectively. A bandwidth of approximately 10 GHz was reported (figure 10.20).

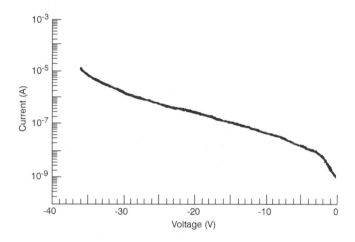

*Figure 10.19. Dark current vs. bias voltage for a Si/SiO_2 interdigitated, resonant cavity photodiode. After J. C. Bean et al., IEEE Photon. Tech. Lett., **9**, 806–808 (1997), copyright ©1997 IEEE.*

Zhu *et al.* [61] have proposed a method for fabricating a SiGe/Si MQW resonant microcavity phototransistor. Figure 10.21 shows a schematic diagram of the double heterojunction phototransistor. The SiGe/Si MQW acts as the base and the absorption layer. A window is opened by etching deeply into the substrate from the back side of the substrate. A four period SiO_2/Si quarter wave stack used as the bottom mirror is deposited on the window and can provide a reflectivity of approximately 98%. Another one period SiO_2/Si quarter wave stack is deposited on the top of the device and provides a reflectivity of about 65%. The cavity is defined by the lower SIMOX substrate and upper SiO_2/Si mirror.

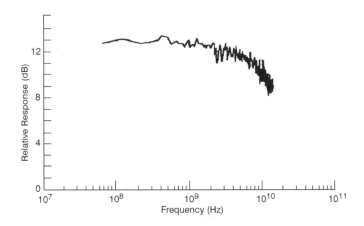

*Figure 10.20. Frequency response of a Si/SiO_2 interdigitated, resonant cavity photodiode. After J. C. Bean et al., IEEE Photon. Tech. Lett., **9**, 806–808 (1997), copyright ©1997 IEEE.*

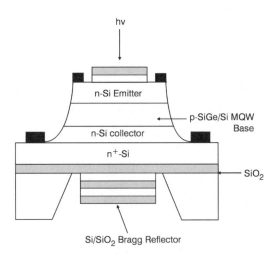

Figure 10.21. Schematic cross-section of SiGe/Si MQW resonant microcavity phototransistor. After Y. Zhu et al., Proc. Electronic Components and Technology Conf., 1199–1204 (1997), copyright ©1997 IEEE.

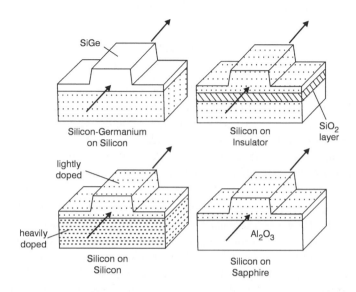

*Figure 10.22. Methods of fabricating optical waveguide in Si technology: (i) SiGe on Si, (ii) Si on insulator, (iii) Si on doped Si, (iv) silicon-on-sapphire. After R. Soref, Proc. IEEE, **81**, 1687–1706 (1993), copyright ©1993 IEEE.*

10.4 Optical Waveguides

Low loss waveguides have been proposed using group-IV alloy films. Light can propagate in four types of group-IV waveguides: lightly doped Si on heavily doped Si [62–66], epitaxial SiGe on Si (bulk or quantum well) [67–74], silicon-on-saphire [69] and silicon-on-insulator [69, 75–84]. In addition to epitaxial SiGe, SiC or SiGeC can be used as waveguide cores. Structurally, the two main waveguide structures are the rib waveguide and the diffused $Si_{1-x}Ge_x$ buried channel. Schematic representations of these structures are shown in figure 10.22.

The addition of Ge to Si increases the refractive index making epitaxial layers of $Si_{1-x}Ge_x$ useful for Si-based waveguide devices. Assuming a linear relationship between the refractive indices of Si and Ge (3.5 and 4.3, respectively, at $\lambda = 1.3$ μm), the refractive indices for small Ge concentrations is given by the relation [68, 73]

$$n_{SiGe} \approx 3.5 + 0.8x \qquad (10.8)$$

It has also been determined empirically that for strained layers [64]

$$n_{SiGe} \approx 3.5 + 0.3x + 0.32x^2 \qquad (10.9)$$

From these relations it is clear that $Si_{1-x}Ge_x$ waveguiding layers can be formed with only a few percent of Ge. The upper limit on x is determined by the fact that the bandgap decreases with increasing Ge content and the bandgap energy of the waveguiding layer needs to be greater than the photon energy of the optical signal. Crystallographic defects like threading dislocations should be kept below 10^4 defects/cm^2 in order to keep losses below 1 dB/cm in SOI and SiGe/Si waveguides.

The relatively high index grading for higher Ge contents allows for a tighter mode. However, the resulting size mismatch with respect to even single mode fibers presents a daunting coupling problem. Thick (several μm) low Ge content waveguides with low losses have been reported [2]. Typically losses are in the 1 dB/cm range. These high numbers for integrated waveguides are obtained because of scattering from the surfaces and interface of these structures. While high from the standpoint of silica fibers, these numbers are quite tolerable for short communication paths on-chip. The ability to grow on relaxed substrates and also obtain very pure materials is expected to play an important role in the further evolution of silicon-based optoelectronic structures. These waveguides can be integrated with other passive components like directional couplers, Mach–Zehnder interferometers and modulators. A detector integrated with a waveguide is shown in figure 10.23.

$Si_{1-x}Ge_x$ rib waveguide structures have been fabricated from epitaxial layers grown by MBE and by CVD. A diffusion technique, similar to that employed for Ti:LiNbO$_3$, has also been developed. This approach has the advantage of utilizing conventional processing on commercial Si substrates without the added cost and complication of epitaxial growth. The diffusion process creates a refractive index profile that is relatively well-matched to a single-mode optical fiber (see figure 10.24).

Losses of 0.5 dB/cm (for transverse electric modes) and 0.6 dB/cm (for transverse magnetic modes) at 1.32 μm have been reported in chemical vapor deposited $Si_{0.99}Ge_{0.01}$ ribs on Si [70]. The propagation loss in a polarization independent single-mode rib made from Ge-indiffused Si has been reported to be 0.3 dB/cm at 1.3 and 1.55 μm. In single-mode SOI/SIMOX ribs, the reported propagation loss was approximately 0.4 dB/cm for polarization independent 1.3 and 1.55 μm infrared radiation [76].

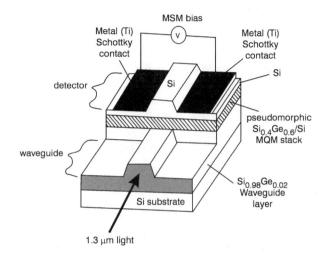

Figure 10.23. Detector integrated with waveguide. After R. Soref, Proc. IEEE, **81**, *1687–1706 (1993), copyright ©1993 IEEE.*

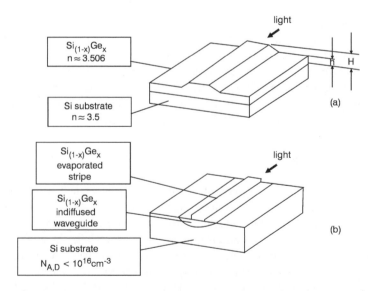

Figure 10.24. (a) Rib waveguide in a SiGe heterostructure and (b) SiGe-indiffused channel waveguide. After B. Schuppert et al., IEEE J. Lightwave Tech., **14**, *2311–2323 (1996), copyright ©1996 IEEE.*

10.4.1 Directional Couplers

Directional couplers are essential components for optical modulators and switches. An SiGe/Si directional coupler was reported by Mayer *et al.* [71]. The coupler consisted of two adjacent $Si_{0.96}Ge_{0.04}$ rib waveguides. The interaction length was varied between 3 and 7 mm and the spacing between the waveguides was varied from 1.5 to 3 μm. For an inter-waveguide gap of 3 μm the coupling coefficient was 3.9 cm^{-1} and the corresponding coupling length was 4.6 mm. Ge-indiffused couplers have also been fabricated and characterized for different coupling lengths and waveguide spacings [73].

10.4.2 Modulators

Modulation of light propagating in semiconductor waveguides is achieved by changing the index of refraction. A common technique is to employ electro-optic or acousto-optic effects in titanates and niobates. Because of the central symmetry of Si and SiGe alloys, the linear electro-optic effect is negligible. Modulation of the guided light can, however, be achieved by changing the index of refraction through the dependence on free-carrier density. Primarily free-carrier injection and the thermo-optic effect have been used to realize modulation and switching; the speed of these devices is much lower than that of III–V compound electro-optic devices. Variation of the electro-reflectance of SiGe superlattice structures has also been used for achieving modulation. The modulator depends on the optical properties of excitons confined by a quantum well structure. It is known that the absorption edge for confined electrons is sensitive to the local electric field. For SiGe systems the useful range of modulation wavelength lies near 1.5–1.6 μm.

10.5 SiGe LEDs

Electroluminescence is the generation of light by electrical means and has a narrow spectrum of width a few hundred angstroms to a fraction of an angstrom. When minority carriers are injected into a semiconductor p-n junction, injection electroluminescence may take place through radiative recombination under favorable conditions. The study of isoelectronic doping in Si has led to luminescence using carbon, sulfur, erbium, and indium. Practical electroluminescence devices are rare-earth-doped Si and $Si_{1-x}Ge_x$. Erbium has emerged as the favorite due to its relatively narrow emission near 1.5 μm, an important wavelength for communication applications.

Optical recombination in silicon-based materials requires the presence of a phonon to conserve momentum. Photoluminescence in the wavelength range 1.2–2.7 μm has been reported by Houghton *et al.* [85] for thick (100–200 nm) SiGe alloys and SiGe/Si strained layer superlattices. The photoluminescence and electroluminescence peaks were found to shift consistently with Ge fraction and were about 120 meV below the established bandgap for the composition. Properly chosen post-growth annealing enhanced luminescence efficiency by two orders of magnitude. Doping with isoelectronic impurities is the basis for visible light emitting diodes made from GaP. Electroluminescence from sulfur impurities in p-n junctions formed in epitaxial Si [86] has been observed in the 1.25–1.5 μm range and persisted up to 150 K. The external quantum efficiency was about 0.2%–0.5%.

For applications in temperature-stable optical sources such as light emitting diodes and semiconductor lasers, rare earth metal doped semiconductors are commonly used. Among the various rare earth ions, Er^{3+} seems to be the most useful element since it emits luminescence at 1.54 μm which corresponds to the minimum absorption region of the silica-based optical fiber. The Er-doped Si-based LED is particularly interesting. Although Si is widely used in the semiconductor industry, its band related luminescence is weak due to the indirect bandgap. In a SiGe-based host, the Ge mole fraction can be altered to change the host bandgap energy and thus change the energy mismatch. The SiGe host can also be integrated with the Si substrate easily. Furthermore, if it is desired to realize an Er-doped semiconductor laser on a Si substrate, it is necessary to dope Er in a quantum well structure such as Si/SiGe:Er/Si.

Photoluminescence (PL) properties of Er-doped SiGe prepared by MBE and ion implantation have both been reported [87]. It was shown that a reasonably strong PL signal of the Er-doped SiGe can be observed at room temperature, and the activation energy of an ion-implanted $Si_{0.83}Ge_{0.17}$:Er sample annealed at 850 °C for 20 min is 130 meV which is slightly smaller than the activation energy of Er-doped Si.

Er-doped SiGe light emitting diodes were fabricated by implanting Er^{3+} ions into SiGe epi-layers. By injecting minority carriers into the diodes, Er^{3+} related emission was observed in the 1.54 μm region at 77 K [88]. Figure 10.25 shows the current–voltage characteristics of one of the SiGe:Er diodes measured at 77 K. Typical reverse breakdown voltages of these diodes were between 10 and 14 V. In the forward bias region, the ideality factor is found to be 1.84 which indicates that the forward current in this diode is dominated by space charge recombination at low forward bias. At high forward bias, the series resistance limits the current as shown in figure 10.25.

Figure 10.26a shows a typical EL spectrum of SiGe:Er LEDs measured at 77 K. The post implantation annealing condition of this sample was 850 °C

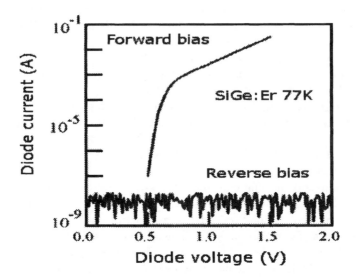

*Figure 10.25. The current–voltage characteristics of a typical $Si_{0.87}Ge_{0.13}{:}Er$ diode measured at 77 K. After S.-J. Chang et al., J. Appl. Phys., **83**, 1426–1428 (1998).*

for 20 min. The injection current was 50 mA. The PL spectrum of the same sample measured at 77 K is shown in figure 10.26b for comparison. In the PL experiment, an argon 514.5 nm ion laser was used as the excitation source. It can be seen that there is great similarity between the EL spectrum shown in figure 10.26a and the PL spectrum shown in figure 10.26b although the background level is higher and the Er-related luminescence signal is 50 times weaker in figure 10.26a. The similarity observed suggests that the Er centers emitting the luminescence signal at 1.54 μm are the same for the EL and PL processes.

10.6   Solar Cells

Amorphous hydrogenated Si has a bandgap lying in the visible region and has been used as the basic material for solar photovoltaic cells and low-cost detectors for visible LEDs. Amorphous hydrogenated silicon–germanium–carbon alloys have extended the efficacy and range of the above applications. Amorphous SiGeC:H is typically grown by plasma enhanced CVD using silane,

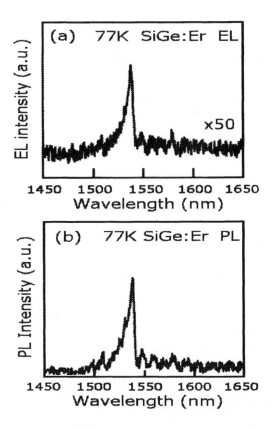

Figure 10.26. (a) EL and (b) PL spectra of a typical $Si_{0.87}Ge_{0.13}$:Er LED measured at 77 K. This sample was annealed at 850 °C for 20 min. After S.-J. Chang et al., J. Appl. Phys., **83**, 1426–1428 (1998).

germane and methane. The composition is controlled as usual by varying the fraction of Ge and carbon containing compounds and hydrogen dilution. The growth of a:SiGe:H has been extensively studied for use in low-cost yet efficient tandem solar cells. The cost is low because amorphous SiGeC can be grown even on glass substrates.

Incorporation of Ge and carbon into amorphous Si enables a useful variation of bandgap. Phototransistors employing heterojunctions of Si/SiGe/SiC have been reported to provide a low-cost solution to detection problems when the speed requirement is within a few tens of megahertz.

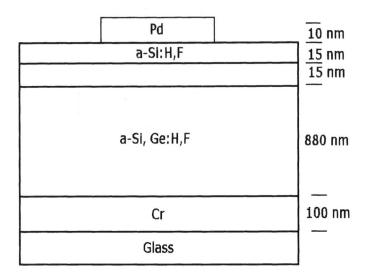

Figure 10.27. Amorphous SiGe:H thin film photodetector array: single cell, bandgap 1.34 eV and speed 25 MHz. After D. S. Shen et al., IEEE Electron Dev. Lett., 13, 5–7 (1992), copyright © 1992 IEEE.

The ease with which hetero-wells and barriers can be fabricated makes these detector configurations quite attractive.

Hydrogenated amorphous SiGe:H has been used for implementing solar cells and infrared phototransistors. Figure 10.27 shows the structure of a single cell of an a:SiGe:H photodetector array [89]. The a-Si, Ge:H, F alloy was deposited at low temperature in an rf plasma enhanced CVD system. The film had an optical bandgap of 1.34 eV with optical absorption for 750–830 nm light between 8×10^3 and 3×10^4 cm^{-1}. The transit time is about 0.1 μs, thus permitting detection of light pulses from commercial medium frequency sources. Implementation of amorphous Si phototransistors on glass substrates has been reported [90]. An amorphous silicon-SiC-SiGe:H phototransistor is shown in figure 10.28. The structure is glass/TCO/n$^+$-i(a-SiGe:H/ a-SiC:H)p$^+$-i-n$^+$ (a-Si:H). The associated band diagram is shown in figure 10.29.

The structure of a Si:H/SiC:H superlattice reach-through APD (RAPD) [91, 92] and the energy band diagram of the SL-RAPD are shown in figures 10.30 and 10.31, respectively. An attractive feature of the amorphous

SiGeC structure is the possibility of integrating a solar cell and a low-cost photodetector array with high speed optoelectronics partially powered by the solar cell.

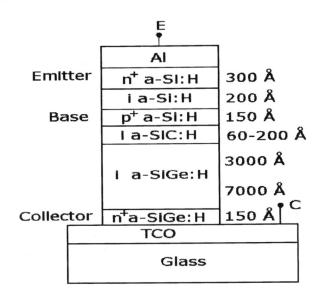

Figure 10.28. a:Si/a:SiGe:H bulk barrier phototransistor with a-SiC:H barrier enhancement layer. After S.-B. Hwang et al., IEEE Trans. Electron. Dev., **40**, *721–726 (1993), copyright ©1993 IEEE.*

10.7 Integrated Optoelectronics

An integrated optical circuit on Si wafers for fiber optical communication requires Si based emitter/receiver devices which can be monolithically integrated on a Si IC chip [93]. By incorporating Si-based optical devices (emitters and detectors) with existing Si-based electronic circuitry on a single "superchip", Si-based optoelectronic integrated circuits will have great cost effectiveness compared to III–V counterparts and with added computational power. A proposed superchip based on Si optoelectronic components and electronic circuits is shown in figure 10.32. A possible realization of interchip and intrachip coupling on a Si substrate via Si/SiGe light emitting and receiving devices can be seen. Figure 10.32 shows schematically a Si based LED distributing a signal via different Si waveguides, transmitting it in free

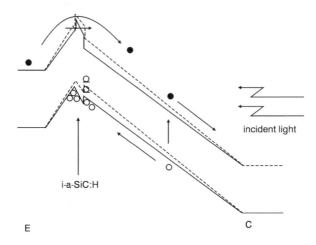

i-a-SiC:H

E   C

*Figure 10.29.   Energy band diagram of an a:Si/a:SiGe:H bulk barrier phototransistor with a-SiC:H barrier enhancement layer. After S.-B. Hwang et al., IEEE Trans. Electron. Dev., **40**, 721–726 (1993), copyright ©1993 IEEE.*

space and collecting it in different waveguides on the next chip. $Si_{1-x}Ge_x$ quantum wells and Si_mGe_n strained layer superlattices (SLS) have attracted a great deal of attention. The original motivation for studying Si_mGe_n SLS was to create a quasi-direct band structure by using the periodicity of the superlattice to fold the Brillouin zone [94].

In addition, other passive optical device components such as modulators and interferometers with SiGe waveguides have been realized on Si substrates [2] which are necessary for an integrated optical and electronic circuit on Si. There have been several reports of Si-CMOS receiver circuits with integrated optical devices [95]. Integrated optical receivers with a CMOS preamplifier and a vertical p-i-n photodetector fabricated using BiCMOS technology have also been reported [96, 97].

For some applications, it is desirable to integrate a photodiode with waveguide devices. Kesan *et al.* [98] showed that SiGe photodiodes can be integrated with SOI rib waveguides. The SLS $Si_{0.4}Ge_{0.6}$ photodiodes had internal and external quantum efficiencies of 50% and 12% at 1.1 μm and 10 V bias, respectively. The RC limited bandwidth was 200 ps.

Splett *et al.* [66] have fabricated an all SiGe circuit consisting of a 2.5 μm thick $Si_{0.98}Ge_{0.02}$ waveguide beneath an SLS SiGe photodiode. The absorbing

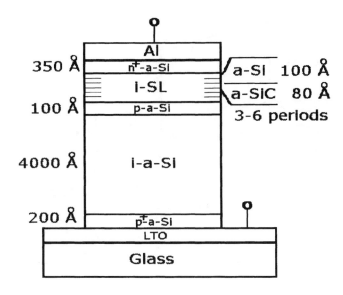

Figure 10.30. The structure of an a-Si:H/SiC:H superlattice reach-through APD. After J. W. Hong et al., IEEE Trans. Electron. Dev., **37**, *1804–1809 (1990), copyright © 1990 IEEE.*

region of the photodiode consisted of 20 periods of 30 nm Si and 5 nm $Si_{0.55}Ge_{0.45}$. The external quantum efficiency for a single-mode fiber input was $\sim 11\%$. Taking into account the losses due to fiber-to-waveguide coupling, attenuation in the waveguide, and waveguide-to-photodiode coupling, it was estimated that the internal quantum efficiency was $\sim 40\%$. The measured time response showed two components: a fast response due to the RC time constant (~ 2 GHz) and a slow response that was caused by hole trapping in the quantum wells (~ 50 ns). It was projected that the slow response could be eliminated by using triangularly shaped barriers to facilitate hole transport out of the wells.

Light emitting devices made from group-IV alloy films are still in infancy. Light emission has been observed in rare-earth-doped Si, SiGe and strained quantum wells of SiGe. Nano-porous Si and Ge, isoelectronic-impurity-doped SiGe, and nanocrystalline Si and Ge films have attracted much attention for light emission. Quasi-direct gap short-period SiGe superlattices, Si quantum wire and SiGe strained quantum well structures are very intriguing areas for light emission [99]. Silicon-based columnar structures have been used to

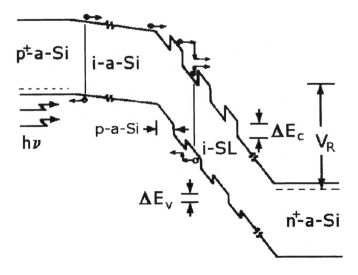

Figure 10.31. The energy band diagram of SL-RAPD. After J. W. Hong et al.,
IEEE Trans. Electron. Dev., **37**, *1804–1809 (1990), copyright © 1990 IEEE.*

demonstrate infrared light radiation [100]. The success could lead the way to
commercial SiGe electronics and more importantly to SiGe/Si optoelectronics
operating at the preferred wavelengths of 1.30 μm and 1.55 μm in optical fiber
communication.

10.8   Summary

Si/SiGe heterostructure based approaches to different photodetectors, emitters, waveguides, modulators and switches have been described. At present,
most of these optoelectronic devices and circuits have not demonstrated performance comparable to state-of-the-art III–V optoelectronics. Applications
of SiGe/Si photodetectors have been well established although problems remain. Since participation of a phonon is involved in indirect bandgap Si/SiGe
materials obtaining luminescence appears to evade an acceptable solution, but
use of porous Si seems promising.

Although in MSMs it is difficult to obtain low dark current, for Si MSMs
this is not a concern because excellent passivation techniques are available.
For SiGe structures, on the other hand, passivation and Schottky contacts

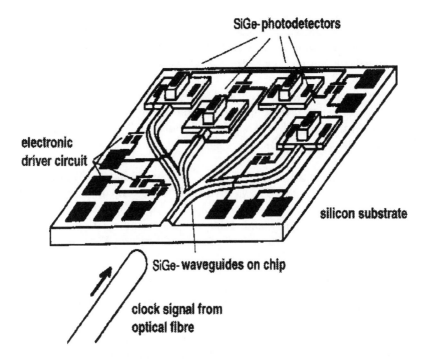

Figure 10.32. Concept of optical clock distribution on a Si IC chip using SiGe waveguides and photodetectors. After H. Presting and U. Konig, Proc. Current Developments of Microelectronics, 139–150 (1999).

may require further development. Although the outlook is very promising in electronics because the commercialization of SiGe-HBTs has started, the pace of progress is slower in SiGe/Si photonics.

10.9 References

1 T. P. Pearsall, "Si-Ge alloys and superlattices for optoelectronics," *Mater. Sci. Eng.*, vol. B9, pp. 225–231, 1991.

2 R. A. Soref, "Silicon based optoelectronics," *IEEE Proc.*, vol. 81, pp. 1687–1706, 1993.

3 R. A. Soref, "Silicon-based group IV heterostructures for optoelectronic applications," *J. Vac. Sci. Technol. A*, vol. 14, pp. 913–918, 1996.

4 O. Gurdal, M.-A. Hasan, M. R. Sardela, J. E. Greene, H. H. Radamson, J. E. Sundgren and G. V. Hansson, "Growth of metastable $Ge_{1-x}Sn_x$/Ge strained layer superlattices on Ge(001)2×1 by temperature-modulated molecular beam epitaxy," *Appl. Phys. Lett.*, vol. 67, pp. 956–958, 1995.

5 J. C. G. de Sande, A. Rodriguez and T. Rodriguez, "Spectroscopic ellipsometry determination of the refractive index of strained $Si_{1-x}Ge_x$ layers in the near-infrared wavelength range (0.9-1.7 μm)," *Appl. Phys. Lett.*, vol. 67, pp. 3402–3404, 1995.

6 S. Janz, J. M. Baribeau, A. Delage, H. Lafontaine, S. Mailhot, R. L. Williams, D. X. Xu, D. M. Bruce, P. E. Jessop and M. Robillard, "Optical properties of pseudomorphic $Si_{1-x}Ge_x$ for Si-based waveguides at the $\lambda = 1300$-nm and 1550-nm telecommunications wavelength bands," *IEEE J. Selected Topics in Quantum Electronics*, vol. 4, pp. 990–996, 1998.

7 R. Strong, D. W. Greve, R. Mishra, M. Weeks and P. Pellegrini, "GeSi infrared detectors," *Thin Solid Films*, vol. 294, pp. 343–346, 1997.

8 A. Splett, B. Schuppert, K. Petermann, E. Kasper, H. Kibbel and H. J. Herjog, "Waveguide/photodetector combination in SiGe for long wavelength operation," in *Dig. Conf. on Integrated Photonic Research*, vol. 10, pp. 122–123, 1992.

9 B. Jalali, L. Naval and A. F. J. Levi, "Si-based receivers for optical data links," *IEEE J. Lightwave Technol.*, vol. 11, pp. 930–934, 1994.

10 B. Jalali, L. Naval, A. F. Levi and P. Watson, "GeSi infrared photodetectors grown by rapid thermal CVD," in *SPIE Proc.*, vol. 1802, pp. 94–107, 1992.

11 S. Murtaza, R. Mayer, M. Rashed, D. Kinosky, C. Maziar, S. Banerjee, C. Campbell, J. C. Bean and L. J. Peticolas, "Room temperature electroabsorption in Ge_xSi_{1-x} PIN photodiode," *IEEE Trans. Electron Dev.*, vol. 41, pp. 2297–2300, 1994.

12 F. Huang, X. Zhu, M. O. Tanner and K. L. Wang, "Normal-incidence strained-layer superlattice $Ge_{0.5}Si_{0.5}$/ Si photodiodes near 1.3 μm," *Appl. Phys. Lett.*, vol. 67, pp. 566–568, 1995.

13 H. Temkin, T. P. Pearsall, J. C. Bean, R. A. Logan and S. Luryi, "Ge$_x$Si$_{1-x}$ strained-layer superlattice waveguide photodetectors operating near 1.3 μm," *Appl. Phys. Lett.*, vol. 48, pp. 963–965, 1986.

14 H. Temkin, J. C. Bean, T. P. Pearsall, N. A. Olsson and D. V. Lang, "Ge$_{0.6}$Si$_{0.4}$ rib waveguide avalanche photodetector for 1.3 μm operation," *Appl. Phys. Lett.*, vol. 49, pp. 809–811, 1986.

15 J. Silverman, J. M. Mooney and F. D. Shepherd, "Infrared Video Cameras," *Scientific American*, vol. 266, pp. 78–83, 1992.

16 B.-Y. Tsaur, C. K. Chen and S. A. Marino, "Long-wavelength Ge$_x$Si$_{1-x}$/Si heterojunction infrared detectors and 400×400 element imager arrays," *IEEE Electron Dev. Lett.*, vol. 12, pp. 293–296, 1991.

17 H. Kanaya, F. Hasegawa, E. Yamaka, T. Moriyama and M. Nakajima, "Reduction of Barrier Height of silicide/p-Si$_{1-x}$Ge$_x$ contact for application in an image sensor," *Jap. J. Appl. Phys.*, vol. 28, pp. L544–L546, 1989.

18 X. Xiao, J. C. Sturm, S. R. Parihar, S. A. Lyon, D. Meyerhafer, S. Palfrey and F. V. Shallcross, "Silicide/Strained Si$_{1-x}$Ge$_x$ Schottky-Barrier Infrared Detectors," *IEEE Electron Dev. Lett.*, vol. 14, pp. 199–201, 1993.

19 S. Luryi, A. Kastalsky and J. C. Bean, "New infrared detector on silicon chip," *IEEE Trans. Electron Dev.*, vol. ED-31, p. 1135, 1984.

20 H. Temkin, J. C. Bean, T. P. Pearsall, N. A. Olsson and D. V. Lang, "High photoconductive gain in Ge$_x$Si$_{1-x}$ strained-layer superlattice detectors operating at $\lambda = 1.3$ μm," *Appl. Phys. Lett.*, vol. 49, pp. 155–157, 1986.

21 F. Y. Huang and K. L. Wang, "Normal-incidence epitaxial SiGeC photodetector near 1.3μm wavelength grown on Si substrate," *Appl. Phys. Lett.*, vol. 69, pp. 2330–2332, 1996.

22 B. Schuppert, J. Schmidtchen, A. Splett, U. Fisher, T. Zinke, R. Moosburger and K. Petermann, "Integrated optics in silicon and SiGe heterostructures," *J. Lightwave Technol.*, vol. 14, pp. 2311–2323, 1996.

23 T. Tashiro, T. Tatsumi, M. Sugiyama, T. Hashimoto and T. Morikawa, "A Selective Epitaxial SiGe/Si Planar Photodetector for Si-Based OEIC's," *IEEE Trans. Electron Dev.*, vol. 44, pp. 545–550, 1997.

24 S. B. Samavedam, M. T. Currie, T. A. Langdo and E. A. Fitzgerald, "High-quality germanium photodiodes integrated on silicon substrates using optimized relaxed graded buffers," *Appl. Phys. Lett.*, vol. 73, pp. 2125–2127, 1998.

25 X. Shao, S. L. Rommel, B. A. Orner, H. Feng, M. W. Dashiell, R. T. Troeger, J. Kolodzey, P. R. Berger and T. Laursen, "1.3 micron photoresponsivity in Si-based GeC photodiodes," *Appl. Phys. Lett.*, vol. 72, pp. 1860–1862, 1998.

26 L. Colace, G. Masini, F. Galluzzi, G. Assanto, G. Capellini, L. Di Gaspare, E. Palange and F. Evangelisti, "Metal-semiconductor-metal near infrared light detector based on epitaxial Ge/Si," *Appl. Phys. Lett.*, vol. 72, pp. 3175–3177, 1998.

27 T. P. Pearsall, H. Temkin, J. C. Bean and S. Luryi, "Avalanche Gain in Ge_xSi_{1-x}/Si Infrared Waveguide Detectors," *IEEE Electron Dev. Lett.*, vol. EDL-7, pp. 330–332, 1986.

28 B. C. Roy and N. B. Chakrabarti, "Gain and Frequency Response of Graded-Base Heterojunction Bipolar Phototransistor," *IEEE Trans. Electron. Dev.*, vol. ED-34, pp. 1482–1490, 1987.

29 H. Kamitsuna, "Ultra-wideband monolithic photoreceivers using HBT-compatible HPTs with novel base circuits, and simultaneously integrated with an HBT amplifier," *J. Lightwave Tech.*, vol. 13, pp. 2301–2307, 1995.

30 H. Kamitsuna, "Monolithically integrated high-gain and high-sensitive photoreceivers with tunable filtering functions for subcarrier multiplexed optical/microwave systems," *IEEE Trans. Microwave Th. Tech.*, vol. 43, pp. 2351–2356, 1995.

31 A. K. Sharma, K. A. M. Scott, S. R. J. Brueck, J. C. Zolper and D. R. Myers, "Ion implantation enhanced metal-Si-metal photodetectors," *IEEE Photon. Tech. Lett.*, vol. 6, pp. 635–638, 1994.

32 N. K. Dutta, D. T. Nichols, D. C. Jacobson and G. Livescu, "Fabrication and performance characteristics of high-speed ion-implanted Si metal-semiconductor-metal photodetectors," *Appl. Opt.*, vol. 36, pp. 1180–1184, 1997.

33 M. Y. Liu, S. Y. Chou, S. Alexandrou, C. C. Wang and T. Y. Hsiang, "110 GHz Si MSM photodetectors," *IEEE Trans. Electron Dev.*, vol. 40, pp. 2145–2146, 1993.

34 A. Richter, P. Steiner, F. Kozlowski and W. Lang, "Current-induced light emission from a porous silicon device," *IEEE Electron Dev. Lett.*, vol. 12, pp. 691–692, 1991.

35 C.-C. Wang, S. Alexandrou, D. Jacobs-Perkins and T. Y. Hsiang, "Comparison of the picosecond characteristics of silicon and silicon-on-sapphire metal-semiconductor-metal photodiodes," *Appl. Phys. Lett.*, vol. 64, pp. 3578–3580, 1994.

36 M. Y. Liu, E. Chen and S. Y. Chou, "140 GHz metal-semiconductor-metal photodetectors on silicon-on-insulator substrate with a scaled active layer," *Appl. Phys. Lett.*, vol. 65, pp. 887–889, 1994.

37 F. Yablonovitch and G. D. Cody, "Intensity enhancement in textured optical sheets for solar cells," *IEEE Trans. Electron Dev.*, vol. ED-29, pp. 300–305, 1982.

38 H. C. Lee and B. van Zghbroeck, "A novel high-speed silicon MSM photodetector operating at 830 nm wavelength," *IEEE Electron Dev. Lett.*, vol. 16, pp. 175–177, 1995.

39 S. Y. Chou and Y. Liu, "32 GHz Metal-Semiconductor-Metal Photodetectors on Thick Crystalline Silicon," *Appl. Phys. Lett.*, vol. 61, pp. 1760–1762, 1992.

40 S. Chattopadhyay, *Studies on Optoelectronic Applications of SiGe Alloys.* PhD Thesis, Jadavpur University, 1999.

41 L. Colace, G. Masini and G. Assanto, "Ge-on-Si Approaches to the Detection of Near-Infrared Light," *IEEE J. Quantum Electron.*, vol. 35, pp. 1843–1852, 1999.

42 S. Chattopadhyay, L. K. Bera, S. K. Ray, P. K. Bose, D. Dentel, L. Kubler, J. L. Bischoff and C. K. Maiti, "Determination of Interface State Density of PtSi/strained-Si$_{1-x}$Ge$_x$/Si Schottky Diodes," *J. Mat. Sci.: Electronic Mater.*, vol. 9, pp. 403–407, 1998.

43 H. Hertle, G. Schuberth, E. Gornik and G. Abstreiter, "Intersubband absorption in the conduction band of Si/SiGe multiple quantum wells," *Appl. Phys. Lett.*, vol. 59, pp. 2977–2979, 1991.

44 T. L. Lin and J. Maserjian, "Novel $Si_{1-x}Ge_x$/Si heterojunction internal photoemission long-wavelength infrared detectors," *Appl. Phys. Lett.*, vol. 57, pp. 1422–1424, 1990.

45 J. S. Park, T. L. Lin, E. W. Jones, H. M. Del Castillo and S. D. Gunapala, "Long-wavelength stacked SiGe/Si heterojunction internal photoemission infrared detectors using multiple SiGe/Si layers," *Appl. Phys. Lett.*, vol. 64, pp. 2370–2372, 1994.

46 J. R. Jimenez, X. Xiao, J. C. Sturm and P. W. Pellegrini, "Tunable, long-wavelength PtSi/SiGe/Si Schottky diode infrared detectors," *Appl. Phys. Lett.*, vol. 67, pp. 506–508, 1995.

47 R. People, J. C. Bean, C. G. Bethea, S. K. Sputz and L. J. Peticolas, "Broadband (8-14 μm) normal incidence, pseudomorphic Ge_xSi_{1-x}/Si strained-layer infrared photodetector operating between 20 and 77 K," *Appl. Phys. Lett.*, vol. 61, pp. 1122–1124, 1992.

48 J. S. Park, R. P. G. Karunasiri and K. L. Wang, "Intervalence-subband transition in SiGe/Si multiple quantum wells normal incident detection," *Appl. Phys. Lett.*, vol. 61, pp. 681–683, 1992.

49 R. P. G. Karunasiri, J. S. Park and K. L. Wang, "Normal incidence infrared detector using intervalence-subband transitions in $Si_{1-x}Ge_x$/Si quantum wells," *Appl. Phys. Lett.*, vol. 61, pp. 2434–2436, 1992.

50 R. T. Carline, D. J. Robbins, M. B. Stanaway and W. Y. Leong, "Long-wavelength SiGe/Si resonant cavity infrared detector using a bonded silicon-on-oxide reflector," *Appl. Phys. Lett.*, vol. 68, pp. 544–546, 1996.

51 P. Kruck, M. Helm, T. Fromherz, G. Bauer, J. F. Nutzel and G. Abstreiter, "Medium-wavelength, normal-incidence, p-type Si/SiGe quantum well infrared photodetector with background limited performance up to 85 K," *Appl. Phys. Lett.*, vol. 69, pp. 3372–3374, 1996.

52 H. Presting, "Near and mid infrared silicon/germanium based photodetection," *Thin Solid Films*, vol. 321, pp. 186–195, 1998.

53 D. J. Robbins, M. B. Stanaway, W. Y. Leong, J. L. Glasper and C. Pickering, "$Si_{1-x}Ge_x$ quantum well infrared photodetectors," *J. Mater. Sci.: Mater. Electron.*, vol. 6, pp. 363–367, 1995.

54 M. Ghioni, A. Laciata, G. Ripamonti and S. Cova, "All-silicon avalanche photodiode sensitive at 1.3 μm with picosecond time resolution," *IEEE J. Quantum Electron.*, vol. 28, pp. 2678–2681, 1992.

55 A. Vonsovici, L. Vescan, R. Apetz, A. Koster and K. Schmidt, "Room temperature photocurrent spectroscopy of SiGe/Si p-i-n photodiodes grown by selective epitaxy," *IEEE Trans. Electron Dev.*, vol. 45, pp. 538–542, 1998.

56 A. Chin and T. Y. Chang, "Enhancement of quantum efficiency in thin photodiodes through absorptive resonance," *IEEE J. Lightwave Technol.*, vol. 9, pp. 321–323, 1991.

57 S. S. Murtaza, H. Nie, J. C. Campbell, J. C. Bean and L. J. Peticolas, "Short-wavelength, high-speed, Si-based resonant-cavity photodetector," *IEEE Photon. Tech. Lett.*, vol. 8, pp. 927–929, 1996.

58 Y. Ishikawa, N. Shibata and S. Fukatsu, "Epitaxy-ready Si/SiO_2 Bragg reflectors by multiple separation-by-implanted-oxygen," *Appl. Phys. Lett.*, vol. 69, pp. 3881–3883, 1996.

59 D. C. Diaz, C. L. Schow, J. Qi and J. C. Campbell, J. C. Bean and L. J. Peticolas, "Si/SiO_2 resonant cavity photodetector," *Appl. Phys. Lett.*, vol. 69, pp. 2798–2800, 1996.

60 J. C. Bean, J. Qi, C. L. Schow, R. Li, H. Nie, J. Schaub and J. C. Campbell, "High-speed polysilicon resonant-cavity photodiode with SiO_2-Si Bragg reflectors," *IEEE Photon. Tech. Lett.*, vol. 9, pp. 806–808, 1997.

61 Y. Zhu, Q. Yang and Q. Wang, "Resonant cavity SiGe/Si MQW heterojunction phototransistor grown on the SIMOX substrate for 1.3 μm operation," in *Proc. Electronic Components and Technology Conf.*, pp. 1199–1204, 1997.

62 R. A. Soref and J. P. Lorenzo, "Single-crystal - A new material for 1.3 and 1.6 μm integrated-optical components," *Electron. Lett.*, vol. 21, pp. 953–954, 1985.

63 R. A. Soref and J. P. Lorenzo, "Epitaxial silicon guided-wave components for $\lambda = 1.3$ μm," in *OSA Integrated and Guided-Wave Optics Conf. Dig. Papers, Feb. 26, 1986*, pp. 18–19, 1986.

64 R. A. Soref and J. P. Lorenzo, "All-silicon active and passive guided-wave components for $\lambda = 1.3\ \mu$m," *IEEE J. Quantum Electron.*, vol. QE-22, pp. 873–879, 1986.

65 T. G. Brown, P. L. Bradfield, D. G. Hall and R. A. Soref, "Optical emission from impurities within an epitaxial silicon optical waveguide," *Opt. Lett.*, vol. 12, pp. 753–755, 1987.

66 A. Splett and K. Petermann, "Low loss single-mode optical waveguides with large cross-section in standard epitaxial silicon," *Photonics Technology Lett.*, vol. PTL-6, pp. 425–427, 1994.

67 R. A. Soref, F. Namavar and J. P. Lorenzo, "Optical waveguiding in a single-crystal layer of germanium-silicon grown on silicon," in *SPIE Proc.*, vol. 1177, pp. 175–184, 1989.

68 R. A. Soref, F. Namavar and J. P. Lorenzo, "Optical waveguiding in a single crystal layer of germanium-silicon grown on silicon," *Opt. Lett.*, vol. 15, pp. 270–272, 1990.

69 F. Namavar and R. A. Soref, "Optical waveguiding in $\mathrm{Si/Si}_{1-x}\mathrm{Ge}_x\mathrm{/Si}$ heterostructures," *J. Appl. Phys.*, vol. 70, pp. 3370–3372, 1991.

70 S. F. Pesarcik, G. V. Treyz, S. S. Iyer and J. M. Halbout, "Silicon germanium optical waveguides with 0.5 dB/cm losses for singlemode fibre optic systems," *Electronics Lett.*, vol. 28, pp. 159–160, 1992.

71 R. A. Mayer, K. H. Jung, T. Y. Hsieh, D.-L. Kwong and J. C. Campbell, "$\mathrm{Ge}_x\mathrm{Si}_{1-x}$ optical directional coupler," *Appl. Phys. Lett.*, vol. 58, pp. 2744–2745, 1991.

72 Y. M. Liu and P. R. Prucnal, "Deeply etched singlemode GeSi rib waveguides for silicon-based optoelectronic integration," *Electron. Lett.*, vol. 28, pp. 1434–1435, 1992.

73 J. Schmidtchen, B. Schuppert and K. Petermann, "Passive integrated-optical waveguide structures by Ge-diffusion in silicon," *IEEE J. Lightwave Technol.*, vol. 12, pp. 842–848, 1994.

74 A. Splett, J. Schmidtchen, B. Schuppert, K. Petermann, E. Kasper and H. Kibbel, "Low loss optical ridge waveguides in a strained GeSi epitaxial layer grown on silicon," *Electronics Lett.*, vol. 26, pp. 1035–1037, 1990.

75 R. A. Soref and J. P. Lorenzo, "Light-by-light modulation in silicon-on-insulator waveguides," in *Proc. IGWO'89 (OSA Tech. Dig. Series)*, vol. 4, pp. 86–89, 1989.

76 J. Schmidtchen, A. Splett, B. Schuppert, K. Petermann and G. Burbach, "Low-loss singlemode optical waveguides with large cross section in silicon-on-insulator," *Electron. Lett.*, vol. 27, pp. 1486–1488, 1991.

77 R. M. Emmons, B. N. Kurdi and D. G. Hall, "Buried-oxide Silicon-on-Insulator structures I: Optical waveguide characteristics," *IEEE J. Quantum Electron.*, vol. 28, pp. 157–163, 1992.

78 R. M. Emmons and D. G. Hall, "Buried-oxide Silicon-on-Insulator structures II: Waveguide grating coupler," *IEEE J. Quantum Electron.*, vol. 18, pp. 164–173, 1992.

79 R. A. Soref, J. Schmidtchen and K. Petermann, "Large single-mode rib waveguides in GeSi–Si and Si-on-SiO$_2$," *IEEE J. Quantum Electron.*, vol. 27, pp. 1971–1974, 1991.

80 T. Zinke, U. Fischer, A. Splett, B. Schuppert and K. Petermann, "Comparison of optical waveguide losses in silicon-on-insulator," *Electronics Lett.*, vol. 29, pp. 2031–2033, 1993.

81 A. G. Rickman and G. T. Reed, "Silicon-on-insulator optical rib waveguides: loss, mode characteristics, bends and y-junctions," *IEE Proc. Optoelectronics*, vol. 141, pp. 391–393, 1994.

82 P. D. Trinh, S. Yegnarayanan and B. Jalali, "Integrated optical directional couplers in silicon-on-insulator," *Electronics Lett.*, vol. 31, pp. 2097–2098, 1995.

83 A. G. Rickman, G. T. Reed and F. Namavar, "Silicon-on-insulator optical rib waveguide loss and mode characteristics," *IEEE J. Lightwave Technol.*, vol. 12, pp. 1771–1776, 1994.

84 U. Fischer, T. Zinke, J.-R. Kropp, F. Arndt and K. Petermann, "0.1 dB/cm waveguide losses in single-mode SOI rib waveguides," *IEEE Photonics Tech. Lett.*, vol. 8, pp. 647–648, 1996.

85 D. C. Houghton, J. P. Noel and N. L. Rowell, "Electroluminescence and photoluminescence from SiGe alloys grown on (100) silicon by MBE," *Mater. Sci. Eng.*, vol. B9, pp. 237–244, 1991.

86 P. L. Bradfield, T. G. Brown and D. G. Hall, "Electroluminescence from sulfur impurities in a p-n junction formed in epitaxial silicon," *Appl. Phys. Lett.*, vol. 55, pp. 100–102, 1989.

87 M. Q. Huda, J. H. Evans-Freeman, A. R. Peaker, D. C. Houghton and A. Nejim, "Luminescence from erbium implanted silicon–germanium quantum wells," *J. Vac. Sci. Technol. B*, vol. 16, pp. 2928–2933, 1998.

88 S.-J. Chang, D. K. Nayak and Y. Shiraki, "1.54 μm electroluminescence from erbium-doped SiGe light emitting diodes," *J. Appl. Phys.*, vol. 83, pp. 1426–1428, 1998.

89 D. S. Shen, J. P. Conde, V. Chu, S. Aljishi, J. Z. Liu and S. Wagner, "Amorphous Silicon–Germanium Thin-Film Photodetector Array," *IEEE Electron Dev. Lett.*, vol. 13, pp. 5–7, 1992.

90 B. S. Wu, C.-Y. Chang, Y.-Y. Fang and R. H. Lee, "Amorphous Silicon Phototransistor on Glass Substrate," *IEEE Trans. Electron. Dev.*, vol. ED-32, pp. 2192–2196, 1985.

91 J. W. Hong, W.-L. Laih, Y.-W. Chen, Y.-K. Fang, C.-Y. Chang and J. Gong, "Optical and Noise Characteristics of Amorphous Si/SiC Superlattice Reach-Through Avalanche Photodiodes," *IEEE Trans. Electron. Dev.*, vol. 37, pp. 1804–1809, 1990.

92 S.-B. Hwang, Y. K. Fang, K.-H. Chen, C.-R. Liu, J.-D. Hwang and M.-H. Chou, "An a-Si:H/a-Si, Ge:H Bulk Barrier Phototransistor with a-SiC:H Barrier Enhancement Layer for High-Gain IR Optical Detector," *IEEE Trans. Electron. Dev.*, vol. 40, pp. 721–726, 1993.

93 U. Hilleringham and K. Goser, "Optoelectronic system integration on silicon: Waveguides, photodetectors, and VLSI CMOS circuits on one chip," *IEEE Trans. Electron. Dev.*, vol. 42, pp. 841–846, 1995.

94 T. P. Pearsall, J. Bevk, L. C. Feldman, J. M. Bonar, J. P. Mannaerts and A. Ourmazd, "Structurally induced optical transitions in Ge-Si superlattices," *Phys. Rev. Lett.*, vol. 58, pp. 729–732, 1987.

95 M. Steyaert, M. Ingels, J. Crols and G. van der Plas, "A CMOS 240 Mb/s optical receiver with a transimpedance-bandwidth of 18 THzΩ," in *IEEE ISSCC Tech. Dig.*, pp. 182–183, 1994.

96 P. J.-W. Lim, A. Y. C. Tzeng, H. L. Chuang and S. A. St. Onge, "A 3.3-V monolithic photodetector/CMOS-preamplifier for 531 Mb/s optical data link applications," in *IEEE ISSCC Tech. Dig.*, pp. 96–97, 1993.

97 D. M. Kuchta and C. J. Mahon, "Mode selective loss penalties in VCSEL optical fiber transmission links," *IEEE Photon. Technol. Lett.*, vol. 6, pp. 288–290, 1994.

98 V. P. Kesan, P. G. May, E. Bassous and S. S. Iyer, "Integrated waveguide-photodetector using Si/SiGe multiple quantum wells for long wavelength applications," in *IEEE IEDM Tech. Dig.*, pp. 637–640, 1990.

99 S. Fukatsu, N. Usami and Y. Shiraki, "High-temperature operation of strained $Si_{0.65}Ge_{0.35}$/Si (111) p-type multiple-quantum well light-emitting diode grown by solid source Si molecular beam epitaxy," *Appl. Phys. Lett.*, vol. 63, pp. 967–969, 1993.

100 G. V. Hansson, W. X. Ni, K. B. Joelsson and I. A. Buyanova, "Silicon-based structures for IR light emission," *Physica Scripta*, vol. T69, pp. 60–64, 1997.

Index